Handbook of Experimental Pharmacology

Volume 215

For further volumes:
http://www.springer.com/series/164

Erich Gulbins • Irina Petrache
Editors

Sphingolipids: Basic Science and Drug Development

Springer

Editors
Erich Gulbins
Department of Molecular Biology
University of Duisburg-Essen
Essen
Germany

Irina Petrache
Division of Pulmonary, Allergy,
Critical Care and Occupational Medicine
Department of Medicine
Indiana University School of Medicine
Indianapolis, IN, USA

ISSN 0171-2004 ISSN 1865-0325 (electronic)
ISBN 978-3-7091-1367-7 ISBN 978-3-7091-1368-4 (eBook)
DOI 10.1007/978-3-7091-1368-4
Springer Heidelberg New York Dordrecht London

Library of Congress Control Number: 2013935698

Printed on acid-free paper

Springer is part of Springer Science+Business Media (www.springer.com)

Preface

Until the late 1980s, sphingolipids were believed to represent structural components of the plasma membrane, whose function was to provide a protective barrier to the cell. This picture dramatically changed within the last years. Sphingolipids are now recognized signals for fundamental cellular processes, such as proliferation, survival, cell death, adhesion, migration, angiogenesis, and embryogenesis. The explosion of knowledge regarding sphingolipids was facilitated by biochemical studies of their signaling properties, the cloning of enzymes of the sphingolipid metabolism, development of genetic models for determining their physiologic roles, and the establishment of biochemical, biophysical, and optical methods for their detection and quantitation. The next step in the evolution of sphingolipids will be the transfer of basic insights into the biochemistry and cell biology of human diseases. The recent success of the sphingolipid drug, fingolimod, a sphingosine-1-phosphate agonist, which rapidly became a therapy for multiple sclerosis, exemplifies the potential of targeting sphingolipids for the treatment of human disorders. The aim of our two volumes in this series—*Sphingolipds: Basic Science and Drug Development* and *Sphingolipids in Disease*—is to define the state of the art of sphingolipid biology and to present preclinical developments and early clinical applications of this fascinating class of lipids.

Essen, Germany — Erich Gulbins
Indianapolis, IN, USA — Irina Petrache

Contents

Part II Drug Development

Part I
Basic Mechanisms

The Genetics of Sphingolipid Hydrolases and Sphingolipid Storage Diseases

Edward H. Schuchman and Calogera M. Simonaro

Contents

Abstract The relationship of sphingolipids with human disease first arose from the study of sphingolipid storage diseases over 50 years ago. Most of these disorders are due to inherited deficiencies of specific sphingolipid hydrolases, although a small number also result from defects in sphingolipid transport or activator proteins. Due to the primary protein deficiencies sphingolipids and other macromolecules accumulate in cells and tissues of affected patients, leading to a diverse presentation of clinical abnormalities. Over 25 sphingolipid storage diseases have been described to date. Most of the genes have been isolated, disease-causing mutations have been identified, the recombinant proteins have been produced and characterized, and animal models exist for most of the human diseases. Since most sphingolipid hydrolases are enriched within the endosomal/lysosomal system, macromolecules first accumulate within these compartments. However, these abnormalities rapidly spread to other compartments and cause a wide range of cellular dysfunction.

E.H. Schuchman (✉) • C.M. Simonaro
Department of Genetics and Genomic Sciences, Mount Sinai School of Medicine, New York, NY 10029, USA
e-mail: edward.schuchman@mssm.edu

E. Gulbins and I. Petrache (eds.), *Sphingolipids: Basic Science and Drug Development*, Handbook of Experimental Pharmacology 215, DOI 10.1007/978-3-7091-1368-4_1,

This review focuses on the genetics of sphingolipid storage diseases and related hydrolytic enzymes with an emphasis on the relationship between genetic mutations and human disease.

Keywords Sphingolipids • Hydrolases • Lysosomes • Genes • Mutations

1 Introduction and Overview

Sphingolipids are a diverse group of over 100 bioactive lipids involved in many aspects of cellular function (see Chen et al. 2012; Furuya et al. 2011; Petrache et al. 2011; Podbielska et al. 2012 for recent reviews). They all share a common sphingosine (2-amino-4-octadecene-1,3-diol) backbone, which may be linked via an amide bond to a fatty acid (forming ceramide), or phosphorlyated to form sphingosine-1-phosphate. Ceramide itself also may be modified to form glycosphingolipids, sphingomyelin, or ceramide-1-phosphate.

The metabolism of sphingolipids is complex and carefully regulated (e.g., Mullen et al. 2012; Wennekes et al. 2009; Worgall 2011). Abnormal metabolism can have profound effects on cellular function, leading to enhanced cell death, proliferation, and/or abnormal cellular differentiation. In general, the synthesis and breakdown of sphingolipids occur in distinct cellular compartments. For example, numerous enzymes contribute to the de novo synthesis of sphingolipids, mostly found in the endoplasmic reticulum and Golgi apparatus. In contrast, the breakdown of sphingolipids occurs primarily within endosomes and lysosomes by a series of hydrolytic enzymes that function optimally at acidic pH. Hydrolysis of sphingolipids also may occur in non-endosomal/lysosomal compartments, for example by enzymes located at the plasma membrane or within mitochondria. Overall, over 100 enzymes participate in the synthesis or breakdown of sphingolipids.

Historically, most information concerning the importance of sphingolipids in human disease stems from studies of the sphingolipid storage diseases (Schulze and Sandhoff 2011). These disorders are each due to deficient function of one or more of the hydrolytic enzymes found within the endosomal/lysosomal system, resulting in sphingolipid accumulation (Cox and Cachón-González 2012). Each of these diseases is inherited as a Mendelian trait (recessive or X-linked), the cDNAs/genes encoding the respective enzymes have been isolated, and many mutations causing the human diseases have been identified. In addition, genetic abnormalities in several of these enzymes/genes have now been linked to more common diseases (e.g., Parkinson's disease and beta-glucocerebrosidase) (Young et al. 2012). Table 1 lists the sphingolipid storage diseases and the respective enzymes involved.

This review focuses on the genetics of sphingolipid storage diseases and related hydrolytic enzymes. Each of the diseases/enzymes is briefly reviewed, with an emphasis on the importance of genetic mutations in disease.

Table 1 Genetics of human sphingolipid hydrolases and diseases

Sphingolipid hydrolase/ activator	Disease	Primary accumulating lipid	Gene designation	Gene location
Acid ceramidase	Farber disease	Ceramide	ASAH1	8p22
Acid sphingomyelinase	Niemann-Pick (types A and B)	Sphingomyelin	SMPD1	11p15.4-p15.1
Beta glucocerebrosidase	Gaucher disease	Glucosylceramide	GBA1	1q21
Beta galactocerebrosidase	Krabbe disease	Galactosylceramide	GALC	14q31
Arylsulfatase A	Metachromatic leukodystrophy	Sulfatide	ARSA	22q13.31-qter
Alpha galactosidase A	Fabry disease	Globotriosylceramide	GLA	Xq22.1
Beta galactosidase	GM1 gangliosidosis	GM1 ganglioside	GLB1	3p22.31
Hexosaminidase	GM2 gangliosidosis			
Hexosaminidase A	Tay-Sachs disease	GM2 ganglioside	HEXA	15q24
Hexosaminidase B	Sandhoff disease	GM2 ganglioside	HEXB	5q13
GM2 activator	Tay-Sachs A/B variant	GM2 ganglioside	GMA2	5q31.1
Prosaposin	Variable	Variable	PSAP	10q21-q22

2 Acid Ceramidase Deficiency: Farber Disease

Patients with Farber disease (also known as Farber lipogranulomatosis) usually present within the first few months of life with a triad of clinical manifestations, including painful and progressively deformed joints, subcutaneous nodules, particularly near joints, and progressive hoarseness due to laryngeal involvement (Farber et al. 1957). About half of the patients have neurological deficits (Ahmad et al. 2009). As with all of the sphingolipid storage disorders, there is considerable clinical heterogeneity among Farber disease patients, and later onset, attenuated cases also have been described. Farber disease is the least common of the sphingolipid storage diseases, likely because severe mutations in the acid ceramidase (AC) gene (*ASAH1*) lead to embryonic lethality (see below).

Acid ceramidase (E.C. 3.5.1.23) was first identified and purified in the 1960s. It was recognized as the enzyme deficient in Farber disease in the 1970s (Sugita et al. 1972). The full-length cDNAs and genes encoding AC have been isolated from humans, mice, and other species. The human cDNA encodes a 395 amino acid precursor polypeptide, and the gene is located on chromosomal region 8p22. Of note, the newly synthesized AC precursor is inactive, and only assumes an active form after *N*-glycosylation and the internal cleavage of a peptide bond, resulting in alpha and beta subunits. Interestingly, this cleavage event is carried out by AC itself, and it is therefore considered a "self-regulating" enzyme. In addition, AC associates with other lipid hydrolases within lysosomes and other cell compartments, including acid sphingomyelinase (see below). The importance of this multienzyme complex on AC activity, as well as on sphingolipid metabolism in general, is not clearly understood. Finally, full AC activity, at least in lysosomes, depends on the expression of a sphingolipid activator protein (SAP-D), which is expressed by a distinct gene (see Sect. 10, below).

Acid ceramidase is an "amidase" that hydrolyzes the amide bond between sphingosine and the fatty acid in ceramide. Ceramides are a heterogenous group of lipids defined by the length of their fatty acid chains. While the substrate specificity of AC is not entirely clear, ceramides with fatty acid chains shorter than 12 carbons cannot be efficiently hydrolyzed by the enzyme in vitro. In vivo, it is likely that AC can recognize and cleave ceramides of varying chain lengths. The ceramides that accumulate in Farber disease patients are primarily found within lysosomes, leading to the original designation of AC as a lysosomal enzyme. AC also has maximal activity at lysosomal (acidic) pH, although it has been localized to sites other than the lysosomes (e.g., the spindle in oocytes) as well. It is therefore not entirely clear to what extent the biological activity of AC depends on the acidic environment, and in fact a "reverse" ceramidase reaction has been attributed to the enzyme at more neutral pH (Okino et al. 2003). Numerous in vitro assays have been developed for AC, and some can be used to diagnose Farber disease in cultured cells from suspected patients, or prenatally using aminiocytes or cultured chorionic villi. However, due to the very hydrophobic nature of ceramides, it remains one of the more difficult diagnostic enzyme assays to carry out.

To date, only about 100 Farber disease patients have been reported in the literature (Park and Schuchman 2006). Several cases of hydrops fetalis also have been reported. Complete inactivation of the AC gene in mice leads to very early embryonic lethality (four-cell stage), suggesting that severe mutations in the human *ASAH1* gene will lead to embryonic lethality as well (Li et al. 2002). Of note, mouse embryos lacking AC can be rescued by the addition of recombinant AC (obtained from overexpressing Chinese hamster ovary [CHO] cells) to the culture media. Moreover, addition of recombinant AC to the culture media during in vitro fertilization (mouse, bovine) enhances embryo production and quality, resulting in more live births (Eliyahu et al. 2010).

Thus, AC is an essential enzyme required for very early mammalian development. One explanation for the lethal phenotype in mice is that in the absence of this enzymatic activity, early-stage embryos (four cells in mice) undergo apoptosis due to ceramide accumulation. It should be recognized, however, that an additional consequence of such AC "loss of function" during embryogenesis is reduced production of sphingosine, which in turn is converted to S1P, an important mitogenic/proliferative lipid. Thus, AC mutations also may reduce S1P production, contributing to the lethality. Indeed, embryos lacking AC can be partially rescued by S1P as well.

As noted above, the first relationship between AC and human disease arose from studies on Farber disease. To date over 20 mutations have been described in the *ASAH1* gene leading to Farber disease (Park and Schuchman 2006). Most are point mutations, although several splice variants leading to small deletions have been described as well. To date, no patients have been found homozygous for a complete loss-of-function mutation, likely because such individuals do not survive embryonic development. Farber disease is inherited as an autosomal recessive disorder, and to develop the disorder mutations must be inherited on both *ASAH1* alleles. No reports of clinical findings in the parents of affected patients (obligate carriers) have been published; however such individuals are not usually followed by clinicians and follow-up analysis is lacking.

Recently, a "knock-in" mouse model of Farber disease has been reported where a specific human mutation was expressed (add ref). Preliminary analysis of these animals indicates that they develop a Farber disease-like phenotype. In addition, a conditional AC knockout mouse has been developed where AC can be inactivated at various stages of development or in specific tissues.

Other than Farber disease, which is due to the deficiency of AC, constitutive overexpression of the AC gene also occurs in many types of cancer (Park et al. 2005; Beckham et al. 2012). This may promote tumorigenesis and/or metastasis due to overproduction of S1P, or may lead to chemoresistance due to an enhanced ability to hydrolyze pro-apoptotic ceramide (Flowers et al. 2012; Furuya et al. 2011). The molecular basis of AC overexpression in cancer has not been studied extensively, but at least in some cases it appears to be due to elevated transcription. Gain-of-function mutations also could explain this finding, although to date these have not been described. In addition, a recent report showed that the *ASAH1* gene was down-regulated in a Chinese population of schizophrenic patients, and was associated with two single nucleotide polymorphisms in this gene (Zhang et al. 2012).

Finally, it must be noted that at least six other ceramidase activities have been described in man, and at least four other genes have been cloned (Mao and Obeid 2008). These ceramidases are defined by their unique pH, subcellular location, and other biological properties. These enzymes also play important roles in sphingolipid metabolism and/or signaling, and are likely to contribute to human disease pathogenesis. The biology of the other ceramidases has been reviewed extensively elsewhere (e.g., Mao and Obeid 2008).

3 Acid Sphingomyelinase Deficiency: Types A and B Niemann–Pick Disease

Types A and B Niemann–Pick Disease (NPD) result from mutations in the gene (*SMPD1*) encoding acid sphingomyelinase (E.C. 3.1.4.12) (McGovern and Schuchman 2009; Schuchman 2009). Patients with Type A NPD present in early infancy with hepatosplenomegaly and failure to thrive, and develop a rapidly degenerative neurological phenotype that leads to death by 3 years of age. In contrast, patients with Type B NPD have a later onset disease, and usually lack neurological findings. Such individuals also frequently present with hepatosplenomegaly, and may have pulmonary involvement as well. Dyslipidemia (high LDL-C, low HDL-C, high triglycerides) also is common. Patients with Type B

NPD may survive into adulthood, although their life span is generally shortened by the disease. The cause of death is unknown, but may occur from complications related to liver disease, trauma-induced bleeding episodes due to thrombocytopenia, early-onset cardiovascular disease, or pulmonary disease.

Acid sphingomyelinase (ASM) was found to be deficient in Type A and B NPD patients in the late 1960s. Although the enzyme had been discovered several years earlier in rat tissues, it was not substantially purified until the late 1980s. The cDNAs and genes encoding human and mouse ASM were cloned in the early 1990s. The human gene is located on chromosome 11p15.1. Of interest, among genes encoding lysosomal enzymes, it is the only one for which genomic imprinting has been described. This form of genetic regulation is generally reserved for genes with important developmental roles.

Over 100 mutations have been found in the *SMPD1* gene leading to Types A or B NPD (http://www.hgmd.org). Of note, several of these mutations are common within specific regions or ethnic populations, allowing genetic screening programs to be established. Homozygosity for these mutations is required to develop NPD, and it is inherited as an autosomal recessive trait. However, heterozygous carriers of *SMPD1* mutations may develop dyslipidemia and late-onset features of the disorder. In part, the development of a phenotype in the carrier individuals may be related to their imprinting status (i.e., whether the mutations are inherited from the mother or the father) (Simonaro et al. 2006). Unlike acid ceramidase and several other sphingolipid hydrolases, ASM does not require an exogenous activator protein to achieve full function. Thus, mutations in the sphingolipid activator protein gene (see below) do not affect ASM activity or cause Types A and B NPD. Of interest, the lysosomal lipid, bis(monacyl) phosphate, may activate the enzyme in vitro and in vivo.

In addition to ASM, several other sphingomyelinases (nonacidic) exist in mammalian cells (Clarke et al. 2011). These are defined by their unique subcellular location or other biochemical properties. At least four human genes have been isolated encoding these sphingomyelinases. Definitive proof that mutations in the *SMPD1* gene were solely responsible for Types A and B NPD came from studies in the knockout mouse model (ASMKO), where neutral and other sphingomyelinase activities were normal, despite the fact that these animals completely lacked ASM and developed NPD-like pathology and clinical features (Horinouchi et al. 1995). In addition, these early studies in the ASMKO mice showed that the previously identified zinc-activated, serum (secreted) form of ASM was encoded by the same *SMPD1* gene as lysosomal ASM (Schissel et al. 1998). It is now known that ASM is, in fact, a zinc-requiring enzyme. However, the lysosomal form of the enzyme has high levels of zinc bound, and thus does not require exogenous zinc when measured in vitro. In contrast, the serum form of the enzyme requires exogenous zinc in the assay systems routinely used.

Another early finding in the ASMKO mouse was that ceramide, the product of sphingomyelin hydrolysis by ASM, also was elevated in tissues of these animals. While initially a surprising result, it is likely that the elevated ceramide derives from breakdown of the accumulating sphingomyelin in NPD tissues by other sphingomyelinases. Sphingomyelin levels may be up to 30-fold above normal in the ASMKO mice, and ceramide two- to fivefold above normal. Most recently, sphingosine levels also were found to be high in tissues from ASMKO mice, probably due to breakdown of the accumulating ceramide by ceramidases (Schuchman et al., unpublished observations). Of interest, the degree of sphingosine accumulation in these animals (>20-fold above normal) cannot be explained by ceramide alone, perhaps suggesting some defect in converting the sphingosine to S1P.

The ASMKO mice develop a phenotype intermediate between Types A and B NPD. They die prematurely at ~6–8 months, secondary to neurological disease. They have been extensively used to study the pathobiology of this disease, and to develop/evaluate new therapies. In particular, enzyme replacement therapy (ERT) using recombinant ASM produced in CHO cells has been developed in the mouse model (Miranda et al. 2000), and a phase 1 clinical trial in adult patients with Type B NPD was recently completed (http://clinicaltrials.gov). A phase 2 trial is planned for 2012. Of interest, in the ASMKO mice dose-related toxicity was observed (which was not seen in healthy animals) using recombinant ASM, presumably due to rapid breakdown of the accumulating sphingomyelin to ceramide. This hypothesis was confirmed by pretreating animals with low dose of enzyme ("debulking" the accumulating sphingomyelin), which prevented any toxicity. In the patients treated by ERT, transient changes in several inflammatory biomarkers were observed, corresponding to dose and serum levels of ceramide.

Beginning in the late 1990s, the ASMKO mice also were extensively used to investigate the role of ASM in cell signaling. This is the subject of several reviews, and is not focused on in this chapter. However, it is now clear that ASM plays an important role in stress-induced ceramide generation and apoptosis, and that this occurs through hydrolysis of sphingomyelin at the cell surface within lipid rafts (Zeidan and Hannun 2010). How ASM, which has an acidic pH optimum in vitro and functions within lysosomes, participates in these processes has been the subject of much debate. It is now known that after the induction of stress or developmental signals, ASM may be rapidly translocated to lipid rafts at the cell surface. The mechanism underlying this rapid translocation is not entirely clear, but phosphorylation within the ER/Golgi may be responsible in part (Zeidan et al. 2008). It is also known that although ASM requires an acidic pH for optimal activity in vitro, in vivo it is capable of efficiently hydrolyzing sphingomyelin at physiological pH. This highlights the complex biology of this and other sphingolipid hydrolases, particularly in terms of pH, subcellular location, and other properties.

These observations on the ASMKO mice have also highlighted the important role of ASM in human disease, beyond NPD (e.g., Becker et al. 2010; Bikman and Summers 2011; Truman et al. 2011). Elevated ASM activity has now been detected in numerous common diseases, including diabetes, Alzheimer disease, cardiovascular diseases, pulmonary diseases, depression, and others. These will also be reviewed elsewhere and not discussed further here. At the present time it is not known if the elevated ASM found in these diseases is due to an activating genetic mutation, enhanced transcription, and/or posttranslational modification of the enzyme.

4 Beta-Glucocerebrosidase Deficiency: Gaucher Disease

Gaucher disease, due to the deficient activity of acid beta-glucocerebrosidase (also known as beta-glucosidase [E.C. 3.2.1.45]), is the most common sphingolipid storage disorder (Campbell and Choy 2012; Lee et al. 2012). There are three clinical subtypes of Gaucher disease, including an early-onset, neurological form (Type 2), an intermediate form (Type 3), and a late-onset, non-neurological form (Type 1) (Lo et al. 2012; Pastores 2010). All Gaucher patients present with hepatosplenomegaly, but the progression of the disease and survival are markedly different. For example, Type 3 children may only survive until 2–3 years of age, while Type 1 patients may survive into adulthood. If left untreated (see below), the hematological and bone lesions in Type 1 patients progress with age, seriously limiting quality of life. A massively enlarged spleen and frequent bone fractures may become common, requiring splenectomy, joint replacement, or other surgical interventions (Mikosch and Hughes 2010; Goker-Alpan 2011). Bleeding also is a common presenting feature of Gaucher disease due to thrombocytopenia. Pulmonary involvement may similarly occur. Of note, neoplastic disorders, particularly of the blood-forming organs, are more common in Gaucher disease patients than in the general population. These include chronic lymphocytic leukemia, multiple myeloma, and Hodgkin and non-Hodgkin lymphoma. The molecular basis of this increased cancer risk remains unknown (Choy and Campbell 2011).

All Gaucher patients are characterized by the presence of "Gaucher" cells in various organs, lipid-filled cells derived from the monocyte/macrophage system. Gaucher cells have a distinctive appearance by light microscopy, and can be readily distinguished from cells of other sphingolipid storage disorders. Gaucher cells are distributed throughout the body, wherever macrophages reside. The largest numbers are found in the spleen, sinusoids of the liver, bone marrow, and lymph nodes.

The major accumulating lipid in Gaucher disease is *N*-acyl-sphingosyl-1-0-β-D-glucoside (glucosylceramide, also known as glucosylcerebroside, GC). GC is widely distributed in mammalian tissues in small quantities as a metabolic intermediate in the synthesis and degradation of complex glycosphingolipids, such as gangliosides or globoside. The deacylated form of GC, glucosylsphingosine, is not normally found in tissues, but also accumulates in Gaucher disease cells. GC levels in plasma may be 2- to 20-fold above normal.

The major site of GC turnover in cells is the lysosome. Beta-glucocerebrosidase is normally found in this organelle and functions optimally at an acidic pH. As with the other lipid hydrolases, in vitro detection of this enzymatic activity can be difficult, and there have been various factors (lipids, protein, sugars) shown to influence the assay systems. In vivo, sphingolipid activator-derived peptides, particularly saposin (SAP) C, appear to be required for full enzymatic activity (see below). Nonacidic glucocerebrosidases also have been described and are encoded by distinct genes. Whether these activities can be modifying factors for the Gaucher phenotype has been the subject of much discussion, but remains unclear.

The gene encoding beta-glucocerebrosidase (*GBA1*) has been mapped to chromosome 1q21. It is ~7 kb and contains 11 exons. Of note, a 5 kb "pseudogene" is located immediately downstream from the *GBA1* gene. This pseudogene has maintained a high degree of homology with the functional gene and is transcribed. However, the nucleotide changes within the pseudogene sequence prevent the occurrence of a long open reading frame, precluding the possibility of expressing an active beta-glucocerebrosidase (Brown et al. 2006). Full-length cDNAs from many species also have been obtained that can be used to express the active enzyme.

Numerous mutations have been found in the *GBA1* gene causing Gaucher disease (Hruska et al. 2008). As described for Niemann–Pick disease above, several common mutations are found in specific populations, and DNA-based screening is now routinely carried out in the Ashkenazi Jewish population (Scott et al. 2010). Estimates of the Gaucher disease carrier frequency in this population range from ~1:10 to 1:25. Some of the mutations also can be used to predict the occurrence of severe or mild Gaucher phenotypes, assisting in family planning and genetic counseling.

Recently, mutations in the *GBA1* gene also have been associated with the occurrence of Parkinson's disease, although the precise relationship between the *GBA1* gene, glucocerebrosidase activity, and Parkinson's disease remains uncertain (Kinghorn 2011). One of the first clues suggesting this linkage came from the small number of Gaucher patients who exhibited Parkinson-like features, while stronger associations later came from Gaucher relatives who were carriers for *GBA1* mutations. Currently, heterozygous *GBA1* mutations are considered to be one of the most common genetic risk factor for Parkinson's disease (Anheim et al. 2012).

A complete knockout mouse model for Gaucher was constructed and had an early neonatal lethal phenotype, dying within 34 days after birth. These animals had massive GC storage, and while the cause of death was not entirely clear, skin abnormalities were prevalent. Several knock-in models also have been constructed, and some exhibited longer survival and later onset GC storage in macrophages, more closely resembling Gaucher disease patients (Farfel-Becker et al. 2011; Sardi et al. 2012).

Gaucher disease was the first sphingolipid storage disease for which ERT was developed in the early 1990s, first using beta glucocerebrosidase purified from placentae (Ceredase), and subsequently using recombinant enzyme produced in CHO cells (Cerezyme) (Hughes and Pastores 2010; Lachmann 2011). For non-neurological patients, ERT was life changing, resulting in reduction in spleen size and correction of the hematological abnormalities. Bone disease has been much less responsive to ERT, and remains a challenge in the treatment of Gaucher disease, as does treatment of the brain disease (Hughes and Pastores 2010). Another treatment strategy currently available for Gaucher patients uses a small molecule inhibitor of GC synthesis (Miglustat) to slow the accumulation of this lipid (Abian et al. 2011; Benito et al. 2011). The target of this inhibitor is the enzyme glucosylceramide synthase.

Of note, treatment of Gaucher patients either by ERT or with Miglustat should elevate ceramide, and could potentially result in ceramide-mediated toxicity. To date, however, such toxicity has not been observed. There are several factors that might explain this observation. First, compared to sphingomyelin, GC is generally found in small quantities in cells, and is not a major structural component of cell membranes. Thus, accumulation of GC in Gaucher cells is more exclusively found in the lysosomes than sphingomyelin in Niemann–Pick cells, where substantial amounts of the accumulating lipid also are found in the plasma membrane. Thus, ceramide-mediated raft reorganization and signaling are less likely to become activated in Gaucher disease compared to NPD. In addition, ceramide produced by ASM following ERT in NPD is likely to remain within the cell membrane, while the ceramide produced following ERT with Cerezyme or by inhibiting GC synthesis with Miglustat may be found in different cellular compartments, and therefore more readily accessible for biosynthetic pathways. Lastly, the ERT dose used to treat Gaucher patients (1.6 mg/kg, every 2 weeks) may be below the threshold needed to induce ceramide-mediated toxicity.

5 Galactocerebrosidase Deficiency: Krabbe Disease/Globoid Cell Leukodystrophy

Infantile globoid cell leukodystrophy (GLD), also known as Krabbe disease, is a rapidly progressive, fatal neurological disease of infants (Kohlschütter and Eichler 2011). The disease usually begins between 3 and 6 months of life with symptoms that may include irritability or hypersensitivity to stimuli. Patients rapidly exhibit severe mental and motor deterioration, and rarely survive beyond the second year (Duffner et al. 2011). Clinical manifestations in classical metachromatic leukodystrophy (MLD) patients are limited to the nervous system. Blindness and deafness are common, and peripheral neuropathy is almost always detectable. Patients with later onset forms of the disease, including adult-onset forms, may present with blindness, spastic paraparesis, and dementia (Malandrini et al. 2013; Perlman and Mars 2012).

The presence of numerous multinucleated globoid cells, almost total loss of myelin and oligodendrocytes, and astrocytic gliosis in the white matter are the morphologic basis of the disease. "Globoid cells" in Krabbe disease are enlarged macrophages that contain undigested galacotsylceramide and other lipids. Demyelination, axonal degeneration, fibrosis, and histiocytic infiltration are also common in the peripheral nervous system. Consistent with the myelin loss, the white matter is severely depleted of all lipids, particularly glycolipids. Under normal circumstances, galactosylceramide is present almost exclusively in the myelin sheath. The ratio of galactosylceramide to sulfatide (3-0 sulfogalactosylceramide) is abnormally high in MLD.

The genetic cause of Krabbe disease is mutations in the gene encoding galactocerebrosidase beta-galactosidase (*GALC*; E.C. 3.2.1.46) (see below). This lysosomal enzyme normally degrades galactosylceramide to ceramide and galactose. It is postulated that accumulation of a toxic metabolite, psychosine (galactosylsphingosine), which is also a substrate for the missing enzyme, leads to early destruction of the oligodendrocytes (see below). The total brain content of galactosylceramide is not increased in patients with Krabbe disease, supporting this hypothesis. It should be noted that mammalian tissues contain two genetically distinct lysosomal beta-galactosidases, GALC and GM1-ganglioside beta-galactosidase (see Sect. 8, below). These two enzymes have overlapping substrate specificities in vitro, which can present complications when attempting to make the enzymatic diagnosis of Krabbe disease (Sakai 2009).

The human *GALC* gene has been mapped to chromosome 14q24.3-32.1, and the genomic and/or full-length cDNA sequences have been cloned from multiple species. The human cDNA encodes a 669 amino acid polypeptide with six potential *N*-glycosylation sites. The fully glycosylated precursor polypeptide is ~85 kDa, and is processed into ~50 and 30 kDa subunits. To date, over 70 mutations in the *GALC* gene have been identified in infantile and late-onset GALC-deficient patients (http://www.hgmd.org). Most of these mutations are "private," occurring in individual families, but some may be commonly found in specific regions or ethnicities. Among these, the most common mutation occurs in patients of European ancestry, and deletes a large segment of the *GALC* gene. Several polymorphisms also have been found in the *GALC* gene, and some of these may alter amino acids and thus potentially affect the function of the GALC polypeptide, although not enough to cause GLD.

In general, carriers of *GALC* mutations are considered clinically normal, although one interesting, recent report did show an association between the risk for primary open-angle glaucoma and a heterozygous deletion of the *GALC* gene (Liu et al. 2011). Another interesting recent paper also showed that GALC contributed to the maintenance of a functional hematopoietic stem cell niche, suggesting that genetic alterations in the GALC gene may be associated with hematological conditions as well. Indeed, hematopoietic stem cells retrieved from Twitcher mice (GALC deficient, see below) had defects in their frequency, proliferation, clonogenic potential, and engraftment.

As noted above, in 1972 Miyatake and Suzuki formulated a hypothesis to explain the unusually rapid and complete destruction of oligodendrocytes in GLD. It was already known at that time that psychosine (galactosyl sphingosine), due to its free amino group, was highly cytotoxic, and that this lipid served as a substrate for GALC, but not GM1 beta-galactosidase. In the central nervous system, psychosine is primarily formed within oligodendrocytes, the site of myelin synthesis, and this may explain the selective destruction of these cells in GLD brains. Indeed, in the white matter of brains from Krabbe patients, psychosine levels may be elevated up to 100-fold, while galactosylceramide levels are either normal or only elevated slightly.

GALC deficiency occurs in a number of naturally occurring animal models, including mice, sheep, several breeds of dogs, and the rhesus monkey. The mouse model in particular (also known as the "Twitcher" mouse) has been exploited for numerous therapeutic studies (Luzi et al. 2001). Affected mice develop clinical signs at ~20 days, with stunted growth, twitching, and hind leg weakness. By 40 days they reach a near-terminal stage. Twitcher mice have been treated by ERT, gene therapies, BMT, and other modalities (Neri et al. 2011; Reddy et al. 2011; Lin et al. 2011). BMT in particular may have positive therapeutic effects; however all of the treated mice eventually die of characteristic pathological findings. Thus, this and the other available therapies may slow the disease progress to some degree, but not prevent it.

Since 2006, New York State in the USA has screened all newborns for the occurrence of Krabbe disease. Pilot screening programs are now under way in other states and parts of the world. This screening is accomplished by GALC activity assays on dried blood spots, followed by DNA analyses. The rationale for this screening program is that the early identification of infants with Krabbe disease will permit early BMT and other therapeutic interventions, improving the quality of life (Duffner et al. 2012). However, since BMT has not been curative in the animal models, screening for severe neurological disorders such as this remains controversial.

6 Arylsulfatase A Deficiency: Metachromatic Leukodystrophy

MLD is an autosomal recessively inherited disorder in which the desulfation of 3-0-sulfogalactosylsphingosine or 3-0-sulfogalactosyl glycolipids is deficient. Among the accumulating glycolipids in MLD are sulfatide (galactosyceramide-3-sulfate), lysosulfatide (i.e., similar to sulfatide but lacking the fatty acid in the ceramide moiety), lactosylceramide-3-sulfate, and others. These sulfated glycolipids normally occur in the myelin sheaths in the central and peripheral nervous systems, and to a lesser extent in visceral organs.

The clinical onset and severity of MLD are highly variable (Renaud 2012). The late infantile form is usually recognized in the second year of life and fatal in early childhood. The juvenile form presents between age 4 and puberty; the adult form may occur at any age after puberty. In both of the later onset variants, gait disturbance and mental regression are the earliest signs. In the childhood variants, other common signs are blindness, loss of speech, quadriparesis, peripheral neuropathy, and seizures (Kehrer et al. 2011).

In 1963, Austin et al. reported the deficiency of arylsulfatase A (ASA, E.C. 3.1.6.8) activity in MLD, also known as sulfatide sulfatase, and mutations in this gene (*ARSA*) account for the vast majority of MLD patients. There also are two other minor, genetic causes of MLD. One is due to mutations in the gene (*PSAP*, see Sect. 10, below) encoding the sphingolipid activator protein, saposin B. Mutations in the saposin B encoding region of the *PSAP* gene lead to deficient ASA activity, and a clinical picture indistinguishable from MLD. In fact, these observations in MLD patients provided the first in vivo proof that such sphingolipid activator proteins were required for in vivo activity of this and other sphingolipid hydrolases. In addition to ASA and SAP B mutations, patients with multiple sulfatase deficiency, due to mutations in the gene encoding sulfatase-modifying factor 1 (*SUMF-1*) (see below), also have a reduction in ASA and other sulfatase activities, and develop features of MLD.

As with the other sphingolipid hydrolases, full-length genomic and cDNA sequences encoding ASA have been isolated from several species. The human *ARSA* gene has been mapped to chromosomal region 22q13. It is a remarkably simple gene, and encompasses only 3.2 kb divided into eight exons. As noted above, almost all cases of MLD are due to mutations in this gene, except for the rare cases of *PSAP* or *SUMF-1* mutations. The human *ARSA* gene encodes a 507 amino acid polypeptide precursor that undergoes posttranslational modifications much like other lysosomal enzymes (e.g., *N*-glycosylation). Also, as with all known eukaryotic sulfatases, a formylglycine residue is substituted for cysteine 69 in human ASA. The formylglycine residue is generated by oxidation of the thiol group to an aldehyde. Mutations in the gene encoding the cysteine-modifying enzyme, SUMF-1, lead to the synthesis of inactive sulfatases, and are the cause for multiple sulfatase deficiency (Gieselmann et al. 1994).

Over 100 MLD-causing mutations have been described in the *ARSA* gene. The most common mutation in patients with late-infantile MLD is a G-to-A transition that eliminates the donor splice site at the start of intron 2 (Ługowska et al. 2011). This causes a loss of all ASA enzymatic activity, producing a severe, early-onset phenotype. In contrast, in adults with MLD the most frequent mutation is a C-to-T transition that results in substitution of a leucine for proline in amino acid residue 426. Patients with this mutation have a small amount of residual enzyme activity and therefore exhibit late-onset MLD. Common mutations in specific populations also have been described.

Notably, there also are two mutations that cause "pseudodeficiency" of ASA activity. The term "pseudodeficiency" reflects the fact that these mutations cause severe reduction of ASA activity in vitro, but do not result in MLD (Tinsa et al. 2010). One of these mutations alters an *N*-glycosylation site, but this has no effect on enzyme function. The other affects a polyadenylation site in the mRNA. The pseudodeficiency alleles may be found in ~3–8 % of the normal population, and therefore pose a diagnostic challenge since the artificial in vitro assay systems may reveal very low ASA activity, even though the individual is highly unlikely to develop MLD.

Alterations in the *ARSA* gene also have been linked with the occurrence of several common diseases. For example, as early as 1991, Kappler et al. found that among a small set of patients with multiple sclerosis (MS), there was a higher incidence of the ASA pseudodeficiency alleles than in the general population (Baronica et al. 2011). In addition, variants of the ASA protein and/or gene have been associated with depression and alcoholism. MLD itself also has been associated with several psychiatric disorders, and there is at least one report where polymorphisms in the genes encoding the serotonin and dopamine pathways affected the MLD phenotype (Kumperscak et al. 2008).

Other than humans, naturally occurring mutations causing MLD do not occur in other species. ASA knockout mice have been constructed that exhibit a complete loss of ASA activity and sulfatide storage multiple organs (Geiselmann et al. 1998). While these animals exhibit some neurological deficits, they do not develop the predicted severe phenotype of MLD and live a near-normal life span. They also do not develop demyelinating disease. Thus, these observations in the mouse suggest the existence of compensating pathways in this species that do not exist in humans, perhaps providing a link to novel therapies. Several therapies have been evaluated in the MLD knockout mouse model, including BMT, ERT, and gene therapies (Matthes et al. 2012; Stroobants et al. 2011). BMT also has been extensively used in MLD patients. While it is not effective in the infantile cases, BMT is often recommended for the later onset cases (Solders et al. 1998; Gassas et al. 2011). However, the data supporting an improved clinical outcome in the late-onset MLD cases are fragmented and incomplete, and it is unclear whether the benefit outweighs the potential risks.

7 Alpha Galactosidase A Deficiency: Fabry Disease

Fabry disease in inherited as an X-linked recessive trait, the only such X-linked disease among the sphingolipid storage disorders (Tarabuso 2011; Toyooka 2011). Therefore, all males ("hemizygotes") carrying mutations in the Fabry gene (*GLA*, see below) develop symptoms of the disease, although the severity may vary depending on the nature of the individual mutation. Female heterozygotes who inherit one copy of the mutant gene on one of their X chromosomes also may exhibit features of the disease depending on the pattern of X-inactivation ("lyonization") that occurs in their individual tissues.

The gene responsible for Fabry disease encodes an enzyme, alpha-galactosidase A (E.C. 3.2.1.22), that is required to hydrolyze glycosphingolipids with terminal alpha galactosyl moieties, predominately globotriosylceramide, and to a lesser extent, galabiosylceramide and blood group B substances. Affected males have extensive deposition of these lipids in body fluids and in the lysosomes of endothelial, perithelial, and smooth muscle cells of blood vessels. Deposition also occurs in ganglion cells, and in many cell types in the heart, kidney, eyes, and most other tissues (Aerts et al. 2011).

Clinical manifestations in classically affected hemizygotes with no detectable alpha galactosidase A activity include the onset during childhood or adolescence of pain and paresthesias in the extremities, angiokeratomas in the skin and mucous membranes, and hypohidrosis. Corneal and lenticular opacities also are early findings. With increasing age, proteinuria, hyposthenuria, and lymphedema appear. Severe renal impairment leads to hypertension and uremia. Death usually occurs from renal failure or from cardiac or cerebrovascular disease. Atypical variant males with residual alpha galactosidase A activity may be asymptomatic or have late-onset disease. Heterozygous females may have an attenuated form of the disease depending on the pattern of X-inactivation. They usually are asymptomatic, although rarely present with clinical disease as severe as hemizygous males.

As noted above, in Fabry disease glycosphingolipids with terminal alpha-D-galactosyl residues are not properly broken down due to the alpha galactosidase A enzyme defect, and these lipids therefore accumulate in cells. The predominant accumulating lipid is globotriaosylceramide (Gal(α1-4)Gal(α1-4)Glc(β1-1′)Cer). Since alpha galactosidase A is predominately found in lysosomes, the major site of lipid accumulation is this organelle. Elevated globotriaosylceramide is found in most tissues, but predominantly in kidney.

Other minor accumulating lipids include galabiosylceramide (Galα1-4)(Gal(β1-1′)Cer) and the blood group B and P glycosphingolipids (Linthorst et al. 2004). In human erythrocytes there are two neutral glycosphingolipids with terminal alpha galactosyl residues that inhibit blood group B-specific hemagglutination. The structure of these blood group B glycosphingolipids has been determined, and they are found at high levels in Fabry patients. This raises an interesting and unique point about Fabry disease. Fabry hemizygotes and heterozygotes who have blood group B and AB accumulate four glycosphingolipids, while those who are blood group A or O only accumulate two (globotriaosylceramide and galabiosylceramide). A fifth neutral glycosphingolipid that can accumulate in Fabry is the P1 blood group antigen, which also has terminal alpha galactosyl residues.

In man, there are actually two alpha galactosidases (alpha galactosidases A and B). Fabry disease is caused by mutations in the gene (*GLA*) encoding the "A" type (see below) (Tomasic et al. 2010). However, when measuring alpha galactosidase activity in blood from Fabry patients in vitro, classic hemizygotes may exhibit up to 25 % of normal activity due to the presence of the "B" form. Alpha galactosidase B is encoded by a distinct gene (*NAGA*), and is now known to be an alpha-*N*-acetylgalactosaminidase that recognizes these moieties in glycoproteins, complex carbohydrates, and other molecules. Because the artificial substrates routinely used to measure alpha galactosidase activity do not readily distinguish these enzymes, substantial misdiagnosis can occur by simple enzyme testing.

The human gene encoding alpha galactosidase A resides on chromosomal region Xq22. It encompasses ~12 kb and contains seven exons. The full-length human cDNA contains a 1,290 bp open reading frame that encodes an unglycosylated precursor polypeptide of ~48 kDa. As with most lysosomal hydrolases, there are several *N*-glycosylation sites in the predicted polypeptide sequence and the oligosaccharide chains undergo mannose-6-phosphate modifications, which facilitates targeting of the enzyme to lysosomes. The mature, fully glycosylated alpha galactosidase A in lysosomes is ~51 kDa. Genomic and cDNA sequences encoding alpha galactosidases have now been isolated from many species. An unusual feature of most human alpha galactosidase A cDNAs is the lack of a 3′ untranslated region. The polyadenylation signal sequence is often actually found in the coding region, 12 bp from the termination codon, followed by a poly(A) tract.

The full-length cDNA encoding human alpha galactosidase A had been used to express and purify the recombinant enzyme from CHO cells. This enzyme was extensively characterized and used in the Fabry disease mouse model to evaluate ERT (see below). Based on these results and after extensive clinical testing, a recombinant enzyme drug (Fabrazyme) was approved for use in human Fabry disease patients in 2003. This represents the second sphingolipid storage disorder for which ERT became available (Gaucher disease was the first). ERT in Fabry patients reduces pain, proteinuria, and endothelial cell storage of glycolipids, and improves the quality of life for Fabry patients (Rozenfeld and Neumann 2011). Since no naturally occurring animal model of Fabry disease has been found, the preclinical evaluation of ERT was performed in a knockout mouse model. These animals exhibit age-dependent glycosphingolipid storage, but there are only a few clinical findings of Fabry disease.

Numerous mutations have been found in the *GLA* gene causing Fabry disease, spread throughout the enzyme-coding region (Schaefer et al. 2005). Several regulatory (promoter) and splice site mutations also have been found. As noted above, the presence of one *GLA* mutation in hemizygous males will cause disease, but the severity will depend on the effect of this mutation on the residual alpha galactosidase A activity. In contrast, in female heterozygotes, because of X-inactivation, only some may develop disease (Bouwman et al. 2012). It should be noted that such "Fabry" females carrying single mutations will be essentially "mosaics" for alpha galactosidase A, with some cells expressing the mutant allele and others the normal allele (due to random inactivation of one X chromosome). Because the enzyme can be released from cells at low levels (as with all lysosomal hydrolases), and then taken up by neighboring cells, "cross-correction" of the lipid storage is possible. Aside from Fabry disease, there has been some evidence that mutations in the *GLA* gene also may be a previously unrecognized cause of stroke. Indeed, newborn screening programs have revealed that in some populations the incidence of *GLA* mutations is unexpectedly high, suggesting that this gene/disease association should be studied further as a potential risk factor for this and other common disorders of the vasculature (Spáčil et al. 2011).

8 Beta Galactosidase Deficiency: GM1 Gangliosidosis

An inherited deficiency of lysosomal acid beta galactosidase (E.C. 3.2.1.23) is expressed as two clinically distinct diseases, GM1 gangliosidosis and Morquio B disease (Brunetti-Pierri and Scaglia 2008; Morita et al. 2009; Caciotti et al. 2011). GM1 gangliosidosis affects both neurological and somatic tissues, occurring mainly in early infancy (type 1). Developmental arrest is usually observed a few months after birth, followed by progressive neurologic deterioration and generalized rigospasticity with sensorimotor and psycho/intellectual dysfunction. As with many of the other neurodegenerative lysosomal storage diseases, cherry-red maculae are common, as is facial dysmorphism, hepatosplenomegaly, and generalized skeletal dysplasia. Later onset, juvenile and adult forms of beta galactosidase deficiency also have been described (types 2 and 3, respectively). They are observed as progressive neurologic diseases in childhood or in young adults. Dysmorphic skeletal changes are less prominent or absent in these clinical forms. A protracted course, mainly presenting as dystonia, is the major neurological manifestation in adults with GM1 gangliosidosis.

Morquio B disease is clinically expressed as a generalized skeletal dysplasia with corneal clouding, resulting in short stature, pectus carinatum, platyspondylia, odontoid hypoplasia, kyphoscoliosis, and genu valgum. There is no central nervous system involvement, although spinal cord compression may occur at late stages of the disease. Intelligence is normal and hepatosplenomegaly is not present.

In man, two lysosomal enzymes are known to be responsible for the hydrolysis of terminal beta-linked galactose residues at acidic pH in various glycoconjugates. One is the enzyme being discussed here, beta galactosidase (E.C. 3.2.1.23), and the other is galactocerebrosidase (E.C. 3.2.1.46), whose primary substrates include galactosylceramide, galactosylsphingosine, and other lipids. This is the enzyme deficient in globoid cell leukodystrophy (Krabbe disease), and was discussed in the section above. Beta galactosidase activity is severely deficient in cells and tissues from patients with both GM1 gangliosidosis and Morquio B disease, and both diseases are now known to be due to genetic mutations in the same gene (*GLB1*) (Hofer et al. 2009). The primary substrates for beta galactosidase include galactose-containing oligosaccharides (e.g., keratan sulfate), and GM1 ganglioside. The asialo derivative of GM1, GA1 ganglioside, also may be a substrate for this enzyme.

As expected, tissues from patients with beta galactosidase deficiency exhibit a broad spectrum of accumulating galactose-containing macromolecules, including GM1, GA1, and keratan sulfate. Gangliosides are normal components of plasma membranes, concentrated in neuronal membranes, especially in the regions of nerve endings and dendrites. GM1 is the major ganglioside in the brains of vertebrates. Gangliosides display a broad spectrum of interactions, and may act as binding molecules for toxins and hormones, and also are involved in cell differentiation, cell–cell interactions, and cell signaling (see below).

In the brains of patients with GM1 gangliosidosis, storage of GM1 is the most prominent observation (Nada et al. 2011). The accumulating GM1 has the same fatty acid composition, sugar composition and sequence, and glycosidic linkages as normal GM1. Visceral organs also show storage of GM1 ganglioside, but to a lesser degree. Also to a lesser degree, GA1 ganglioside may accumulate in the brain of GM1 gangliosidosis patients, as do glycosylceramide, lactosylceramide, and GM2 ganglioside. White matter shows chemical manifestations of myelin breakdown, including low proteolipid protein, low total lipid, and the presence of esterified cholesterol (Steenweg et al. 2010).

Other than gangliosides, the other major storage molecule in beta galactosidase-deficient patients is keratan sulfate (Ohto et al. 2012). This glycosaminoglycan normally exists in a proteoglycan linked with chondrotin sulfate. It is found primarily within connective tissues, particularly cartilage, explaining the skeletal dysplasia typical of many beta galactosidase-deficient patients. After proteolysis and release from the proteoglycan, free keratan sulfate chains are hydrolyzed by a series of exoenzymes, including beta galactosidase.

It should be noted that another protein, referred to as protective protein/cathepsin A (PPCA), has been associated with both beta galactosidase and another lysosomal hydrolase, neuraminidase, and stabilizes and in some cases activates the enzymes (Tatano et al. 2006). A genetic deficiency of this protective protein results in combined deficiency of beta galactosidase and neuraminidase (galactosialidosis). The gene encoding PPCA is distinct from both of these enzymes, and is not discussed further here.

The gene encoding beta galactosidase (*GLB1*) is located in chromosomal region 3p21.33. Genes encoding this enzyme also have been isolated from mice and several other species. The full-length human *GLB1* cDNA encodes a 677 amino acid polypeptide with beta galactosidase activity. Mutations in the *GLB1* gene are responsible for both GM1 gangliosidosis and Morquio B disease. As with other lysosomal diseases, most are point mutations altering single amino acids, although small deletions, splice site mutations, and other alterations also have been identified. To date, no correlation has been made between the location or type of *GLB1* mutations and the occurrence of GM1 gangliosidosis or Morquio B disease.

The pathogenesis of beta galactosidase deficiency is complex. GM1 ganglioside stimulates neurite outgrowth and affects neuronal differentiation, and enhances the action of nerve growth factor (van der Voorn et al. 2004). Golgi and electron microscopic studies of cortical neurons from this and several other neurological lysosomal diseases exhibit large neural processes (meganeurites), which may be explained by GM1 accumulation. The extent of meganeurite development is related to the onset, severity, and clinical course of the disease. In other studies, significant changes in neuronal connectivity were found in the cerebral cortex of animals with beta galactosidase deficiency, and cholinergic function was altered as well. Phosphoinositol-specific phospholipase C and adenyl cyclase activities were altered in the membranes of cerebral cortex from GM1 gangliosidosis cats. In addition, the beta subunit of cholera toxin, which binds specifically to GM1 in the outer leaflet of the cell membrane, was found to induce an increase of calcium and manganese influx in N18 cells, and this has been associated with the activation of an L-type voltage-dependent calcium channel. This channel has important implications for neural development, and its reliance on GM1 ganglioside is also likely related to the pathogenesis of GM1 storage. These and other data suggest that various morphologic and metabolic aberrations occur in the brains of GM1 gangliosidosis patients and animals, and can be linked to the biology of GM1.

Naturally occurring animal models of GM1 gangliosidosis have been reported in cats, dogs, sheep, and calves. The cat model in particular has been studied in some detail, and is a neurologic disorder with clinical, morphological, biochemical, and genetic similarities to human GM1 gangliosidosis (Claro et al. 1991). Affected kittens appear normal at birth, but tremors of the head and hind limbs are first noted at 2–3 months of age, followed by generalized dysmetria and spastic quadriplegia. GM1 ganglioside and other galactose-containing oligosaccharides accumulate in the tissues of affected animals. A genetic knockout of beta galactosidase activity also has been generated in mice by gene targeting (Itoh et al. 2001). Progressive and severe neurodegeneration occurs in these animals, and GM1 and GA1 gangliosides accumulate in the brain. Unlike patients, GA1 ganglioside may accumulate in beta galactosidase deficiency mice to a greater extent than GM1, which may be explained by the fact that there is a unique neuraminidase in mice that converts GM1 to GA1. Skeletal dysplasia, which is characteristic of patients with beta galactosidase deficiency, does not occur in the mice. This also is due to the unique metabolism of keratan sulfate in mice compared to humans.

9 Hexosaminidase A and B Deficiency: GM2 Gangliosidoses

The GM2 gangliosidoses are a group of genetic disorders caused by excessive accumulation of GM2 ganglioside and related glycolipids in lysosomes, mainly of neuronal cells (Maegawa 2012). The general cause of GM2 gangliosidosis is deficient activity of the enzyme beta hexosaminidase (E.C. 3.2.1.30). The enzymatic hydrolysis of GM2 ganglioside by this enzyme requires that it be complexed with a substrate-specific cofactor, the GM2 activator. There are two isozymes of beta hexosaminidase, Hex A, which is composed of alpha and beta subunits, and Hex B, which is composed of two beta subunits. Hex A and B can only act on the GM2/GM2 activator complex.

Defects in any of the three genes can lead to GM2 gangliosidosis, *HEXA*, which encodes the alpha subunit of Hex A, *HEXB*, which encodes the beta subunit of Hex A and Hex B, or *GM2A*, which encodes the GM2 activator protein (Cordeiro et al. 2000). Three clinical forms of GM2 gangliosidosis occur from mutations in these genes: Tay–Sachs disease and variants, resulting from mutations of the *HEXA* gene, and resulting in deficient activity of Hex A; Sandhoff disease and variants, resulting from mutations of the *HEXB* gene, and resulting in deficient activity of both Hex A and Hex B; and GM2 activator deficiency, caused by mutations in the *GM2A* gene and characterized by normal Hex A and Hex B, but the inability to form a functional ganglioside GM2/GM2 activator complex.

GM2 gangliosidosis can present with a wide spectrum of severity. These range from infantile-onset, rapidly progressive neurodegenerative disease culminating in death before the age of 4 years (classical Tay–Sachs and Sandhoff diseases and GM2 activator deficiency) to adult-onset, slowly progressive neurological conditions compatible with long survival with little or no effect on intellect (Bley et al. 2011; Smith et al. 2012; Jain et al. 2010; Osher et al. 2011). The clinical phenotypes of the acute, infantile form of any of the three genetic GM2 gangliosidosis (*HEXA*, *HEXB*, or *GM2A* mutations) are essentially indistinguishable. Affected infants generally appear normal at birth. The earliest signs of disease are often mild motor weakness, beginning at 3–5 months of age. An exaggerated startle response is also commonly observed at this early stage as well. Soon after, regression and loss of already acquired mental and motor skills become obvious. The disease is rapidly progressive and leads to death in early childhood.

As noted above, later onset forms of GM2 gangliosidosis also may occur. The clinical phenotype varies widely among these patients, and onset can occur at any time from the late infantile period to adults. In general, in the later onset cases involvement of the deeper brain structures is more prominent compared to the overwhelming generalized gray matter involvement in the infantile form. Manifestations include dystonia, other extrapyramidal signs such as ataxia, choreoathetoid movements, signs of spinocerebellar degeneration, and ALS-like motor neuron involvement. Psychotic manifestations are not uncommon as well. It should also be noted that in a small number of later onset cases, mental capacity is well preserved, although severe dysarthria often masks the preserved intelligence.

Like most lysosomal glycosidases, beta hexosaminidase hydrolyzes a broad spectrum of substrates. It is specific for only the terminal nonreducing sugar (GlcNac or GalNac) in beta linkage. The primary substrates for Hex A and Hex B are glycoproteins, oligosaccharides, glycosaminoglycans, and glycolipids, including the ganglioside GM2 when complexed with the GM2 activator protein. It should be noted that a minor isoform composed of two alpha subunits (Hex S) also has been identified, but the biochemical and clinical significance of this isoform remains unclear. Both the alpha and beta subunits of beta hexosaminidase possess an active site, although dimerization is required for activity. The substrate specificity of the two subunits differs however; the beta subunit prefers neutral, water-soluble substrates, while the alpha subunit also hydrolyzes negatively charged substrates such as GM2 gangliosides or glycosaminoglycans. The fact that mutations in the *HEXA* gene, in which the Hex B isozyme (composed of two beta subunits) is normal, lead to accumulated GM2 ganglioside demonstrates that the preferred substrate for the alpha subunit is GM2 (see below). However, as noted above, the Hex A isozyme can only hydrolyze GM2 if complexed with the GM2 activator protein.

Mutations in the alpha subunit encoding the *HEXA* gene are therefore primarily characterized by accumulation of GM2 ganglioside. Indeed, GM2 ganglioside is the primary neuronal storage material in all three genetic causes, hence their grouping as GM2 gangliosidoses. *HEXA* mutations are responsible for classical Tay–Sachs disease and the later onset Tay–Sachs variants. The pathology of classical Tay–Sachs disease has been described extensively (add ref). The brain is atrophic during the early stage, but usually increases over time until death. Histological analysis shows classical neuronal storage disease in essentially all neurons of the CNS and the peripheral nervous system as well, including swollen retinal ganglion cells. The membranous storage bodies characteristic of Tay–Sachs disease upon electron microscopic examination are composed of cholesterol, phospholipid, and GM2 ganglioside. Increased apoptotic neurons also have been observed. The neuronal pathology in the acute or subacute forms of Tay–Sachs disease is generally less severe than that in the infantile form, and tends to be more prominent in the hypothalamus, cerebellum, brain stem, and spinal cord.

In the Sandhoff disease variants, due to mutations in the *HEXB* gene, the pathological and clinical findings are very similar to those of Tay–Sachs (Maegawa 2012). However, yellowing of the cerebral cortex and deeper structures has been documented in Sandhoff disease, possibly due to accumulated asialoganglioside. Also, additional accumulation of sphingoglycolipids with a terminal hexosamine residue and fragments of undigested glycoprotein in systemic organs has been found as well. Patients with *GM2A* mutations are indistinguishable from those with *HEXA* or *HEXB* mutations.

The *HEXA* gene, encoding the alpha subunit, is located within the chromosomal region 15q23–q24. It is ~35 kb in length and contains 14 exons. The *HEXB* gene, encoding the beta subunit, is located at chromosomal region 5q13 and is ~45 kb in length that is divided into 14 exons. Analysis of these genes strongly suggests that they arose from a common ancestor. They share a very common exon structure, and in both genes the promoter activity resides within ~150 bp of the initiating ATG codon. Overall, the alpha and beta subunits share ~57 % amino acid identity. The *GM2A* gene maps to chromosomal region 5q31.1–31.3. In addition, an evolutionarily related pseudogene, *GM2AP*, maps to chromosome 3. The *GM2A* gene is ~16 kb and contains four exons.

Numerous mutations have been described in the *HEXA* gene responsible for Tay–Sachs disease and its variants. Infantile Tay–Sachs disease occurs most commonly among individuals of Ashkenazi Jewish ancestry (carrier frequency ~1:25), and there are two mutations that are the most frequent; one is a 4 bp insertion in exon 11, and the second is a donor splice site mutation in intron 11 (Frisch et al. 2004). Both of these mutations result in severely deficient or absent *HEXA* mRNA expression, leading to absent Hex A and Hex S activities. Another common HEXA mutation causing infantile Tay–Sachs disease occurs in French Canadian patients, and is due to a deletion that extends from ~2 kb upstream of the 5′ end of the gene into intron 1 (De Braekeleer et al. 1992; Martin et al. 2007). This also results in the absence of Hex A mRNA and protein. Many other mutations in the *HEXA* gene have been described causing classical Tay–Sachs disease, effecting protein processing, catalytic activity, and/or mRNA processing. In addition, mutations causing the later onset forms are known, and generally result in expression of *HEXA* mRNA and residual enzyme activity. An important *HEXA* mutation also is known as the B1 variant, in which there is normal activity towards the artificial substrates generally used to measure beta hexosaminidase activity, but no activity of the mutant enzyme towards GM2 ganglioside. Individuals with this mutation pose a unique diagnostic challenge since they will appear enzymatically normal, but develop severe disease. Many mutations in the *HEXB* gene causing Sandhoff disease and its variants also have been described. A much small number of mutations in the *GM2A* gene are known. As noted above, the fact that mutations in this gene caused a severe clinical disease provided the first proof of the physiological significance of the GM2/GM2 activator complex (Peleg et al. 1995; Tanaka et al. 2003).

Naturally occurring mutations causing GM2 gangliosidosis have been described in dogs, cats, and pigs (Yamato et al. 2002; Kosanke et al. 1979; Bradbury et al. 2009).

In all species, abundant neuronal storage is the major finding. Visceral storage is only found in the feline model however. Meganurite formation, which has been observed in human GM2 gangliosidosis patients, occurs in the dog and cat models. A number of groups also have independently generated murine models of Tay–Sachs, Sandhoff, and GM2 activator deficiency in mice using gene targeting strategies. In general, knockout of the *HEXA* gene in mice (Tay–Sachs disease) results in a much milder neurological disease than predicted from the human cases (Gulinello et al. 1994). Mutations in the *HEXB* gene in mice result in a more severe phenotype, with the onset of clinical disease at ~3 months associated with excessive neuronal GM2 storage (Yamanaka et al. 2008). Mutations in the *GM2A* gene of mice also resulted in a much milder disease than predicted (Phaneuf et al. 1996).

In part, the distinction between the mouse models and the human disorder is due to the distinct degradation of GM2 ganglioside by these species (Baek et al. 2009). In humans, GM2 is degraded nearly exclusively by Hex A in collaboration with the activator protein to yield the ganglioside GM3. In contrast, in mice GM2 can be degraded by two different pathways. One is identical to the human pathway, and the second is essentially unique to the mouse and is initiated by sialidase acting on GM2 to yield GA2 ganglioside. The GA2 is then degraded by either Hex A with activator protein or to a lesser extent Hex B. This explains the mild phenotype due to alpha subunit mutations (*HEXA*), since the Hex B can act on the GA2 produced by sialidase. In contrast, *HEXB* mutations result in deficiency of both enzymes, and a complete block of degradation.

10 Sphingolipid Activator Proteins

As discussed in the sections above, the lysosomal degradation of sphingolipids requires small nonenzymatic glycoproteins, referred to as "sphingolipid activator proteins" (SAPs) (Sandhoff and Kolter 1998; Kolter and Sandhoff 2010). These include the saposin proteins (SAP A–D, and GM2 activator protein). There are two genes responsible for these proteins. One is *GMA2*, which encodes the GM2 activator protein and is located on chromosome 5 (Wendeler et al. 2003), and the other is *PSAP*, located on chromosome 10 and encoding the saposin precursor protein and the individual SAP A–D, which are derived from proteolytic cleavage of the precursor (Misasi et al. 2009). The GM2 activator protein and resulting mutations (responsible for the "A/B" Tay–Sachs variant) were discussed in the section above and will not be discussed further here.

The *PSAP* gene encodes the SAP precursor protein (prosaposin), with a total of 524 amino acids and five *N*-glycosylation sites (Misasi et al. 2009). There are four homologous domains located within the precursor protein, each of ~80 amino acids. A major portion of the newly synthesized precursor is exported to the cell surface and then imported into the lysosomal compartment, where it is processed to the mature SAP A–D proteins. Notably, unlike most lysosomal proteins the prosaposin protein is transported to the lysosome via its interaction with the alternative receptor, sortilin (Wähe et al. 2010; Yuan and Morales 2011). The occurrence of the non-processed SAP precursor in body fluids and its neurotrophic properties indicates that its function may not be restricted to being the precursor of the individual SAPs. For example, prosaposin has been found in milk, serum, and seminal fluid, and treatment of cell lines with PSAP increased cell survival and was anti-apoptotic. Indeed, serum prosapsin levels are increased in patients with advanced prostate cancer. Of interest is whether these effects of PSAP on cell survival are direct effects of the protein itself, or rather the effects of the individual processed SAPs on the activation of the sphingolipid hydrolases.

In vitro, SAP A stimulates beta glucocerebrosidase (the enzyme deficient in Gaucher disease) and beta galactocerebrosidase (the enzyme deficient in Krabbe disease) activities in the presence of detergents. SAP B is a nonspecific glycolipid-binding protein that stimulates the hydrolysis of ~20 or more glycolipids by different enzymes in vitro, including the hydrolysis of sulfatide by ASA (the enzyme deficient in MLD). SAP C stimulates the in vitro activities of glucosyl- and galactosylcerebrosidases, as well as sphingomyelin by ASM (the enzyme deficient in Types A and B Niemann–Pick disease) (Kölzer et al. 2004). SAP D is required for the degradation of ceramide by AC (the enzyme deficient in Farber disease) (Linke et al. 2001).

Analysis of mutations in the genes encoding the *PSAP* protein has provided important insights regarding the role of the SAP proteins for the in vivo hydrolysis of sphingolipids, and their relationship to human disease. Of note, a SAP precursor deficiency was found in a child who died at 16 weeks and who was homozygous for a mutation in the initiation codon of the *PSAP* gene. The complete absence of the precursor protein and four resultant SAPs led to a generalized accumulation of ceramide, glucosylceramide, galactosylceramide, sulfatide, lactosylceramide, digalactosylceramide, and other sphingolipids. The symptoms of the disease resembled those of type 2 Gaucher disease. Other point mutations within the region of the *PSAP* gene encoding SAP B caused accumulation of sulfatide and some other sphingolipids, and a clinical course resembling that of juvenile MLD. Mutations in the SAP C region only led to accumulation of glucosylceramide, similar to juvenile Gaucher disease (Vaccaro et al. 2010). No sphingomyelin storage was observed in individuals with SAP C mutations, and this has been explained by the fact that there is a SAP-like domain within ASM that compensates for the loss of SAP C in these patients. Patients with mutations in the SAP A region of the *PSAP* gene resemble those with Krabbe disease (GLD), and accumulate galactosylcerebroside within the CNS and peripheral nervous system (Grossi et al. 2008). There are no human patients with SAP D mutations, but the requirement by AC for SAP D has been confirmed in vivo in mutant mice, which accumulate ceramides with alpha hydroxylated fatty acids mainly in the brain and kidney, and develop Purkinje cell loss and defects in the urinary system (Oya et al. 1998; Tohyama et al. 1999).

A mouse model of SAP precursor deficiency also has been constructed by gene targeting, and the clinical and biochemical features resembled those of the human disease (Matsuda et al. 2007). Mice developed massive sphingolipid storage and neurological disease by ~20 days, and died by ~35–38 days. Several knock-in models expressing specific human mutations also have been constructed, confirming the in vivo importance of these proteins (Sun et al. 2008, 2010).

References

Abian O, Alfonso P, Velazquez-Campoy A, Giraldo P, Pocovi M, Sancho J (2011) Therapeutic strategies for Gaucher disease: miglustat (NB-DNJ) as a pharmacological chaperone for glucocerebrosidase and the different thermostability of velaglucerase alfa and imiglucerase. Mol Pharm 8(6):2390–2397

Aerts JM, Boot RG, van Eijk M, Groener J, Bijl N, Lombardo E, Bietrix FM, Dekker N, Groen AK, Ottenhoff R, van Roomen C, Aten J, Serlie M, Langeveld M, Wennekes T, Overkleeft HS (2011) Glycosphingolipids and insulin resistance. Adv Exp Med Biol 721:99–119

Ahmad A, Mazhar AU, Anwar M (2009) Farber disease: a rare neurodegenerative disorder. J Coll Physicians Surg Pak 19(1):67–68

Anheim M, Elbaz A, Lesage S, Durr A, Condroyer C, Viallet F, Pollak P, Bonaïti B, Bonaïti-Pellié C, Brice A (2012) Penetrance of Parkinson disease in glucocerebrosidase gene mutation carriers. Neurology 78(6):417–420

Baek RC, Martin DR, Cox NR, Seyfried TN (2009) Comparative analysis of brain lipids in mice, cats, and humans with Sandhoff disease. Lipids 44(3):197–205

Baronica KB, Mlinac K, Ozretić D, Vladić A, Bognar SK (2011) Arylsulfatase a gene polymorphisms in relapsing remitting multiple sclerosis: genotype-phenotype correlation and estimation of disease progression. Coll Antropol 35(Suppl 1):11–16

Becker KA, Grassmé H, Zhang Y, Gulbins E (2010) Ceramide in Pseudomonas aeruginosa infections and cystic fibrosis. Cell Physiol Biochem 26(1):57–66

Beckham TH, Lu P, Cheng JC, Zhao D, Turner LS, Zhang X, Hoffman S, Armeson KE, Liu A, Marrison T, Hannun YA, Liu X (2012) Acid ceramidase-mediated production of sphingosine 1-phosphate promotes prostate cancer invasion through upregulation of cathepsin B. Int J Cancer 131:2034–2043. doi:10.1002/ijc.27480

Benito JM, García Fernández JM, Ortiz Mellet C (2011) Pharmacological chaperone therapy for Gaucher disease: a patent review. Expert Opin Ther Pat 21(6):885–903

Bikman BT, Summers SA (2011) Sphingolipids and hepatic steatosis. Adv Exp Med Biol 721: 87–97

Bley AE, Giannikopoulos OA, Hayden D, Kubilus K, Tifft CJ, Eichler FS (2011) Natural history of infantile G(M2) gangliosidosis. Pediatrics 128(5):e1233–e1241

Bouwman MG, Rombach SM, Schenk E, Sweeb A, Wijburg FA, Hollak CE, Linthorst GE (2012) Prevalence of symptoms in female Fabry disease patients: a case-control survey. J Inherit Metab Dis 35:891–898

Bradbury AM, Morrison NE, Hwang M, Cox NR, Baker HJ, Martin DR (2009) Neurodegenerative lysosomal storage disease in European Burmese cats with hexosaminidase beta-subunit deficiency. Mol Genet Metab 97(1):53–59

Brown JT, Lahey C, Laosinchai-Wolf W, Hadd AG (2006) Polymorphisms in the glucocerebrosidase gene and pseudogene urge caution in clinical analysis of Gaucher disease allele c.1448T > C (L444P). BMC Med Genet 7:69

Brunetti-Pierri N, Scaglia F (2008) GM1 gangliosidosis: review of clinical, molecular, and therapeutic aspects. Mol Genet Metab 94(4):391–396

Caciotti A, Garman SC, Rivera-Colón Y, Procopio E, Catarzi S, Ferri L, Guido C, Martelli P, Parini R, Antuzzi D, Battini R, Sibilio M, Simonati A, Fontana E, Salviati A, Akinci G, Cereda C, Dionisi-Vici C, Deodato F, d'Amico A, d'Azzo A, Bertini E, Filocamo M, Scarpa M, di Rocco M, Tifft CJ, Ciani F, Gasperini S, Pasquini E, Guerrini R, Donati MA, Morrone A (2011) GM1 gangliosidosis and Morquio B disease: an update on genetic alterations and clinical findings. Biochim Biophys Acta 1812(7):782–790

Campbell TN, Choy FY (2012) Gaucher disease and the synucleinopathies: refining the relationship. Orphanet J Rare Dis 7:12

Chen H, Tran JT, Brush RS, Saadi A, Rahman AK, Yu M, Yasumura D, Matthes MT, Ahern K, Yang H, LaVail MM, Mandal MN (2012) Ceramide signaling in retinal degeneration. Adv Exp Med Biol 723:553–558

Choy FY, Campbell TN (2011) Gaucher disease and cancer: concept and controversy. Int J Cell Biol 2011:150450

Clarke CJ, Wu BX, Hannun YA (2011) The neutral sphingomyelinase family: identifying biochemical connections. Adv Enzyme Regul 51(1):51–58

Claro E, Wallace MA, Fain JN, Nair BG, Patel TB, Shanker G, Baker HJ (1991) Altered phosphoinositide-specific phospholipase C and adenylyl cyclase in brain cortical membranes of cats with GM1 and GM2 gangliosidosis. Brain Res Mol Brain Res 11(3–4):265–271

Cordeiro P, Hechtman P, Kaplan F (2000) The GM2 gangliosidoses databases: allelic variation at the HEXA, HEXB, and GM2A gene loci. Genet Med 2(6):319–327

Cox TM, Cachón-González MB (2012) The cellular pathology of lysosomal diseases. J Pathol 226(2):241–254

De Braekeleer M, Hechtman P, Andermann E, Kaplan F (1992) The French Canadian Tay-Sachs disease deletion mutation: identification of probable founders. Hum Genet 89(1):83–87

Duffner PK, Barczykowski A, Jalal K, Yan L, Kay DM, Carter RL (2011) Early infantile Krabbe disease: results of the World-Wide Krabbe Registry. Pediatr Neurol 45(3):141–148

Duffner PK, Granger C, Lyon N, Niewczyk P, Barczykowski A, Bauer S (2012) Msall ME (2012) developmental and functional outcomes in children with a positive newborn screen for Krabbe disease: a pilot study of a phone-based interview surveillance technique. J Pediatr 161:258–263

Eliyahu E, Shtraizent N, Martinuzzi K, Barritt J, He X, Wei H, Chaubal S, Copperman AB, Schuchman EH (2010) Acid ceramidase improves the quality of oocytes and embryos and the outcome of in vitro fertilization. FASEB J 24(4):1229–1238

Farber S, Cohen J, Uzman LL (1957) Lipogranulomatosis; a new lipo-glycoprotein storage disease. J Mt Sinai Hosp NY 24(6):816–837

Farfel-Becker T, Vitner EB, Futerman AH (2011) Animal models for Gaucher disease research. Dis Model Mech 4(6):746–752

Flowers M, Fabriás G, Delgado A, Casas J, Abad JL, Cabot MC (2012) C6-Ceramide and targeted inhibition of acid ceramidase induce synergistic decreases in breast cancer cell growth. Breast Cancer Res Treat 133:447–458

Frisch A, Colombo R, Michaelovsky E, Karpati M, Goldman B, Peleg L (2004) Origin and spread of the 1278insTATC mutation causing Tay-Sachs disease in Ashkenazi Jews: genetic drift as a robust and parsimonious hypothesis. Hum Genet 114(4):366–376

Furuya H, Shimizu Y, Kawamori T (2011) Sphingolipids in cancer. Cancer Metastasis Rev 30 (3–4):567–576

Gassas A, Raiman J, White L, Schechter T, Clarke J, Doyle J (2011) Long-term adaptive functioning outcomes of children with inherited metabolic and genetic diseases treated with hematopoietic stem cell transplantation in a single large pediatric center: parents' perspective. J Pediatr Hematol Oncol 33(3):216–220

Geiselmann V, Matzner, U, Hess B, Lullmann-Rauch R, Coenen R, Hartmann D, D'Hooge R, DeDeyn P, Nagels G (1998) Metachromatic leukodystrophy: molecular genetics and an animal model. J. Inherit Metab Dis:564–574

Gieselmann V, Zlotogora J, Harris A, Wenger DA, Morris CP (1994) Molecular genetics of metachromatic leukodystrophy. Hum Mutat 4(4):233–242

Goker-Alpan O (2011) Therapeutic approaches to bone pathology in Gaucher disease: past, present and future. Mol Genet Metab 104(4):438–447

Grossi S, Regis S, Rosano C, Corsolini F, Uziel G, Sessa M, Di Rocco M, Parenti G, Deodato F, Leuzzi V, Biancheri R, Filocamo M (2008) Molecular analysis of ARSA and PSAP genes in twenty-one Italian patients with metachromatic leukodystrophy: identification and functional characterization of 11 novel ARSA alleles. Hum Mutat 29(11):E220–E230

Gulinello M, Chen F, Dobrenis K (1994) Targeted disruption of the Hexa gene results in mice with biochemical and pathologic features of Tay-Sachs disease. Proc Natl Acad Sci U S A 91(21):9975–9979

Hofer D, Paul K, Fantur K, Beck M, Bürger F, Caillaud C, Fumic K, Ledvinova J, Lugowska A, Michelakakis H, Radeva B, Ramaswami U, Plecko B, Paschke E (2009) GM1 gangliosidosis and Morquio B disease: expression analysis of missense mutations affecting the catalytic site of acid beta-galactosidase. Hum Mutat 30(8):1214–1221

Horinouchi K, Erlich S, Perl DP, Ferlinz K, Bisgaier CL, Sandhoff K, Desnick RJ, Stewart CL, Schuchman EH (1995) Acid sphingomyelinase deficient mice: a model of types A and B Niemann-Pick disease. Nat Genet 10(3):288–293

Hruska KS, LaMarca ME, Scott CR, Sidransky E (2008) Gaucher disease: mutation and polymorphism spectrum in the glucocerebrosidase gene (GBA). Hum Mutat 29(5):567–583

Hughes DA, Pastores GM (2010) The pathophysiology of GD - current understanding and rationale for existing and emerging therapeutic approaches. Wien Med Wochenschr 160 (23–24):594–599

Itoh M, Matsuda J, Suzuki O, Ogura A, Oshima A, Tai T, Suzuki Y, Takashima S (2001) Development of lysosomal storage in mice with targeted disruption of the beta-galactosidase gene: a model of human G(M1)-gangliosidosis. Brain Dev 23(6):379–384

Jain A, Kohli A, Sachan D (2010) Infantile Sandhoff's disease with peripheral neuropathy. Pediatr Neurol 42(6):459–461

Kehrer C, Blumenstock G, Gieselmann V, Krägeloh-Mann I, Leukonet G (2011) The natural course of gross motor deterioration in metachromatic leukodystrophy. Dev Med Child Neurol 53(9):850–855

Kinghorn KJ (2011) Pathological looping in the synucleinopathies: investigating the link between Parkinson's disease and Gaucher disease. Dis Model Mech 4(6):713–715

Kohlschütter A, Eichler F (2011) Childhood leukodystrophies: a clinical perspective. Expert Rev Neurother 11(10):1485–1496

Kolter T, Sandhoff K (2010) Lysosomal degradation of membrane lipids. FEBS Lett 584(9): 1700–1712

Kölzer M, Ferlinz K, Bartelsen O, Hoops SL, Lang F, Sandhoff K (2004) Functional characterization of the postulated intramolecular sphingolipid activator protein domain of human acid sphingomyelinase. Biol Chem 385(12):1193–1195

Kosanke SD, Pierce KR, Read WK (1979) Morphogenesis of light and electron microscopic lesions in porcine GM2-gangliosidosis. Vet Pathol 16(1):6–17

Kumperscak HG, Dolzan V, Videtic A, Plesnicar BK (2008) Polymorphisms in genes encoding the serotonin and dopamine pathways in two sisters with metachromatic leukodystrophy. J Int Med Res 36(5):1123–1128

Lachmann RH (2011) Enzyme replacement therapy for lysosomal storage diseases. Curr Opin Pediatr 23(6):588–593

Lee JY, Lee BH, Kim GH, Jung CW, Lee J, Choi JH, Yoo HW (2012) Clinical and genetic characteristics of Gaucher disease according to phenotypic subgroups. Korean J Pediatr 55(2): 48–53

Li CM, Park JH, Simonaro CM, He X, Gordon RE, Friedman AH, Ehleiter D, Paris F, Manova K, Hepbildikler S, Fuks Z, Sandhoff K, Kolesnick R, Schuchman EH (2002) Insertional mutagenesis of the mouse acid ceramidase gene leads to early embryonic lethality in homozygotes and progressive lipid storage disease in heterozygotes. Genomics 79(2):218–224

Lin DS, Hsiao CD, Liau I, Lin SP, Chiang MF, Chuang CK, Wang TJ, Wu TY, Jian YR, Huang SF, Liu HL (2011) CNS-targeted AAV5 gene transfer results in global dispersal of vector and prevention of morphological and function deterioration in CNS of globoid cell leukodystrophy mouse model. Mol Genet Metab 103(4):367–377

Linke T, Wilkening G, Sadeghlar F, Mozcall H, Bernardo K, Schuchman E, Sandhoff K (2001) Interfacial regulation of acid ceramidase activity. Stimulation of ceramide degradation by lysosomal lipids and sphingolipid activator proteins. J Biol Chem 276(8):5760–5768

Linthorst GE, Hollak CE, Donker-Koopman WE, Strigland A, Aerts JM (2004) Enzyme therapy for Fabry disease: neutralizing antibodies toward agalsidase slpha and beta. Kidney Int 66: 1589–95

Liu Y, Gibson J, Wheeler J, Kwee LC, Santiago-Turla CM, Akafo SK, Lichter PR, Gaasterland DE, Moroi SE, Challa P, Herndon LW, Girkin CA, Budenz DL, Richards JE, Allingham RR, Hauser MA (2011) GALC deletions increase the risk of primary open-angle glaucoma: the role of Mendelian variants in complex disease. PLoS One 6(11):e27134

Lo SM, Choi M, Liu J, Jain D, Boot RG, Kallemeijn WW, Aerts JM, Pashankar F, Kupfer GM, Mane S, Lifton RP, Mistry PK (2012) Phenotypic diversity in type 1 Gaucher disease: discovering the genetic basis of Gaucher disease/hematological malignancy phenotype by individual genome analysis. Blood 119:4731–4740

Ługowska A, Ponińska J, Krajewski P, Broda G, Płoski R (2011) Population carrier rates of pathogenic ARSA gene mutations: is metachromatic leukodystrophy underdiagnosed? PLoS One 6(6):e20218

Luzi P, Rafi MA, Zaka M, Curtis M, Vanier MT, Wenger DA (2001) Generation of a mouse with low galactocerebrosidase activity by gene targeting: a new model of globoid cell leukodystrophy (Krabbe disease). Mol Genet Metab 73(3):211–223

Maegawa G (2012) GM2 gangliosidosis: the prototype of lysosomal storage disorders. Dev Med Child Neurol 54(2):104–105

Malandrini A, D'Eramo C, Palmeri S, Gaudiano C, Gambelli S, Sicurelli F, Berti G, Formichi P, Kuqo A, Dotti MT, Federico A (2013) Peripheral neuropathy in late-onset Krabbe disease: report of three cases.Neurol Sci 34:79–83

Mao C, Obeid L (2008) Ceramidases: regulators of cellular responses mediated by ceramide, sphingosine and sphingosine-1-phosphate. Biochim Biophys Acta 1781(9):424–434

Martin DC, Mark BL, Triggs-Raine BL, Natowicz MR (2007) Evaluation of the risk for Tay-Sachs disease in individuals of French Canadian ancestry living in new England. Clin Chem 53(3): 392–398

Matsuda J, Yoneshige A, Suzuki K (2007) The function of sphingolipids in the nervous system: lessons learnt from mouse models of specific sphingolipid activator protein deficiencies. J Neurochem 103(Suppl 1):32–38

Matthes F, Stroobants S, Gerlach D, Wohlenberg C, Wessig C, Fogh J, Gieselmann V, Eckhardt M, D'Hooge R, Matzner U (2012) Efficacy of enzyme replacement therapy in an aggravated mouse model of metachromatic leukodystrophy declines with age. Hum Mol Genet 21:2599–2609

McGovern MM, Schuchman EH (2009) Acid sphingomyelinase defieciency. In: Pagon RA, Bird TD, Dolan CR, Stephens K, Adam MP (eds) Gene Reviews™ [Internet]. University of Washington, Seattle, WA

Mikosch P, Hughes D (2010) An overview on bone manifestations in Gaucher disease. Wien Med Wochenschr 160(23–24):609–624

Miranda SR, He X, Simonaro CM, Gatt S, Dagan A, Desnick RJ, Schuchman EH (2000) Infusion of recombinant human acid sphingomyelinase into niemann-pick disease mice leads to visceral, but not neurological, correction of the pathophysiology. FASEB J 14(13):1988–1995

Misasi R, Hozumi I, Inuzuka T, Capozzi A, Mattei V, Kuramoto Y, Shimeno H, Soeda S, Azuma N, Yamauchi T, Hiraiwa M (2009) Biochemistry and neurobiology of prosaposin: a potential therapeutic neuro-effector. Cent Nerv Syst Agents Med Chem 9(2):119–131

Miyatake T, Suzuki K (1972) Globoid cell leukodystrophy: additional deficiency of psychosine galactosidase. Biochem Biophys Res Commun 48(3):539–543

Morita M, Saito S, Ikeda K, Ohno K, Sugawara K, Suzuki T, Togawa T, Sakuraba H (2009) Structural bases of GM1 gangliosidosis and Morquio B disease. J Hum Genet 54(9):510–515

Mullen TD, Hannun YA, Obeid LM (2012) Ceramide synthases at the centre of sphingolipid metabolism and biology. Biochem J 441(3):789–802

Nada R, Gupta K, Lal SB, Vasishta RK (2011) An autopsy case of infantile GM1 gangliosidosis with adrenal calcification. Metab Brain Dis 26(4):307–310

Neri M, Ricca A, di Girolamo I, Alcala'-Franco B, Cavazzin C, Orlacchio A, Martino S, Naldini L, Gritti A (2011) Neural stem cell gene therapy ameliorates pathology and function in a mouse model of globoid cell leukodystrophy. Stem Cells 29(10):1559–1571

Ohto U, Usui K, Ochi T, Yuki K, Satow Y, Shimizu T (2012) Crystal structure of human β-galactosidase: structural basis of Gm1 gangliosidosis and morquio B diseases. J Biol Chem 287(3):1801–1812

Okino N, He X, Gatt S, Sandhoff K, Schuchman EH (2003) The reverse activity of human acid ceramidase. J Biol Chem 278(32):29948–29953
Osher E, Fattal-Valevski A, Sagie L, Urshanski N, Amir-Levi Y, Katzburg S, Peleg L, Lerman-Sagie T, Zimran A, Elstein D, Navon R, Stern N, Valevski A (2011) Pyrimethamine increases β-hexosaminidase A activity in patients with Late Onset Tay Sachs. Mol Genet Metab 102 (3):356–633
Oya Y, Nakayasu H, Fujita N, Suzuki K, Suzuki K (1998) Pathological study of mice with total deficiency of sphingolipid activator proteins (SAP knockout mice). Acta Neuropathol 96(1): 29–40
Park JH, Eliyahu E, Narla G, DiFeo A, Martignetti JA, Schuchman EH (2005) KLF6 is one transcription factor involved in regulating acid ceramidase gene expression. Biochim Biophys Acta 1732(1–3):82–87
Park JH, Schuchman EH (2006) Acid ceramidase and human disease. Biochim Biophys Acta 1758 (12):2133–2138
Pastores GM (2010) Neuropathic Gaucher disease. Wien Med Wochenschr 160(23–24):605–608
Perlman SJ, Mar S (2012) Leukodystrophies. Adv Exp Med Biol 724:154–171
Peleg L, Meltzer F, Karpati M, Goldman B (1995) GM2 gangliosidosis B1 variant: biochemical and molecular characterization of hexosaminidase A. Biochem Mol Med 54(2):126–132
Petrache I, Petrusca DN, Bowler RP, Kamocki K (2011) Involvement of ceramide in cell death responses in the pulmonary circulation. Proc Am Thorac Soc 8(6):492–496
Phaneuf D, Wakamatsu N, Huang JQ, Borowski A, Peterson AC, Fortunato SR, Ritter G, Igdoura SA, Morales CR, Benoit G, Akerman BR, Leclerc D, Hanai N, Marth JD, Trasler JM, Gravel RA (1996) Dramatically different phenotypes in mouse models of human Tay-Sachs and Sandhoff diseases. Hum Mol Genet 5(1):1–14
Podbielska M, Krotkiewski H, Hogan EL (2012) Signaling and regulatory functions of bioactive sphingolipids as therapeutic targets in multiple sclerosis. Neurochem Res 37:1154–1169
Reddy AS, Kim JH, Hawkins-Salsbury JA, Macauley SL, Tracy ET, Vogler CA, Han X, Song SK, Wozniak DF, Fowler SC, Klein RS, Sands MS (2011) Bone marrow transplantation augments the effect of brain- and spinal cord-directed adeno-associated virus 2/5 gene therapy by altering inflammation in the murine model of globoid-cell leukodystrophy. J Neurosci 31 (27):9945–9957
Renaud DL (2012) Lysosomal disorders associated with leukoencephalopathy. Semin Neurol 32 (1):51–54
Rozenfeld P, Neumann PM (2011) Treatment of Fabry disease: current and emerging strategies. Curr Pharm Biotechnol 12:916–22
Sakai N (2009) Pathogenesis of leukodystrophy for Krabbe disease: molecular mechanism and clinical treatment. Brain Dev 31(7):485–487
Sandhoff K, Kolter T (1998) Processing of sphingolipid activator proteins and the topology of lysosomal digestion. Acta Biochim Pol 45(2):373–384
Sardi SP, Singh P, Cheng SH, Shihabuddin LS, Schlossmacher MG (2012) Mutant GBA1 expression and synucleinopathy risk: first insights from cellular and mouse models. Neurodegener Dis 10:195–202
Schaefer E, Mehta A, Gal A (2005) Genotype and phenotype in Fabry disease; analysis of the Fabry outcome survey. Acta Paediatr Suppl 94:87–92
Schissel SL, Keesler GA, Schuchman EH, Williams KJ, Tabas I (1998) The cellular trafficking and zinc dependence of secretory and lysosomal sphingomyelinase, two products of the acid sphingomyelinase gene. J Biol Chem 273(29):18250–18259
Schuchman EH (2009) The pathogenesis and treatment of acid sphingomyelinase-deficient Niemann-Pick disease. Int J Clin Pharmacol Ther 47(Suppl 1):S48–S57
Schulze H, Sandhoff K (2011) Lysosomal lipid storage diseases. Cold Spring Harb Perspect Biol 3(6):pii: a0084804. doi:10.1101/cschperspect.a004804

Scott SA, Edelmann L, Liu L, Luo M, Desnick RJ, Kornreich R (2010) Experience with carrier screening and prenatal diagnosis for 16 Ashkenazi Jewish genetic diseases. Hum Mutat 31 (11):1240–1250

Simonaro CM, Park JH, Eliyahu E, Shtraizent N, McGovern MM, Schuchman EH (2006) Imprinting at the SMPD1 locus: implications for acid sphingomyelinase-deficient Niemann-Pick disease. Am J Hum Genet 78(5):865–870

Smith NJ, Winstone AM, Stellitano L, Cox TM, Verity CM (2012) GM2 gangliosidosis in a UK study of children with progressive neurodegeneration: 73 cases reviewed. Dev Med Child Neurol 54(2):176–182

Solders G, Celsing G, Hagenfeldt L, Ljungman P, Isberg B, Ringdén O (1998) Improved peripheral nerve conduction, EEG and verbal IQ after bone marrow transplantation for adult metachromatic leukodystrophy. Bone Marrow Transplant 22(11):1119–1122

Spáčil Z, Elliott S, Reeber SL, Gelb MH, Scott CR, Tureček F (2011) Comparative triplex tandem mass spectrometry assays of lysosomal enzyme activities in dried blood spots using fast liquid chromatography: application to newborn screening of Pompe, Fabry, and Hurler diseases. Anal Chem 83(12):4822–4828

Steenweg ME, Vanderver A, Blaser S, Bizzi A, de Koning TJ, Mancini GM, van Wieringen WN, Barkhof F, Wolf NI, van der Knaap MS (2010) Magnetic resonance imaging pattern recognition in hypomyelinating disorders. Brain 133(10):2971–2982

Stroobants S, Gerlach D, Matthes F, Hartmann D, Fogh J, Gieselmann V, D'Hooge R, Matzner U (2011) Intracerebroventricular enzyme infusion corrects central nervous system pathology and dysfunction in a mouse model of metachromatic leukodystrophy. Hum Mol Genet 20(14): 2760–2769

Sugita M, Dulaney JT, Moser HW (1972) Ceramidase deficiency in Farber's disease (lipogranulomatosis). Science 178(4065):1100–1102

Sun Y, Ran H, Zamzow M, Kitatani K, Skelton MR, Williams MT, Vorhees CV, Witte DP, Hannun YA, Grabowski GA (2010) Specific saposin C deficiency: CNS impairment and acid beta-glucosidase effects in the mouse. Hum Mol Genet 19(4):634–647

Sun Y, Witte DP, Ran H, Zamzow M, Barnes S, Cheng H, Han X, Williams MT, Skelton MR, Vorhees CV, Grabowski GA (2008) Neurological deficits and glycosphingolipid accumulation in saposin B deficient mice. Hum Mol Genet 17(15):2345–2356

Tanaka A, Hoang LT, Nishi Y, Maniwa S, Oka M, Yamano T (2003) Different attenuated phenotypes of GM2 gangliosidosis variant B in Japanese patients with HEXA mutations at codon 499, and five novel mutations responsible for infantile acute form. J Hum Genet 48 (11):571–574

Tarabuso AL (2011) Fabry disease. Skinmed 9(3):173–177

Tatano Y, Takeuchi N, Kuwahara J, Sakuraba H, Takahashi T, Takada G, Itoh K (2006) Elastogenesis in cultured dermal fibroblasts from patients with lysosomal beta-galactosidase, protective protein/cathepsin A and neuraminidase-1 deficiencies. J Med Invest 53(1–2):103–112

Tinsa F, Caillaud C, Vanier MT, Bousnina D, Boussetta K, Bousnina S (2010) An unusual homozygous arylsulfatase: a pseudodeficiency in a metachromatic leukodystrophy Tunisian patient. J Child Neurol 25(1):82–86

Tohyama J, Oya Y, Ezoe T, Vanier MT, Nakayasu H, Fujita N, Suzuki K (1999) Ceramide accumulation is associated with increased apoptotic cell death in cultured fibroblasts of sphingolipid activator protein-deficient mouse but not in fibroblasts of patients with Farber disease. J Inherit Metab Dis 22(5):649–662

Tomasic IB, Metcalf MC, Guce AI, Clark NE, Garman SC (2010) Interconversion of the specificities of human lysosomal enzymes associated with Fabry and Schindler diseases. J Biol Chem 285(28):21560–21566

Toyooka K (2011) Fabry disease. Curr Opin Neurol 24(5):463–468

Truman JP, Al Gadban MM, Smith KJ, Hammad SM (2011) Acid sphingomyelinase in macrophage biology. Cell Mol Life Sci 68(20):3293–3305

Vaccaro AM, Motta M, Tatti M, Scarpa S, Masuelli L, Bhat M, Vanier MT, Tylki-Szymanska A, Salvioli R (2010) Saposin C mutations in Gaucher disease patients resulting in lysosomal lipid accumulation, saposin C deficiency, but normal prosaposin processing and sorting. Hum Mol Genet 19(15):2987–2997

van der Voorn JP, Kamphorst W, van der Knaap MS, Powers JM (2004) The leukoencephalopathy of infantile GM1 gangliosidosis: oligodendrocytic loss and axonal dysfunction. Acta Neuropathol 107(6):539–545

Wähe A, Kasmapour B, Schmaderer C, Liebl D, Sandhoff K, Nykjaer A, Griffiths G, Gutierrez MG (2010) Golgi-to-phagosome transport of acid sphingomyelinase and prosaposin is mediated by sortilin. J Cell Sci 23(Pt 14):2502–2511

Wendeler M, Lemm T, Weisgerber J, Hoernschemeyer J, Bartelsen O, Schepers U, Sandhoff K (2003) Expression of recombinant human GM2-activator protein in insect cells: purification and characterization by mass spectrometry. Protein Expr Purif 27(2):259–266

Wennekes T, van den Berg RJ, Boot RG, van der Marel GA, Overkleeft HS, Aerts JM (2009) Glycosphingolipids–nature, function, and pharmacological modulation. Angew Chem Int Ed Engl 48(47):8848–8869

Worgall TS (2011) Sphingolipid synthetic pathways are major regulators of lipid homeostasis. Adv Exp Med Biol 721:139–148

Yamanaka S, Johnson MD, Grinberg A, Westphal H, Crawley JN, Taniike M, Suzuki K, Proia RL (2008) Early deficits in motor coordination and cognitive dysfunction in a mouse model of the neurodegenerative lysosomal storage disorder, Sandhoff disease. Behav Brain Res 193(2): 315–319

Yamato O, Matsuki N, Satoh H, Inaba M, Ono K, Yamasaki M, Maede Y (2002) Sandhoff disease in a golden retriever dog. J Inherit Metab Dis 25(4):319–320

Young SA, Mina JG, Denny PW, Smith TK (2012) Sphingolipid and ceramide homeostasis: potential therapeutic targets. Biochem Res Int 2012:248135

Yuan L, Morales CR (2011) Prosaposin sorting is mediated by oligomerization. Exp Cell Res 317 (17):2456–2467

Zeidan YH, Hannun YA (2010) The acid sphingomyelinase/ceramide pathway: biomedical significance and mechanisms of regulation. Curr Mol Med 10(5):454–466

Zeidan YH, Wu BX, Jenkins RW, Obeid LM, Hannun YA (2008) A novel role for protein kinase C delta-mediated phosphorylation of acid sphingomyelinase in UV light-induced mitochondrial injury. FASEB J 22(1):183–193

Zhang H, Li D, Su Y, Jiang S, Xu Y, Jiang K, Cui D (2012) Identification of the N-acylsphingosine amidohydrolase 1 gene (ASAH1) for susceptibility to schizophrenia in a Han Chinese population. World J Biol Psychiatry 13(2):106–113

Sphingolipids and Membrane Domains: Recent Advances

Salvatore Chiantia and Erwin London

Contents

Abstract There is growing evidence that cell membranes can contain domains with different lipid and protein compositions and with different physical properties. Furthermore, it is increasingly appreciated that sphingolipids play a crucial role in the formation and properties of ordered lipid domains (rafts) in cell membranes. This review describes recent advances in our understanding of ordered membrane domains in both cells and model membranes. In addition, how the structure of sphingolipids influences their ability to participate in the formation of ordered domains, as well as how sphingolipid structure alters ordered domain properties, is described. The diversity of sphingolipid structure is likely to play an important role in modulating the biologically relevant properties of "rafts" in cell membranes.

Keywords Rafts • Domains • Sphingolipids • Ceramide • Cholesterol

S. Chiantia • E. London (✉)
Department of Biochemistry and Cell Biology, Stony Brook University, Stony Brook, NY 11794-5215, USA
e-mail: erwin.london@stonybrook.edu

E. Gulbins and I. Petrache (eds.), *Sphingolipids: Basic Science and Drug Development*, Handbook of Experimental Pharmacology 215, DOI 10.1007/978-3-7091-1368-4_2,

1 Membrane Physical State and Membrane Domain Formation

Phospholipids and sphingolipids can form bilayers that exist in either a tightly packed solid-like (gel) state or a loosely packed fluid (liquid disordered, Ld) state (London 2005). For vesicles composed of a specific lipid, the gel state melts to form the Ld state above a characteristic melting temperature (Tm). In mixtures of high-Tm lipids (e.g., natural sphingolipids, which easily form a tightly packed, ordered state due to having saturated acyl chains and no cis double bonds) and low-Tm lipids (natural phospholipids which generally have one acyl chain with one or more cis double bonds), gel and Ld phases can coexist (Korlach et al. 1999). Although the potential biological importance of lipid phase or lipid domain formation involving coexisting ordered and disordered domains was considered in early studies, the biological implications of this possibility were only first seriously considered when it was proposed that sphingolipid domains might have an important role in sorting of molecules between membranes (Simons and van Meer 1988; van Meer and Simons 1988). Such domains were later named "lipid rafts" (Simons and Ikonen 1997). The idea that sphingolipid domains might exist in cells gained an important impetus when it was found that sphingolipid- and cholesterol-rich detergent-resistant membranes (DRM) could be isolated from mammalian cells (Brown and Rose 1992). When cholesterol is present, the tightly packed liquid ordered (Lo) state tends to form in place of the gel state (London 2005) and, based on model membrane studies, the hypothesis was proposed that lipid domains in cells might be sphingolipid- and cholesterol-rich Lo domains coexisting with Ld domains (Ahmed et al. 1997; Schroeder et al. 1994). It has now been widely shown that in vesicles composed of mixtures of sphingolipids, unsaturated phospholipids, and cholesterol, coexisting Lo and Ld domains can be observed by microscopy at conditions close to physiological (Hammond et al. 2005; Veatch and Keller 2003), and Lo domain formation is the working model for the physical state of "membrane rafts/lipid rafts" in cells (Dietrich et al. 2001a, b; Lingwood et al. 2008; London 2005; Sengupta et al. 2007a, b). Rafts are of biological importance because, by co-clustering membrane lipids and specific proteins in a domain or by segregating them into different domains, specific sets of protein–protein interactions can form or be regulated. In addition, the differences in lipid physical properties and composition in raft and non-raft domains could influence protein function via lipid environment-induced changes in protein conformation.

Raft domains have been proposed to play a crucial role in many important processes that take place within eukaryotic cell membranes, including not only protein and lipid sorting into different membranes but also modulation of signal transduction, especially in the immune system, many types of bacterial and viral infections, including HIV and influenza, and amyloid formation (e.g., Cuadras and Greenberg 2003; Drevot et al. 2002; Gulbins and Kolesnick 2003; Kamiyama et al. 2009; Klemm et al. 2009; Lafont et al. 2002; Lu et al. 2008; Lyman et al. 2008; Manneville et al. 2008; Mukherjee et al. 1999; Murphy et al. 2006; Persaud-Sawin et al. 2009; Rentero et al. 2008; Riethmuller et al. 2006; Scheiffele et al. 1999; Scolari et al. 2009; Simons and Toomre 2000; Sohn et al. 2008a, b; Taylor and Hooper 2007; Williamson et al. 2008; Young et al. 2003, 2005; Zech et al. 2009).

There are many important issues that must be considered to understand the principles and details of domain formation and properties. We refer the interested reader to previous reviews that describe these issues and summarize earlier studies (Feigenson 2006, 2009; Heberle and Feigenson 2011; London 2005; Quinn 2010; Veatch and Keller 2005). In this review we concentrate on topics in lipid domain formation that have seen recent advances and also on studies that specifically focus on sphingolipids.

2 Growing Evidence for Rafts in Cells

Unlike the situation in model membranes, whether "lipid rafts" form in cells has been controversial. It appears that domains are hard to study in living cells due to their small size under basal physiological conditions (Lingwood et al. 2008; Lingwood and Simons 2010; Veatch et al. 2007, 2008). The presence of the cytoskeleton may limit raft size, or even inhibit domain formation (Baumgart et al. 2007; Ehrig et al. 2011). It is important to point out that to be functionally important, domains need not be large or permanent. They only need to be large enough to cluster or segregate proteins, do so for long enough to affect protein activity, and form upon physiological-triggering, e.g., by clustering raft-associating components (Hammond et al. 2005; Lingwood et al. 2008). Indeed, what appear to be much larger domains, or membrane regions enriched in one type of small domains, can be induced to form in cells under activated conditions (see below).

Despite the difficulty of detecting domains under most conditions, recent studies have strengthened the hypothesis that membrane domains do form in cells. Novel fluorescence microscopy probes to visualize membrane order (e.g., using Laurdan or its derivatives, or using di-4-ANEPPDHQ) (Gaus et al. 2003, 2005, 2006a, b; Harder et al. 2007; Kim et al. 2007; Owen et al. 2006, 2007; Zech et al. 2009) detect lipid domains in living cells. The domains these methods detect may also be raft-rich regions in which ordered raft nanodomains are especially abundant rather than single uniform domains. Recent advances in single molecule (Pinaud et al. 2009) and other fluorescence methods also provide additional evidence for raft-like cellular lipid domains (Lenne et al. 2006; Pinaud et al. 2009), including studies using advanced superhigh-resolution light microscopy that suggest the formation of very small domains (Eggeling et al. 2009; Sahl et al. 2010; van Zanten et al. 2010) as did early studies (Varma and Mayor 1998). However, the perturbations arising from the labeled lipids (Zhao et al. 2007) and the high laser powers used for super-resolution microscopy may complicate interpretation (Mueller et al. 2011). Evidence that ordered domains control protein–protein interaction, especially in the immune system, is also growing. Mast cell studies show that domains control kinase function by segregation from phosphatases (Young et al. 2003, 2005). Recent studies using sterol modification, probes of membrane order, and lipidomic analysis of plasma membrane domains all indicate that lipid domain formation accompanies both T cell activation (Rentero et al. 2008; Zech et al. 2009), and B cell activation (Sohn et al. 2008a, b). Studies of membrane budding strongly support the hypothesis that traffic of lipids and proteins to different membranes is also dependent upon domain segregation (Brugger et al. 2000; Klemm et al. 2009; Manneville et al. 2008; Mukherjee et al. 1999) and that in at least one case, viral proteins exploit Ld/Lo domain boundaries to induce viral budding (Rossmann et al. 2010). Other recent papers also support an important role for rafts in membrane sorting of proteins (Klemm et al. 2009; Norambuena and Schwartz 2011; Refaei et al. 2011) while in vivo studies show a role of glycosphingolipids in sorting (Zhang et al. 2011). The difference between Lo vs. Ld bending modulus has been shown to be able to drive membrane segregation (Baumgart et al. 2003; Manneville et al. 2008; Rossmann et al. 2010; Sorre et al. 2009).

The argument that the compositional complexity of natural membranes in terms of their numerous lipid and protein species would prevent domain formation is ruled out by studies detecting large domain formation in two types of plasma membrane preparations, giant membrane vesicles, and plasma membrane spheres (Baumgart et al. 2007; Lingwood et al. 2008; Veatch et al. 2008). It is possible that a partial loss of membrane asymmetry and loss of cytoskeletal connections influence domain size in such preparations (Ehrig et al. 2011; Keller et al. 2009). Finally, raft-like domains have even been detected in a bacterium that contains cholesterol obtained from their hosts (LaRocca et al. 2010). Other bacteria may also have raft-like membrane domains (Lopez and Kolter 2010).

3 Recent Advances in Our Understanding of Raft Formation Principles from Model Membrane Studies: Domain Size

Several important raft properties have been the focus of recent model membrane studies. One is domain size. Although Lo domains can be very large in model membranes, and easily detected by light microscopy, it is clear that in cells they are often very tiny, perhaps on the order of a few to 100 nm. Recent studies have shown that in model membranes with realistic plasma membrane outer leaflet lipid compositions, i.e., sphingomyelin/1-palmitoyl, 2-oleoyl phosphatidylcholine/cholesterol (SM/POPC/cholesterol), Lo domains can also be very small (Pathak and London 2011). Under some conditions, ordered domains that are too small to be detected with FRET pairs having a large (~50 Å) donor-acceptor interaction radius (Ro) can be identified in SM/POPC/cholesterol vesicles. These domains are detected using FRET pairs and short-range quenchers with small (12–25 Å) interaction radii (Pathak and London 2011). These studies indicate that domain sizes with estimated radii as small as ~40 Å form in this mixture at 37 °C.

Why do these tiny domains fail to fuse into large ones? The origin of the size stability of such small "nanodomains" is an active area of research. It has been proposed that domains are small because they exist at conditions close to the critical point of the ternary lipid mixtures investigated (Honerkamp-Smith et al. 2008; Veatch et al. 2007, 2008). Consistent with the predictions of this model, nanodomains gradually decrease in size in SM/POPC/cholesterol mixtures as temperature increases (Pathak and London 2011). It has also been proposed that the presence of molecules that prefer locating at domain edges may contribute to small domain size (Brown and London 1998; Chiantia et al. 2007; Mitchell and Litman 1998). Several studies have shown that the presence of lipids with one saturated and one unsaturated acyl chain (e.g., POPC), which are likely to have some affinity for domain edges, produce smaller domains than lipids with two unsaturated acyl chains (Brewster et al. 2009; Brewster and Safran 2010; de Almeida et al. 2003; Heberle et al. 2010; Pokorny et al. 2006; Schafer and Marrink 2010).

Such nanodomains may not have all the properties of true Lo phases, but retention of tight packing and protein binding specificity similar to that of large Lo domains, combined with a sufficient time persistence (under cellular conditions, e.g., in the presence of proteins), would be sufficient for them to be of biological relevance.

4 Recent Advances in Our Understanding of Raft Formation Principles from Model Membrane Studies: Evidence that Detergent Does not Induce Raft Formation

Ordered domains are detergent insoluble and, as noted above, detergent (e.g., TX-100)-insoluble sphingolipid- and cholesterol-rich ordered membranes (DRM) can be isolated from cells (Brown and Rose 1992; Schroeder et al. 1994). However, the isolation of such domains has not been sufficient to prove that DRM arise from preexisting rafts because detergent could alter domain formation (Brown and London 2000). Many studies have confirmed that when Lo and Ld domains coexist, the DRM arise from the Lo region of the membrane (Ahmed et al. 1997; Dietrich et al. 2001a, b; El Kirat and Morandat 2007; Garner et al. 2008). However, it has been reported by one group that domain formation in SM/POPC/chol vesicles can be stabilized by TX-100, so that they form at higher temperatures only in the presence of TX-100 (Heerklotz 2002; Heerklotz et al. 2003). This observation has been frequently cited as evidence that DRM obtained from cells may be a detergent artifact. However, a reinvestigation of the effect of TX-100 upon domains in SM/POPC/chol vesicles has found that TX-100 does NOT stabilize domain formation, but rather induces the coalescence of preexisting nanodomains into larger domains (Pathak and London 2011). This greatly reduces the concern that domains are an artifact of detergent treatment. Nevertheless, it must be kept in mind that solubilization studies carried out at 4 °C cannot be used to argue that ordered domains are present at 37 °C. Less perturbing methods to obtain domains by physical membrane fragmentation or detergent solubilization at 37 °C should be helpful in this regard (Ayuyan and Cohen 2008; Chen et al. 2009; Drevot et al. 2002; Macdonald and Pike 2005; Morris et al. 2011; Smart et al. 1995; Song et al. 1996).

5 Recent Advances in Our Understanding of Raft Formation Principles from Model Membrane Studies: Role of Lipid Asymmetry in Domain Formation

In the vast majority of studies the model membranes used are symmetric in terms of having identical lipid compositions in each leaflet (monolayer). This limits the ability to extrapolate from model membrane vesicles to real membranes, which have a distinct, if poorly characterized, degree of lipid asymmetry. In the case of the mammalian plasma membrane, it is known that the outer (exoplasmic, exofacial) leaflet is rich in sphingolipids, including sphingomyelin (SM) and glycosphingolipids (GSL), as well as phosphatidylcholine (PC). The inner (cytoplasmic, cytofacial) leaflet is rich in phosphatidylethanolamine (PE) and anionic lipids such as phosphatidylserine (PS) and phosphatidylinositides (Verkleij et al. 1973). Cholesterol is found in both leaflets (in relative amounts that are still disputed) (Mondal et al. 2009). In other membranes, much less is known about asymmetry due to the technical difficulty of asymmetry measurements.

An important asymmetry issue complicating our understanding of membrane domain formation is that inner leaflet lipids have little to no sphingolipid, although this is not entirely certain (van Meer 2011). This raises the question: How could ordered domains form in the inner leaflet? The solution may be that the outer leaflet lipids influence inner leaflet physical properties, i.e., inner and outer leaflet physical states may be coupled (Collins 2008; Kiessling et al. 2009). Coupling could provide a mechanism by which information is transferred across the membrane via lipids. For example, inner leaflet domains induced by outer leaflet domains could act by concentrating cytosolic-surface membrane proteins with a high affinity for ordered domains (e.g., proteins anchored by saturated acyl chains (London 2005; Melkonian et al. 1999)).

Until recently, few methods to prepare suitable asymmetric membranes have been available. Asymmetry has been most readily achieved with planar bilayers (Honerkamp-Smith et al. 2008; Kiessling et al. 2009; Wan et al. 2008). However, asymmetric closed lipid vesicles would have an even wider utility for a variety of applications. Past attempts to make asymmetric lipid vesicles have not come into wide use, perhaps because of limited control over asymmetry, applicability to a limited number of lipids (Everett et al. 1986; Hope and Cullis 1987; Hope et al. 1989; Malewicz et al. 2005; Pagano et al. 1981), and for those methods involving a leaflet-by-leaflet assembly method, unsuitability for membrane protein incorporation, and residual-contaminating oils (Hamada et al. 2008; Hu et al. 2011; Pautot et al. 2003).

Studies (Anderson et al. 2004; Niu and Litman 2002) showing that high concentrations of methyl-β-cyclodextrin (MβCD) can bind phospholipids tightly have opened up new possibilities for preparing asymmetric phospholipid and sphingolipid lipid vesicles, removing the original lipid in the outer leaflet of an acceptor membrane while a new lipid replaces it (Cheng et al. 2009). Asymmetric small unilamellar vesicles (SUV), large unilamellar vesicles (LUV), and giant unilamellar vesicles (GUV) have now been prepared with this lipid exchange approach (Cheng and London 2011; Cheng et al. 2009; Chiantia et al. 2011). Exchange protocols have been used to produce asymmetric vesicles with various lipid compositions, including vesicles containing cholesterol.

Studies of asymmetric planar bilayers and vesicles have shown that leaflets rich in sphingolipids or other high Tm membrane lipids (i.e., with long, linear saturated acyl chains) can indeed induce reorganization/domain formation in the opposite membrane leaflet (Collins and Keller 2008; Kiessling et al. 2006; Wan et al. 2008). This ability depends on experimental conditions including the type of lipid in the leaflet opposite that in the high Tm lipid-rich leaflet (Cheng et al. 2009; Chiantia et al. 2011; Collins and Keller 2008; Kiessling et al. 2006; Wan et al. 2008). The exact physical state of the opposite leaflet (composed of low Tm lipid) is not clear; it may be fully or partly ordered (Cheng and London 2011; Cheng et al. 2009; Chiantia et al. 2011; Kiessling et al. 2006; Wan et al. 2008). This suggests that there is some degree of coupling between the inner and outer lipid physical states. However, this interleaflet coupling is not "strong," as strong coupling would imply that the thermal melting of the bilayer occurs at a Tm that is intermediate between that of the inner and outer leaflet lipids, and this is not what is observed (Cheng and London 2011; Cheng et al. 2009). Instead, the high Tm-lipid-rich leaflet has a Tm similar to that in vesicles composed of pure high Tm lipid. There may be weaker coupling that breaks down as temperature increases (Cheng and London 2011), although further studies are needed to confirm this. In any case, our knowledge of the behavior of asymmetric bilayers is very incomplete. Further progress awaits extension of studies to an even wider variety of lipid compositions, and the development and application of additional assays to define the physical state of the individual leaflets.

The role of membrane proteins and high cholesterol concentrations are other areas that need further investigation. We will not cover these topics here, but simply note that studies of the relationship of protein structure to raft affinity and raft properties in model membrane systems are beginning to yield important conclusions (Baumgart et al. 2007; Coskun et al. 2011; Fastenberg et al. 2003; Johnson et al. 2010; Kaiser et al. 2011; Nelson et al. 2008, 2010; Nikolaus et al. 2010; Sengupta et al. 2008; Shogomori et al. 2005; Tong et al. 2009). Computational methods are another important area seeing important advances (Perlmutter and Sachs 2011) that will not be covered here.

6 Effect of Sphingolipid Structure on Domain Formation

General Properties of Sphingolipids: Model membrane studies have shown that, in general, any lipid with high melting temperature can separate from lipids in an Ld phase to form ordered domains. Sphingolipids are particularly important in this context as the main source of lipids with high Tm that are found in cellular membranes. The tendency of this class of lipids to form ordered domains derives from their peculiar chemical structure, very different from glycerophospholipids. Sphingolipids are composed of a long-chain base and an amide-linked acyl chain. The base, which is the analog of the glycerol backbone attached to a fatty acid on the 1-position portion of a typical glycerophospholipid (e.g., PC), is in most cases a sphingosine molecule. Sphingosine has 18 (or 20) carbon atoms with a trans double bond between the carbon atoms 4 and 5, plus 2 OH groups on carbon atoms 1 and 3 and an amino group on carbon atom 2. Sphingosine variants (e.g., saturated sphinganine or the 4-hydroxylated and saturated phytosphingosine) can also be found in nature (Goni and Alonso 2009). It is worth noting that this double bond should have a much smaller effect on Tm than the double bonds found in fatty acids for two reasons. First, a trans double bond only causes a small kink relative to that formed by a cis double bond. Secondly, it is near one end of the hydrocarbon chain, while double bonds in fatty acyl chains of a membrane lipid interfere with tight packing (as judged by low Tm values) most strongly when they are located near the center of the chain (Barton and Gunstone 1975), as is the case with the cis double bonds of natural fatty acids.

Most sphingolipids have a fatty acid (typically 16–24 carbon atoms long with the exception of skin lipids) N-linked to the sphingosine. This fatty acid is the analog of the 2-position fatty acid of glycerophospholipids. However, unlike the cis double bond containing fatty acids common in glycerophospholipids, this acyl chain is usually fully saturated in eukaryotes and is a major factor imparting a high Tm to sphingolipids. Even in the one common case in which eukaryotic sphingolipids have an unsaturated fatty acid, that fatty acid is usually a 24 carbon chain with just one double bond, which still supports a high Tm value. On the other hand, sphingolipids with very long polyunsaturated acyl chains are present in spermatozoa (Sandhoff 2010), suggestive of some specialized function.

In the discussions below on the effects of fatty acid chain structure and length upon domain-forming properties in model membranes it should be kept in mind that almost all studies to date involve bilayers with symmetric bilayer compositions, i.e., in which the lipid composition of the inner and outer monolayers/leaflets are identical. Such bilayers may have different properties than bilayers with lipid asymmetry, as found in natural membranes, especially in cases in which a lipid molecule has two hydrocarbon chains of unequal lengths (i.e., "chain asymmetry"), which gives rise to the possibility of penetration of the longer chain into the opposite leaflet (interdigitation).

At the level of the polar head of the lipid there are other important structural features which may contribute to a high Tm: while each glycerophospholipid can accept two hydrogen bonds (via its carbonyl groups), sphingolipid hydroxy and amide functional groups can act both as donors and acceptors for hydrogen bonding (Mombelli et al. 2003). This feature, together with the presence of long saturated acyl chains, may confer upon sphingolipids the ability to interact strongly with other (similar) lipids. It should be pointed out that hydrogen bonding between two groups in a membrane does not face as much competition with hydrogen bonding to water as in aqueous solution, and so may be a strong driving force for interactions between lipids.

There are a large variety of sphingolipids that can be found in cellular membranes, ranging from the simplest ceramides to more complex glycosphingolipids. In the next paragraphs, we review recent studies regarding the main sphingolipid species.

Sphingosine: Sphingosine (Sph) is by definition the simplest sphingolipid, but was shown to influence a variety of cellular processes, including cell growth and differentiation, receptor modulation, and cytotoxicity (Hannun and Bell 1989; Merrill and Stevens 1989). It is only a minor component of membranes, and the effects of Sph on membrane organization have received little attention. Work by Mustonen et al. (Mustonen et al. 1993) investigated the effects of Sph on membrane–protein electrostatic interactions, since Sph is positively charged at physiological pH (Lopezgarcia et al. 1995). More recently, Contreras et al. (Contreras et al. 2006) showed that Sph is not able to produce gel domains by itself but can reinforce the existing ones, and due to the generation of solid–fluid interfaces and the consequent packing defects, can also produce permeabilization in lipid vesicles. The anti-apoptotic Sph derivative Sph-1-phosphate was shown to have a stabilizing effect on lipid lamellar structures (versus negatively curved, inverted phases) and to be the only sphingolipid-signaling molecule that can be found dissociated from the cell membranes in the cytosol (Garcia-Pacios et al. 2009).

Ceramide: Ceramide (Cer) is another very simple sphingolipid, constituted by the sphingosine base (sphinganine in the case of dihydroceramide) linked to an acyl chain via its amide group. This molecule can be produced in mammalian cells either via de novo synthesis (via a family of six ceramide synthases) or through hydrolysis of SM phosphocholine group, mediated by sphingomyelinases. In response to specific stimuli, ceramide concentration in physiological contexts can reach 10–20 % of the total lipid content (Cremesti et al. 2002; Hannun 1996). Cer is considered both an important second messenger and a membrane structural component involved in several biological processes, such as cell growth, differentiation, apoptosis, senescence, and bacterial and viral pathogenesis (Bollinger et al. 2005; Kolesnick et al. 2000). It has a high Tm (which can be as high at 90 °C (Shah et al. 1995)) and high hydrophobicity (due to its small polar headgroup). These properties are important reasons Cer is one of the main components of the water-impermeable extracellular matrix of the stratum corneum of the skin (Shah et al. 1995; tenGrotenhuis et al. 1996; Wartewig and Neubert 2007).

Although Cer can exert some of its biological functions via direct interaction with specific target proteins (Grosch et al. 2012), we will focus on its effects on the lateral organization of the plasma membrane. It was observed that Cer accumulation in the plasma membrane leads to the formation of large lipid–protein domains ("platforms") involved, for example, in the internalization of viruses and parasites and in the induction of apoptosis (Gulbins and Grassme 2002). Cer-rich platforms may act in these contexts as sorting locations for membrane receptors, inhibitors, and other membrane components involved in signaling (Gulbins et al. 2004). For example, receptor clustering and trapping in Cer-rich domains have been suggested by experiments performed with the receptors Fcγ II (Shakor et al. 2004) and CD95 and CD40 (Grassme et al. 2001, 2002). Similarly, Cer-rich domains seem to recruit the receptors mediating the internalization of *Neisseria gonorrhoeae* (Grassmé et al. 2007).

The molecular mechanisms behind the formation of such Cer-rich domains and their relation to the SM–cholesterol raft domains were investigated in recent years using model membranes with controlled compositions. These studies showed that long-chain Cer which are symmetric in the sense that the acyl chain and sphingoid base have similar effective hydrocarbon lengths (C16:0, C18:0 N-linked acyl chains) can strongly interact with SM, forming a highly ordered Cer-rich phase (Boulgaropoulos et al. 2011; Chiantia et al. 2006; Megha and London 2004; Silva et al. 2007). The interplay between Cer, SM, and cholesterol can be understood in the context of the "umbrella model" (Huang and Feigenson 1999). Several studies have shown that cholesterol and Cer can compete for the interaction with SM, since both molecules have small headgroups that can be shielded from (unfavorable) interactions with water molecules by lipids with large headgroups like SM, which act as umbrellas (Alanko et al. 2005; Megha and London 2004; Nyholm et al. 2010; Sot et al. 2008). As a result, Cer can readily displace cholesterol from ordered domains, and the reverse is also possible (Megha and London 2004; Silva et al. 2007). Further work showed that, in analogy with Cer-rich platforms observed in cells, Cer homeostasis and the presence of Cer-rich domains affect the lateral organization of membrane proteins in model membranes (Chiantia et al. 2008; Dasgupta et al. 2009; Pabst et al. 2009). Very-long-chain Cer (with saturated C20–C24 N-linked acyl chain) promote in general ordering of the membrane, and formation of gel domains (Pinto et al. 2008, 2011). They also may have interesting functions due to the mismatch between the lengths of the hydrocarbon of the sphingoid base and acyl chain (see below).

Unnatural short-chain Cer (with C2–C12 N-linked acyl chains) are used extensively in in vivo experiments to replace their long-chain analogs. They were shown to have very different effects than long-chain Cer on the lateral organization of the bilayer (Chiantia et al. 2007; Megha et al. 2007; Nybond et al. 2005; Westerlund et al. 2010). Unlike long-chain Cer, these molecules disorder membranes (Gidwani et al. 2003; Sengupta et al. 2007a, b). Nevertheless, short-chain Cer often mimic natural Cer in their functional properties (see, e.g., (Bektas et al. 1998; Kolesnick et al. 2000)). This may reflect their remodeling into natural acyl chain Cer in cells, or effects not dependent upon membrane domain formation, e.g., interaction with specific proteins or functions arising from their "umbrella effect."

Another mystery is the origin of the functional differences between biologically active Cer and dihydroCer, which lacks the sphingoid base double bond. It is not yet clear if this reflects a difference in the domain-forming properties of these molecules, in other biophysical properties, or in functional interactions with proteins. However, functional differences between short-chain Cer and short-chain dihydroCer suggest that the physiological effects of Cer are not exclusively related to lipid domain formation (Simon and Gear 1998).

In addition to the effects upon the formation of lipid–protein domains, Cer can influence the properties of a lipid bilayer due to its intrinsic negative curvature and tendency to form non-lamellar inverted phases. Increased lipid flip-flop, bending, and vesiculation of the membrane can be observed when Cer is produced in one leaflet of the bilayer (Contreras et al. 2005; Holopainen et al. 2000). This may provide a protein-free mechanism for the sorting of the membrane into the different populations of intraluminal vesicles in vivo (Trajkovic et al. 2008). Cer can also affect the properties of a lipid membrane via the formation of transmembrane channels (formed by stacked Cer molecules all parallel to the plane of the bilayer). These have been proposed to allow the passage of certain proteins initiating apoptosis (Colombini 2010; Siskind and Colombini 2000).

Sphingomyelin: SM (i.e., Cer phosphorylcholine) is one of the most abundant phospholipids in eukaryotic membranes and, in particular, an important component of raft domains in the outer leaflet of the plasma membrane together with cholesterol. The stability of such domains depends on SM–SM and SM–cholesterol interactions which, in turn, have been shown to be determined by details of the chemical structure of the SM molecule. Several studies have thoroughly investigated the role of many SM structural properties at the level of the membrane interfacial region, e.g., polar head size (Bjorkbom et al. 2011), specific stereoconfiguration (Ramstedt and Slotte 1999), the presence of the phosphocholine methyl group (Terova et al. 2005), the sphingosine double bond (Kuikka et al. 2001; Vieira et al. 2010), the hydroxylation of the N-linked acyl chain (Ekholm et al. 2011), and the role of the 3-hydroxyl group or the amide-linkage in establishing hydrogen bonds (Bittman et al. 1994; Bjorkbom et al. 2011; Kan et al. 1991). Other studies investigated the molecular requirements at the level of the hydrophobic region of SM that allow it to participate in the formation of ordered membrane domains. For example, it was observed that SMs from different natural sources (chicken egg, bovine milk, and porcine brain), which have different acyl chain composition, differ in their ability to segregate into ordered domains in the bilayer (Filippov et al. 2006). Saturated N-linked acyl chains with length similar to that of the sphingoid base have strong lipid–lipid interactions (Epand and Epand 2004; Jaikishan et al. 2010). Increasing the length of the N-linked acyl chain above that of the base increases (although not dramatically) the lipid packing abilities of SM, but decreases its interactions with cholesterol (Jaikishan and Slotte 2011; Niemela et al. 2006).

GSL: GSL are sphingolipids covalently bound to oligosaccharidic groups of different dimensions and complexity (Pontier and Schweisguth 2012). These molecules can be usually found in the plasma membrane. They are involved in cell–cell interaction and are often receptors for viruses and toxins (Hakomori et al. 1998; Lingwood et al. 2000; Viard et al. 2003). From a structural point of view, all GSL usually share a Cer backbone, i.e., a sphingoid base with an N-linked long saturated acyl chain. The heterogeneity of the lipids belonging to this group derives then mostly from the polar moiety bound to the Cer backbone, including (in mammals) small monosaccharides (e.g., glucosylceramide), charged groups (e.g., sulfatides), and large assembles of polysaccharides and sialic acid (e.g., gangliosides) (Westerlund and Slotte 2009). Very different phosphate-containing GSL-like sphingolipids are found in plants and fungi (Rhome and Del Poeta 2010; Sperling and Heinz 2003).

Of interest, GSL often have a remarkably high Tm and packing density compared to the corresponding acyl chain-matched SMs or PCs (Ruocco et al. 1981; Smaby et al. 1996). For this reason, they are often found in DRMs together with SM and cholesterol, but without necessarily implying that all these lipids co-localize always in the same membrane domains (Arvanitis et al. 2005; Braccia et al. 2003). It was shown in fact that GSLs can also form specialized domains (glycosignaling domains) involved in cell–cell recognition, independently from the presence of cholesterol (Hakomori 2004). Several studies in model membranes have investigated the domain forming abilities of GSL (see, e.g., (Bjorkbom et al. 2010a, b; Blanchette et al. 2006; Lin et al. 2007; Maunula et al. 2007)) concluding that the Cer backbone structure as well as the number of sugar units and presence of charge in the GSL headgroup can influence the partitioning of these lipids between lateral membrane domains. For a more thorough discussion, the reader is referred to the extensive review of Westerlund and Slotte (2009).

It is beyond the scope of this review to describe the vast literature on GSL physical and biological properties. However, one case worthy of specific comment involves the physical properties of the lactosylceramide (LacCer). LacCer is one of the most abundant neutral GSL and is expressed in particular in the lipid membranes of human neutrophils (Kniep and Skubitz 1998). In these cells, LacCer acts as one of the several pattern recognition receptors involved in the detection of infectious microorganisms and can bind specifically to various pathogens (e.g., *Escherichia coli* or *Candida albicans*) (Iwabuchi et al. 2010; Sato et al. 2006; Teneberg et al. 2004). Experiments by Pagano and coworkers addressed caveolae-dependent membrane trafficking of LacCer in human skin fibroblasts, proposing that LacCer with natural stereochemistry participate in the formation of membrane microdomains that endocytose via a caveolar pathway and promote β1 integrin signaling (Singh et al. 2006, 2007). Interestingly, LacCer analogs with unnatural stereochemistry were found not to support, or even to inhibit, these processes. It should be noted that the LacCer molecules used had either labeled or shortened acyl chains, which could perturb domain-forming properties.

Other studies show that acyl chain length is important for some of the biological functions of LacCer. LacCer has been found to mediate several biological processes (including chemotaxis, phagocytosis, and superoxide generation) that depend on the Src kinase Lyn (Iwabuchi et al. 2010). The interaction between LacCer in the outer plasma membrane and the Lyn molecules, which are anchored to the cytoplasmic side of the bilayer, was shown in turn to be dependent on the presence of long *N*-acyl molecular species (i.e., 24:0 and 24:1 LacCer). Antibody-mediated cross-linking of LacCer in neutrophils (which naturally possess C24:0 and C24:1 LacCer acyl chain species) caused the lipid to colocalize in DRMs and co-immunoprecipitate with the activated form of Lyn. If the experiment was repeated on a neutrophil-differentiated cell line (D-HL-60) that possesses only shorter C16–18 acyl chain species of LacCer, neither co-localization or activation of Lyn could be observed, unless the cells were loaded with exogenous long-chain LacCer (Iwabuchi and Nagaoka 2002; Nakayama et al. 2008). Analogously, the presence of C24 LacCer was shown to be necessary for generation of superoxide, for chemotaxis towards ß-glucan or anti-LacCer antibodies, and for CD11b/CD18-mediated phagocytosis of non-opsonized microorganisms (Nakayama et al. 2008). These studies suggest that the long *N*-acyl chain of certain sphingolipids can cross the midplane of the lipid bilayer (i.e., interdigitate), to influence proteins on the cytosolic leaflet in order to couple external stimuli with intracellular signal cascades (Iwabuchi et al. 2010; Sonnino et al. 2009). An interesting question that is yet to be answered is whether this involves coupling between rafts in the outer and inner leaflet.

Acknowledgement This work was supported by NIH Grant GM 099892 to E.L.

References

Ahmed SN, Brown DA, London E (1997) On the origin of sphingolipid/cholesterol-rich detergent-insoluble cell membranes: physiological concentrations of cholesterol and sphingolipid induce formation of a detergent-insoluble, liquid-ordered lipid phase in model membranes. Biochemistry 36:10944–10953

Alanko SMK, Halling KK, Maunula S, Slotte JP, Ramstedt B (2005) Displacement of sterols from sterol/sphingomyelin domains in fluid bilayer membranes by competing molecules. Biochim Biophys Acta 1715:111–121

Anderson TG, Tan A, Ganz P, Seelig J (2004) Calorimetric measurement of phospholipid interaction with methyl-beta-cyclodextrin. Biochemistry 43:2251–2261

Arvanitis DN, Min WX, Gong YP, Heng YM, Boggs JM (2005) Two types of detergent-insoluble, glycosphingolipid/cholesterol-rich membrane domains from isolated myelin. J Neurochem 94:1696–1710

Ayuyan AG, Cohen FS (2008) Raft composition at physiological temperature and pH in the absence of detergents. Biophys J 94:2654–2666

Barton PG, Gunstone FD (1975) Hydrocarbon chain packing and molecular motion in phospholipid bilayers formed from unsaturated lecithins. Synthesis and properties of sixteen positional isomers of 1,2-dioctadecenoyl-sn-glycero-3-phosphorylcholine. J Biol Chem 250: 4470–4476

Baumgart T, Hammond AT, Sengupta P, Hess ST, Holowka DA, Baird BA, Webb WW (2007) Large-scale fluid/fluid phase separation of proteins and lipids in giant plasma membrane vesicles. Proc Natl Acad Sci U S A 104:3165–3170

Baumgart T, Hess ST, Webb WW (2003) Imaging coexisting fluid domains in biomembrane models coupling curvature and line tension. Nature 425:821–824

Bektas M, Dullin Y, Wieder T, Kolter T, Sandhoff K, Brossmer R, Ihrig P, Orfanos CE, Geilen CC (1998) Induction of apoptosis by synthetic ceramide analogues in the human keratinocyte cell line HaCaT. Exp Dermatol 7:342–349

Bittman R, Kasireddy CR, Mattjus P, Slotte JP (1994) Interaction of cholesterol with sphingomyelin in monolayers and vesicles. Biochemistry 33:11776–11781

Bjorkbom A, Ohvo-Rekila H, Kankaanpaa P, Nyholm TKM, Westerlund B, Slotte JP (2010a) Characterization of membrane properties of inositol phosphorylceramide. Biochim Biophys Acta 1798:453–460

Bjorkbom A, Rog T, Kankaanpaa P, Lindroos D, Kaszuba K, Kurita M, Yamaguchi S, Yamamoto T, Jaikishan S, Paavolainen L, Paivarinne J, Nyholm TKM, Katsumura S, Vattulainen I, Slotte JP (2011) N- and O-methylation of sphingomyelin markedly affects its membrane properties and interactions with cholesterol. Biochim Biophys Acta 1808:1179–1186

Bjorkbom A, Rog T, Kaszuba K, Kurita M, Yamaguchi S, Lonnfors M, Nyholm TK, Vattulainen I, Katsumura S, Slotte JP (2010b) Effect of sphingomyelin headgroup size on molecular properties and interactions with cholesterol. Biophys J 99:3300–3308

Blanchette CD, Lin WC, Ratto TV, Longo ML (2006) Galactosylceramide domain microstructure: impact of cholesterol and nucleation/growth conditions. Biophys J 90:4466–4478

Bollinger CR, Teichgraber V, Gulbins E (2005) Ceramide-enriched membrane domains. Biochim Biophys Acta 1746:284–294

Boulgaropoulos B, Arsov Z, Laggner P, Pabst G (2011) Stable and unstable lipid domains in ceramide-containing membranes. Biophys J 100:2160–2168

Braccia A, Villani M, Immerdal L, Niels-Christiansen LL, Nystrom BT, Hansen GH, Danielsen EM (2003) Microvillar membrane microdomains exist at physiological temperature - role of galectin-4 as lipid raft stabilizer revealed by "superrafts". J Biol Chem 278:15679–15684

Brewster R, Pincus PA, Safran SA (2009) Hybrid lipids as a biological surface-active component. Biophys J 97:1087–1094

Brewster R, Safran SA (2010) Line active hybrid lipids determine domain size in phase separation of saturated and unsaturated lipids. Biophys J 98:L21–L23

Brown DA, London E (1998) Structure and origin of ordered lipid domains in biological membranes. J Membr Biol 164:103–114

Brown DA, London E (2000) Structure and function of sphingolipid- and cholesterol-rich membrane rafts. J Biol Chem 275:17221–17224

Brown DA, Rose JK (1992) Sorting of GPI-anchored proteins to glycolipid-enriched membrane subdomains during transport to the apical cell surface. Cell 68:533–544

Brugger B, Sandhoff R, Wegehingel S, Gorgas K, Malsam J, Helms JB, Lehmann WD, Nickel W, Wieland FT (2000) Evidence for segregation of sphingomyelin and cholesterol during formation of COPI-coated vesicles. J Cell Biol 151:507–518

Chen X, Jen A, Warley A, Lawrence MJ, Quinn PJ, Morris RJ (2009) Isolation at physiological temperature of detergent-resistant membranes with properties expected of lipid rafts: the influence of buffer composition. Biochem J 417:525–533

Cheng HT, London E (2011) Preparation and properties of asymmetric large unilamellar vesicles: interleaflet coupling in asymmetric vesicles is dependent on temperature but not curvature. Biophys J 100:2671–2678

Cheng HT, Megha, London E (2009) Preparation and properties of asymmetric vesicles that mimic cell membranes: effect upon lipid raft formation and transmembrane helix orientation. J Biol Chem 284:6079–6092

Chiantia S, Kahya N, Ries J, Schwille P (2006) Effects of ceramide on liquid-ordered domains investigated by simultaneous AFM and FCS. Biophys J 90:4500–4508

Chiantia S, Kahya N, Schwille P (2007) Raft domain reorganization driven by short- and long-chain ceramide: a combined AFM and FCS study. Langmuir 23:7659–7665

Chiantia S, Ries J, Chwastek G, Carrer D, Li Z, Bittman R, Schwille P (2008) Role of ceramide in membrane protein organization investigated by combined AFM and FCS. Biochim Biophys Acta 1778:1356–1364

Chiantia S, Schwille P, Klymchenko AS, London E (2011) Asymmetric GUVs prepared by MbetaCD-mediated lipid exchange: an FCS study. Biophys J 100:L1–L3

Collins MD (2008) Interleaflet coupling mechanisms in bilayers of lipids and cholesterol. Biophys J 94:L32–L34

Collins MD, Keller SL (2008) Tuning lipid mixtures to induce or suppress domain formation across leaflets of unsupported asymmetric bilayers. Proc Natl Acad Sci U S A 105:124–128

Colombini M (2010) Ceramide channels and their role in mitochondria-mediated apoptosis. Biochim Biophys Acta 1797:1239–1244

Contreras FX, Basanez G, Alonso A, Herrmann A, Goni FM (2005) Asymmetric addition of ceramides but not dihydroceramides promotes transbilayer (flip-flop) lipid motion in membranes. Biophys J 88:348–359

Contreras FX, Sot J, Alonso A, Goni FM (2006) Sphingosine increases the permeability of model and cell membranes. Biophys J 90:4085–4092

Coskun U, Grzybek M, Drechsel D, Simons K (2011) Regulation of human EGF receptor by lipids. Proc Natl Acad Sci U S A 108:9044–9048

Cremesti AE, Goni FM, Kolesnick R (2002) Role of sphingomyelinase and ceramide in modulating rafts: do biophysical properties determine biologic outcome? FEBS Lett 531: 47–53

Cuadras MA, Greenberg HB (2003) Rotavirus infectious particles use lipid rafts during replication for transport to the cell surface in vitro and in vivo. Virology 313:308–321

Dasgupta U, Bamba T, Chiantia S, Karim P, Abou Tayoun AN, Yonamine I, Rawat SS, Rao RP, Nagashima K, Fukusaki E, Puri V, Dolph PJ, Schwille P, Acharya JK, Acharya U (2009) Ceramide kinase regulates phospholipase C and phosphatidylinositol 4, 5, bisphosphate in phototransduction. Proc Natl Acad Sci U S A 106:20063–20068

de Almeida RFM, Fedorov A, Prieto M (2003) Sphingomyelin/phosphatidylcholine/cholesterol phase diagram: boundaries and composition of lipid rafts. Biophys J 85:2406–2416

Dietrich C, Bagatolli LA, Volovyk ZN, Thompson NL, Levi M, Jacobson K, Gratton E (2001a) Lipid rafts reconstituted in model membranes. Biophys J 80:1417–1428

Dietrich C, Volovyk ZN, Levi M, Thompson NL, Jacobson K (2001b) Partitioning of Thy-1, GM1, and cross-linked phospholipid analogs into lipid rafts reconstituted in supported model membrane monolayers. Proc Natl Acad Sci U S A 98:10642–10647

Drevot P, Langlet C, Guo XJ, Bernard AM, Colard O, Chauvin JP, Lasserre R, He HT (2002) TCR signal initiation machinery is pre-assembled and activated in a subset of membrane rafts. EMBO J 21:1899–1908

Eggeling C, Ringemann C, Medda R, Schwarzmann G, Sandhoff K, Polyakova S, Belov VN, Hein B, von Middendorff C, Schonle A, Hell SW (2009) Direct observation of the nanoscale dynamics of membrane lipids in a living cell. Nature 457:1159–1162

Ehrig J, Petrov EP, Schwille P (2011) Near-critical fluctuations and cytoskeleton-assisted phase separation lead to subdiffusion in cell membranes. Biophys J 100:80–89

Ekholm O, Jaikishan S, Lonnfors M, Nyholm TKM, Slotte JP (2011) Membrane bilayer properties of sphingomyelins with amide-linked 2- or 3-hydroxylated fatty acids. Biochim Biophys Acta 1808:727–732

El Kirat K, Morandat S (2007) Cholesterol modulation of membrane resistance to Triton X-100 explored by atomic force microscopy. Biochim Biophys Acta 1768:2300–2309

Epand RM, Epand RF (2004) Non-raft forming sphingomyelin-cholesterol mixtures. Chem Phys Lipids 132:37–46

Everett J, Zlotnick A, Tennyson J, Holloway PW (1986) Fluorescence quenching of cytochrome b5 in vesicles with an asymmetric transbilayer distribution of brominated phosphatidylcholine. J Biol Chem 261:6725–6729

Fastenberg ME, Shogomori H, Xu X, Brown DA, London E (2003) Exclusion of a transmembrane-type peptide from ordered-lipid domains (rafts) detected by fluorescence quenching: extension of quenching analysis to account for the effects of domain size and domain boundaries. Biochemistry 42:12376–12390

Feigenson GW (2006) Phase behavior of lipid mixtures. Nat Chem Biol 2:560–563

Feigenson GW (2009) Phase diagrams and lipid domains in multicomponent lipid bilayer mixtures. Biochim Biophys Acta 1788:47–52

Filippov A, Oradd G, Lindblom G (2006) Sphingomyelin structure influences the lateral diffusion and raft formation in lipid bilayers. Biophys J 90:2086–2092

Garcia-Pacios M, Collado MI, Busto JV, Sot J, Alonso A, Arrondo JLR, Goni FM (2009) Sphingosine-1-phosphate as an amphipathic metabolite: its properties in aqueous and membrane environments. Biophys J 97:1398–1407

Garner AE, Smith DA, Hooper NM (2008) Visualization of detergent solubilization of membranes: implications for the isolation of rafts. Biophys J 94:1326–1340

Gaus K, Chklovskaia E, de St F, Groth B, Jessup W, Harder T (2005) Condensation of the plasma membrane at the site of T lymphocyte activation. J Cell Biol 171:121–131
Gaus K, Gratton E, Kable EP, Jones AS, Gelissen I, Kritharides L, Jessup W (2003) Visualizing lipid structure and raft domains in living cells with two-photon microscopy. Proc Natl Acad Sci U S A 100:15554–15559
Gaus K, Le Lay S, Balasubramanian N, Schwartz MA (2006a) Integrin-mediated adhesion regulates membrane order. J Cell Biol 174:725–734
Gaus K, Zech T, Harder T (2006b) Visualizing membrane microdomains by Laurdan 2-photon microscopy. Mol Membr Biol 23:41–48
Gidwani A, Brown HA, Holowka D, Baird B (2003) Disruption of lipid order by short-chain ceramides correlates with inhibition of phospholipase D and downstream signaling by FcepsilonRI. J Cell Sci 116:3177–3187
Goni FM, Alonso A (2009) Effects of ceramide and other simple sphingolipids on membrane lateral structure. Biochim Biophys Acta 1788:169–177
Grassme H, Jekle A, Riehle A, Schwarz H, Berger J, Sandhoff K, Kolesnick R, Gulbins E (2001) CD95 signaling via ceramide-rich membrane rafts. J Biol Chem 276:20589–20596
Grassme H, Jendrossek V, Bock J, Riehle A, Gulbins E (2002) Ceramide-rich membrane rafts mediate CD40 clustering. J Immunol 168:298–307
Grassmé H, Riethmüller J, Gulbins E (2007) Prog Lipid Res. Biological aspects of ceramide-enriched membrane domains 46:161–170.
Grosch S, Schiffmann S, Geisslinger G (2012) Chain length-specific properties of ceramides. Prog Lipid Res 51:50–62
Gulbins E, Dreschers S, Wilker B, Grassme H (2004) Ceramide, membrane rafts and infections. J Mol Med 82:357–363
Gulbins E, Grassme H (2002) Ceramide and cell death receptor clustering. Biochim Biophys Acta 1585:139–145
Gulbins E, Kolesnick R (2003) Raft ceramide in molecular medicine. Oncogene 22:7070–7077
Hakomori S (2004) Glycosynapses: microdomains controlling carbohydrate-dependent cell adhesion and signaling. An Acad Bras Cienc 76:553–572
Hakomori S, Handa K, Iwabuchi K, Yamamura S, Prinetti A (1998) New insights in glycosphingolipid function: "glycosignaling domain", a cell surface assembly of glycosphingolipids with signal transducer molecules, involved in cell adhesion coupled with signaling. Glycobiology 8:Xi–Xviii
Hamada T, Miura Y, Komatsu Y, Kishimoto Y, Vestergaard M, Takagi M (2008) Construction of asymmetric cell-sized lipid vesicles from lipid-coated water-in-oil microdroplets. J Phys Chem B 112:14678–14681
Hammond AT, Heberle FA, Baumgart T, Holowka D, Baird B, Feigenson GW (2005) Crosslinking a lipid raft component triggers liquid ordered-liquid disordered phase separation in model plasma membranes. Proc Natl Acad Sci U S A 102:6320–6325
Hannun YA (1996) Functions of ceramide in coordinating cellular responses to stress. Science 274:1855–1859
Hannun YA, Bell RM (1989) Functions of sphingolipids and sphingolipid breakdown products in cellular regulation. Science 243:500–507
Harder T, Rentero C, Zech T, Gaus K (2007) Plasma membrane segregation during T cell activation: probing the order of domains. Curr Opin Immunol 19:470–475
Heberle FA, Feigenson GW (2011) Phase separation in lipid membranes. Cold Spring Harb Perspect Biol 3:1–13
Heberle FA, Wu J, Goh SL, Petruzielo RS, Feigenson GW (2010) Comparison of three ternary lipid bilayer mixtures: FRET and ESR reveal nanodomains. Biophys J 99:3309–3318
Heerklotz H (2002) Triton promotes domain formation in lipid raft mixtures. Biophys J 83: 2693–2701
Heerklotz H, Szadkowska H, Anderson T, Seelig J (2003) The sensitivity of lipid domains to small perturbations demonstrated by the effect of Triton. J Mol Biol 329:793–799

Holopainen JM, Angelova MI, Kinnunen PKJ (2000) Vectorial budding of vesicles by asymmetrical enzymatic formation of ceramide in giant liposomes. Biophys J 78:830–838

Honerkamp-Smith AR, Cicuta P, Collins MD, Veatch SL, den Nijs M, Schick M, Keller SL (2008) Line tensions, correlation lengths, and critical exponents in lipid membranes near critical points. Biophys J 95:236–246

Hope MJ, Cullis PR (1987) Lipid asymmetry induced by transmembrane pH gradients in large unilamellar vesicles. J Biol Chem 262:4360–4366

Hope MJ, Redelmeier TE, Wong KF, Rodrigueza W, Cullis PR (1989) Phospholipid asymmetry in large unilamellar vesicles induced by transmembrane pH gradients. Biochemistry 28:4181–4187

Hu PC, Li S, Malmstadt N (2011) Microfluidic fabrication of asymmetric giant lipid vesicles. ACS Appl Mater Interfaces 3:1434–1440

Huang JY, Feigenson GW (1999) A microscopic interaction model of maximum solubility of cholesterol in lipid bilayers. Biophys J 76:2142–2157

Iwabuchi K, Nagaoka I (2002) Lactosylceramide-enriched glycosphingolipid signaling domain mediates superoxide generation from human neutrophils. Blood 100:1454–1464

Iwabuchi K, Nakayama H, Iwahara C, Takamori K (2010) Significance of glycosphingolipid fatty acid chain length on membrane microdomain-mediated signal transduction. FEBS Lett 584: 1642–1652

Jaikishan S, Bjorkbom A, Slotte JP (2010) Sphingomyelin analogs with branched N-acyl chains: the position of branching dramatically affects acyl chain order and sterol interactions in bilayer membranes. Biochim Biophys Acta 1798:1987–1994

Jaikishan S, Slotte JP (2011) Effect of hydrophobic mismatch and interdigitation on sterol/ sphingomyelin interaction in ternary bilayer membranes. Biochim Biophys Acta 1808: 1940–1945

Johnson SA, Stinson BM, Go MS, Carmona LM, Reminick JI, Fang X, Baumgart T (2010) Temperature-dependent phase behavior and protein partitioning in giant plasma membrane vesicles. Biochim Biophys Acta 1798:1427–1435

Kaiser HJ, Orlowski A, Rog T, Nyholm TKM, Chai WG, Feizi T, Lingwood D, Vattulainen I, Simons K (2011) Lateral sorting in model membranes by cholesterol-mediated hydrophobic matching. Proc Natl Acad Sci U S A 108:16628–16633

Kamiyama H, Yoshii H, Tanaka Y, Sato H, Yamamoto N, Kubo Y (2009) Raft localization of CXCR4 is primarily required for X4-tropic human immunodeficiency virus type 1 infection. Virology 386:23–31

Kan CC, Ruan ZS, Bittman R (1991) Interaction of cholesterol with sphingomyelin in bilayer-membranes - evidence that the hydroxy group of sphingomyelin does not modulate the rate of cholesterol exchange between vesicles. Biochemistry 30:7759–7766

Keller H, Lorizate M, Schwille P (2009) PI(4,5)P(2) degradation promotes the formation of cytoskeleton-free model membrane systems. Chemphyschem 10:2805–2812

Kiessling V, Crane JM, Tamm LK (2006) Transbilayer effects of raft-like lipid domains in asymmetric planar bilayers measured by single molecule tracking. Biophys J 91:3313–3326

Kiessling V, Wan C, Tamm LK (2009) Domain coupling in asymmetric lipid bilayers. Biochim Biophys Acta 1788:64–71

Kim HM, Choo HJ, Jung SY, Ko YG, Park WH, Jeon SJ, Kim CH, Joo T, Cho BR (2007) A two-photon fluorescent probe for lipid raft imaging: C-laurdan. Chembiochem 8:553–559

Klemm RW, Ejsing CS, Surma MA, Kaiser HJ, Gerl MJ, Sampaio JL, de Robillard Q, Ferguson C, Proszynski TJ, Shevchenko A, Simons K (2009) Segregation of sphingolipids and sterols during formation of secretory vesicles at the trans-Golgi network. J Cell Biol 185:601–612

Kniep B, Skubitz KM (1998) Subcellular localization of glycosphingolipids in human neutrophils. J Leukoc Biol 63:83–88

Kolesnick RN, Goni FM, Alonso A (2000) Compartmentalization of ceramide signaling: physical foundations and biological effects. J Cell Physiol 184:285–300

Korlach J, Schwille P, Webb WW, Feigenson GW (1999) Characterization of lipid bilayer phases by confocal microscopy and fluorescence correlation spectroscopy. Proc Natl Acad Sci U S A 96:8461–8466

Kuikka M, Ramstedt B, Ohvo-Rekila H, Tuuf J, Slotte JP (2001) Membrane properties of D-erythro-N-acyl sphingomyelins and their corresponding dihydro species. Biophys J 80: 2327–2337

Lafont F, Tran Van Nhieu G, Hanada K, Sansonetti P, van der Goot FG (2002) Initial steps of Shigella infection depend on the cholesterol/sphingolipid raft-mediated CD44-IpaB interaction. EMBO J 21:4449–4457

LaRocca TJ, Crowley JT, Cusack BJ, Pathak P, Benach J, London E, Garcia-Monco JC, Benach JL (2010) Cholesterol lipids of borrelia burgdorferi form lipid rafts and are required for the bactericidal activity of a complement-independent antibody. Cell Host Microbe 8:331–342

Lenne PF, Wawrezinieck L, Conchonaud F, Wurtz O, Boned A, Guo XJ, Rigneault H, He HT, Marguet D (2006) Dynamic molecular confinement in the plasma membrane by microdomains and the cytoskeleton meshwork. EMBO J 25:3245–3256

Lin WC, Blanchette CD, Longo ML (2007) Fluid-phase chain unsaturation controlling domain microstructure and phase in ternary lipid bilayers containing GalCer and cholesterol. Biophys J 92:2831–2841

Lingwood CA, Boyd B, Nutikka A (2000) Analysis of interactions between glycosphingolipids and microbial toxins. Methods Enzymol 312:459–473

Lingwood D, Ries J, Schwille P, Simons K (2008) Plasma membranes are poised for activation of raft phase coalescence at physiological temperature. Proc Natl Acad Sci U S A 105: 10005–10010

Lingwood D, Simons K (2010) Lipid rafts as a membrane-organizing principle. Science 327:46–50

London E (2005) How principles of domain formation in model membranes may explain ambiguities concerning lipid raft formation in cells. Biochim Biophys Acta 1746:203–220

Lopez D, Kolter R (2010) Functional microdomains in bacterial membranes. Genes Dev 24: 1893–1902

Lopezgarcia F, Villalain J, Gomezfernandez JC (1995) Effect of sphingosine and stearylamine on the interaction of phosphatidylserine with calcium - a study using Dsc, Ft-Ir and Ca-45(2+)-binding. Biochim Biophys Acta 1236:279–288

Lu Y, Liu DX, Tam JP (2008) Lipid rafts are involved in SARS-CoV entry into Vero E6 cells. Biochem Biophys Res Commun 369:344–349

Lyman MG, Curanovic D, Enquist LW (2008) Targeting of pseudorabies virus structural proteins to axons requires association of the viral Us9 protein with lipid rafts. PLoS Pathog 4:e1000065

Macdonald JL, Pike LJ (2005) A simplified method for the preparation of detergent-free lipid rafts. J Lipid Res 46:1061–1067

Malewicz B, Valiyaveettil JT, Jacob K, Byun HS, Mattjus P, Baumann WJ, Bittman R, Brown RE (2005) The 3-hydroxy group and 4,5-trans double bond of sphingomyelin are essential for modulation of galactosylceramide transmembrane asymmetry. Biophys J 88:2670–2680

Manneville JB, Casella JF, Ambroggio E, Gounon P, Bertherat J, Bassereau P, Cartaud J, Antonny B, Goud B (2008) COPI coat assembly occurs on liquid-disordered domains and the associated membrane deformations are limited by membrane tension. Proc Natl Acad Sci U S A 105: 16946–16951

Maunula S, Bjorkqvist YJE, Slotte JP, Ramstedt B (2007) Differences in the domain forming properties of N-palmitoylated neutral glycosphingolipids in bilayer. Biochim Biophys Acta 1768:336–345

Megha, London E (2004) Ceramide selectively displaces cholesterol from ordered lipid domains (rafts) - implications for lipid raft structure and function. J Biol Chem 279:9997–10004

Megha SP, Kolter T, Bittman R, London E (2007) Effect of ceramide N-acyl chain and polar headgroup structure on the properties of ordered lipid domains (lipid rafts). Biochim Biophys Acta 1768:2205–2212

Melkonian KA, Ostermeyer AG, Chen JZ, Roth MG, Brown DA (1999) Role of lipid modifications in targeting proteins to detergent-resistant membrane rafts. Many raft proteins are acylated, while few are prenylated. J Biol Chem 274:3910–3917
Merrill AH Jr, Stevens VL (1989) Modulation of protein kinase C and diverse cell functions by sphingosine–a pharmacologically interesting compound linking sphingolipids and signal transduction. Biochim Biophys Acta 1010:131–139
Mitchell DC, Litman BJ (1998) Effect of cholesterol on molecular order and dynamics in highly polyunsaturated phospholipid bilayers. Biophys J 75:896–908
Mombelli E, Morris R, Taylor W, Fraternali F (2003) Hydrogen-bonding propensities of sphingomyelin in solution and in a bilayer assembly: a molecular dynamics study. Biophys J 84:1507–1517
Mondal M, Mesmin B, Mukherjee S, Maxfield FR (2009) Sterols are mainly in the cytoplasmic leaflet of the plasma membrane and the endocytic recycling compartment in CHO cells. Mol Biol Cell 20:581–588
Morris RJ, Jen A, Warley A (2011) Isolation of nano-meso scale detergent resistant membrane that has properties expected of lipid 'rafts'. J Neurochem 116:671–677
Mueller V, Ringemann C, Honigmann A, Schwarzmann G, Medda R, Leutenegger M, Polyakova S, Belov VN, Hell SW, Eggeling C (2011) STED nanoscopy reveals molecular details of cholesterol- and cytoskeleton-modulated lipid interactions in living cells. Biophys J 101:1651–1660
Mukherjee S, Soe TT, Maxfield FR (1999) Endocytic sorting of lipid analogues differing solely in the chemistry of their hydrophobic tails. J Cell Biol 144:1271–1284
Murphy SC, Hiller NL, Harrison T, Lomasney JW, Mohandas N, Haldar K (2006) Lipid rafts and malaria parasite infection of erythrocytes. Mol Membr Biol 23:81–88
Mustonen P, Lehtonen J, Koiv A, Kinnunen PKJ (1993) Effects of sphingosine on peripheral membrane interactions - comparison of adriamycin, cytochrome-C, and phospholipase-A2. Biochemistry 32:5373–5380
Nakayama H, Yoshizaki F, Prinetti A, Sonnino S, Mauri L, Takamori K, Ogawa H, Iwabuchi K (2008) Lyn-coupled LacCer-enriched lipid rafts are required for CD11b/CD18-mediated neutrophil phagocytosis of nonopsonized microorganisms. J Leukoc Biol 83:728–741
Nelson LD, Chiantia S, London E (2010) Perfringolysin O association with ordered lipid domains: implications for transmembrane protein raft affinity. Biophys J 99:3255–3263
Nelson LD, Johnson AE, London E (2008) How interaction of perfringolysin O with membranes is controlled by sterol structure, lipid structure, and physiological low pH: insights into the origin of perfringolysin O-lipid raft interaction. J Biol Chem 283:4632–4642
Niemela PS, Hyvonen MT, Vattulainen I (2006) Influence of chain length and unsaturation on sphingomyelin bilayers. Biophys J 90:851–863
Nikolaus J, Scolari S, Bayraktarov E, Jungnick N, Engel S, Pia Plazzo A, Stockl M, Volkmer R, Veit M, Herrmann A (2010) Hemagglutinin of influenza virus partitions into the nonraft domain of model membranes. Biophys J 99:489–498
Niu SL, Litman BJ (2002) Determination of membrane cholesterol partition coefficient using a lipid vesicle-cyclodextrin binary system: effect of phospholipid acyl chain unsaturation and headgroup composition. Biophys J 83:3408–3415
Norambuena A, Schwartz MA (2011) Effects of integrin-mediated cell adhesion on plasma membrane lipid raft components and signaling. Mol Biol Cell 22:3456–3464
Nybond S, Bjorkqvist YJE, Ramstedt B, Slotte JP (2005) Acyl chain length affects ceramide action on sterol/sphingomyelin-rich domains. Biochim Biophys Acta 1718:61–66
Nyholm TKM, Grandell PM, Westerlund B, Slotte JP (2010) Sterol affinity for bilayer membranes is affected by their ceramide content and the ceramide chain length. Biochim Biophys Acta 1798:1008–1013
Owen DM, Lanigan PM, Dunsby C, Munro I, Grant D, Neil MA, French PM, Magee AI (2006) Fluorescence lifetime imaging provides enhanced contrast when imaging the phase-sensitive dye di-4-ANEPPDHQ in model membranes and live cells. Biophys J 90:L80–L82
Owen DM, Neil MA, French PM, Magee AI (2007) Optical techniques for imaging membrane lipid microdomains in living cells. Semin Cell Dev Biol 18:591–598

Pabst G, Boulgaropoulos B, Gander E, Sarangi BR, Amenitsch H, Raghunathan VA, Laggner P (2009) Effect of ceramide on nonraft proteins. J Membr Biol 231:125–132
Pagano RE, Martin OC, Schroit AJ, Struck DK (1981) Formation of asymmetric phospholipid membranes via spontaneous transfer of fluorescent lipid analogues between vesicle populations. Biochemistry 20:4920–4927
Pathak P, London E (2011) Measurement of lipid nanodomain (raft) formation and size in sphingomyelin/POPC/cholesterol vesicles shows TX-100 and transmembrane helices increase domain size by coalescing preexisting nanodomains but do not induce domain formation. Biophys J 101:2417–2425
Pautot S, Frisken BJ, Weitz DA (2003) Engineering asymmetric vesicles. Proc Natl Acad Sci U S A 100:10718–10721
Perlmutter JD, Sachs JN (2011) Interleaflet interaction and asymmetry in phase separated lipid bilayers: molecular dynamics simulations. J Am Chem Soc 133:6563–6577
Persaud-Sawin DA, Banach L, Harry GJ (2009) Raft aggregation with specific receptor recruitment is required for microglial phagocytosis of Abeta42. Glia 57:320–335
Pinaud F, Michalet X, Iyer G, Margeat E, Moore HP, Weiss S (2009) Dynamic partitioning of a glycosyl-phosphatidylinositol-anchored protein in glycosphingolipid-rich microdomains imaged by single-quantum dot tracking. Traffic 10:691–712
Pinto SN, Silva LC, de Almeida RFM, Prieto M (2008) Membrane domain formation, interdigitation, and morphological alterations induced by the very long chain asymmetric C24: 1 ceramide. Biophys J 95:2867–2879
Pinto SN, Silva LC, Futerman AH, Prieto M (2011) Effect of ceramide structure on membrane biophysical properties: the role of acyl chain length and unsaturation. Biochim Biophys Acta 1808:2753–2760
Pokorny A, Yandek LE, Elegbede AI, Hinderliter A, Almeida PF (2006) Temperature and composition dependence of the interaction of delta-lysin with ternary mixtures of sphingomyelin/cholesterol/POPC. Biophys J 91:2184–2197
Pontier SM, Schweisguth F (2012) Glycosphingolipids in signaling and development: from liposomes to model organisms. Dev Dyn 241:92–106
Quinn PJ (2010) A lipid matrix model of membrane raft structure. Prog Lipid Res 49:390–406
Ramstedt B, Slotte JP (1999) Comparison of the biophysical properties of racemic and d-erythro-N-acyl sphingomyelins. Biophys J 77:1498–1506
Refaei M, Leventis R, Silvius JR (2011) Assessment of the roles of ordered lipid microdomains in post-endocytic trafficking of glycosyl-phosphatidylinositol-anchored proteins in mammalian fibroblasts. Traffic 12:1012–1024
Rentero C, Zech T, Quinn CM, Engelhardt K, Williamson D, Grewal T, Jessup W, Harder T, Gaus K (2008) Functional implications of plasma membrane condensation for T cell activation. PLoS One 3:e2262
Rhome R, Del Poeta M (2010) Sphingolipid signaling in fungal pathogens. Adv Exp Med Biol 688:232–237
Riethmuller J, Riehle A, Grassme H, Gulbins E (2006) Membrane rafts in host-pathogen interactions. Biochim Biophys Acta 1758:2139–2147
Rossmann JS, Jing X, Leser GP, Lamb RA (2010) Influenza virus M2 protein mediates ESCRT-independent membrane secretion. Cell 142:902–913
Ruocco M, Atkinson D, Small DM, Skarjune R, Oldfield E, Shipley GG (1981) X-ray-diffraction and calorimetric study of anhydrous and hydrated N-palmitoylgalactosylceramide. Biophys J 20:5957–5966
Sahl SJ, Leutenegger M, Hilbert M, Hell SW, Eggeling C (2010) Fast molecular tracking maps nanoscale dynamics of plasma membrane lipids. Proc Natl Acad Sci U S A 107:6829–6834
Sandhoff R (2010) Very long chain sphingolipids: tissue expression, function and synthesis. FEBS Lett 584:1907–1913
Sato T, Iwabuchi K, Nagaoka I, Adachi Y, Ohno N, Tamura H, Seyama K, Fukuchi Y, Nakayama H, Yoshizaki F, Takamori K, Ogawa H (2006) Induction of human neutrophil chemotaxis by

Candida albicans-derived beta-1,6-long glycoside side-chain-branched beta-glucan. J Leukoc Biol 80:204–211
Schafer LV, Marrink SJ (2010) Partitioning of lipids at domain boundaries in model membranes. Biophys J 99:L91–L93
Scheiffele P, Rietveld A, Wilk T, Simons K (1999) Influenza viruses select ordered lipid domains during budding from the plasma membrane. J Biol Chem 274:2038–2044
Schroeder R, London E, Brown D (1994) Interactions between saturated acyl chains confer detergent resistance on lipids and glycosylphosphatidylinositol (GPI)-anchored proteins: GPI-anchored proteins in liposomes and cells show similar behavior. Proc Natl Acad Sci U S A 91:12130–12134
Scolari S, Engel S, Krebs N, Plazzo AP, De Almeida RF, Prieto M, Veit M, Herrmann A (2009) Lateral distribution of the transmembrane domain of influenza virus hemagglutinin revealed by time-resolved fluorescence imaging. J Biol Chem 284:15708–15716
Sengupta P, Baird B, Holowka D (2007a) Lipid rafts, fluid/fluid phase separation, and their relevance to plasma membrane structure and function. Semin Cell Dev Biol 18:583–590
Sengupta P, Hammond A, Holowka D, Baird B (2008) Structural determinants for partitioning of lipids and proteins between coexisting fluid phases in giant plasma membrane vesicles. Biochim Biophys Acta 1778:20–32
Sengupta P, Holowka D, Baird B (2007b) Fluorescence resonance energy transfer between lipid probes detects nanoscopic heterogeneity in the plasma membrane of live cells. Biophys J 92: 3564–3574
Shah J, Atienza JM, Duclos RI, Rawlings AV, Dong ZX, Shipley GG (1995) Structural and thermotropic properties of synthetic C16-0 (palmitoyl) ceramide - effect of hydration. J Lipid Res 36:1936–1944
Shakor ABA, Kwiatkowska K, Sobota A (2004) Cell surface ceramide generation precedes and controls Fc gamma RII clustering and phosphorylation in rafts. J Biol Chem 279:36778–36787
Shogomori H, Hammond AT, Ostermeyer-Fay AG, Barr DJ, Feigenson GW, London E, Brown DA (2005) Palmitoylation and intracellular domain interactions both contribute to raft targeting of linker for activation of T cells. J Biol Chem 280:18931–18942
Silva LC, de Almeida RF, Castro BM, Fedorov A, Prieto M (2007) Ceramide-domain formation and collapse in lipid rafts: membrane reorganization by an apoptotic lipid. Biophys J 92: 502–516
Simon CG Jr, Gear AR (1998) Membrane-destabilizing properties of C2-ceramide may be responsible for its ability to inhibit platelet aggregation. Biochemistry 37:2059–2069
Simons K, Ikonen E (1997) Functional rafts in cell membranes. Nature 387:569–572
Simons K, Toomre D (2000) Lipid rafts and signal transduction. Nat Rev Mol Cell Biol 1:31–39
Simons K, van Meer G (1988) Lipid sorting in epithelial cells. Biochemistry 27:6197–6202
Singh RD, Holicky EL, Cheng ZJ, Kim SY, Wheatley CL, Marks DL, Bittman R, Pagano RE (2007) Inhibition of caveolar uptake, SV40 infection, and beta 1-integrin signaling by a nonnatural glycosphingolipid stereoisomer. J Cell Biol 176:895–901
Singh RD, Liu YD, Wheatley CL, Holicky EL, Makino A, Marks DL, Kobayashi T, Subramaniam G, Bittman R, Pagano RE (2006) Caveolar endocytosis and microdomain association of a glycosphingolipid analog is dependent on its sphingosine stereochemistry. J Biol Chem 281: 30660–30668
Siskind LJ, Colombini M (2000) The lipids C-2- and C-16-ceramide form large stable channels - implications for apoptosis. J Biol Chem 275:38640–38644
Smaby JM, Kulkarni VS, Momsen M, Brown RE (1996) The interfacial elastic packing interactions of galactosylceramides, sphingomyelins, and phosphatidylcholines. Biophys J 70:868–877
Smart EJ, Ying YS, Mineo C, Anderson RG (1995) A detergent-free method for purifying caveolae membrane from tissue culture cells. Proc Natl Acad Sci U S A 92:10104–10108
Sohn HW, Pierce SK, Tzeng SJ (2008a) Live cell imaging reveals that the inhibitory FcgammaRIIB destabilizes B cell receptor membrane-lipid interactions and blocks immune synapse formation. J Immunol 180:793–799

Sohn HW, Tolar P, Pierce SK (2008b) Membrane heterogeneities in the formation of B cell receptor-Lyn kinase microclusters and the immune synapse. J Cell Biol 182:367–379
Song KS, Li S, Okamoto T, Quilliam LA, Sargiacomo M, Lisanti MP (1996) Co-purification and direct interaction of Ras with caveolin, an integral membrane protein of caveolae microdomains. Detergent-free purification of caveolae microdomains. J Biol Chem 271: 9690–9697
Sonnino S, Prinetti A, Nakayama H, Yangida M, Ogawa H, Iwabuchi K (2009) Role of very long fatty acid-containing glycosphingolipids in membrane organization and cell signaling: the model of lactosylceramide in neutrophils. Glycoconj J 26:615–621
Sorre B, Callan-Jones A, Manneville JB, Nassoy P, Joanny JF, Prost J, Goud B, Bassereau P (2009) Curvature-driven lipid sorting needs proximity to a demixing point and is aided by proteins. Proc Natl Acad Sci U S A 106:5622–5626
Sot J, Ibarguren M, Busto JV, Montes LR, Goni FM, Alonso A (2008) Cholesterol displacement by ceramide in sphingomyelin-containing liquid-ordered domains, and generation of gel regions in giant lipidic vesicles. FEBS Lett 582:3230–3236
Sperling P, Heinz E (2003) Plant sphingolipids: structural diversity, biosynthesis, first genes and functions. Biochim Biophys Acta 1632:1–15
Taylor DR, Hooper NM (2007) Role of lipid rafts in the processing of the pathogenic prion and Alzheimer's amyloid-beta proteins. Semin Cell Dev Biol 18:638–648
Teneberg S, Angstrom J, Ljungh A (2004) Carbohydrate recognition by enterohemorrhagic Escherichia coli: characterization of a novel glycosphingolipid from cat small intestine. Glycobiology 14:187–196
tenGrotenhuis E, Demel RA, Ponec M, Boer DR, vanMiltenburg JC, Bouwstra JA (1996) Phase behavior of stratum corneum lipids in mixed Langmuir-Blodgett monolayers. Biophys J 71: 1389–1399
Terova B, Heczko R, Slotte JP (2005) On the importance of the phosphocholine methyl groups for sphingomyelin/cholesterol interactions in membranes: a study with ceramide phosphoethanolamine. Biophys J 88:2661–2669
Tong J, Briggs MM, Mlaver D, Vidal A, McIntosh TJ (2009) Sorting of lens aquaporins and connexins into raft and nonraft bilayers: role of protein homo-oligomerization. Biophys J 97: 2493–2502
Trajkovic K et al (2008) Ceramide triggers budding of exosome vesicles into multivesicular endosomes. Science 319:1244–1247
van Meer G (2011) Dynamic transbilayer lipid asymmetry. Cold Spring Harb Perspect Biol 3: a004671
van Meer G, Simons K (1988) Lipid polarity and sorting in epithelial cells. J Cell Biochem 36: 51–58
van Zanten TS, Gomez J, Manzo C, Cambi A, Buceta J, Reigada R, Garcia-Parajo MF (2010) Direct mapping of nanoscale compositional connectivity on intact cell membranes. Proc Natl Acad Sci U S A 107:15437–15442
Varma R, Mayor S (1998) GPI-anchored proteins are organized in submicron domains at the cell surface. Nature 394:798–801
Veatch SL, Cicuta P, Sengupta P, Honerkamp-Smith A, Holowka D, Baird B (2008) Critical fluctuations in plasma membrane vesicles. ACS Chem Biol 3:287–293
Veatch SL, Keller SL (2003) Separation of liquid phases in giant vesicles of ternary mixtures of phospholipids and cholesterol. Biophys J 85:3074–3083
Veatch SL, Keller SL (2005) Seeing spots: complex phase behavior in simple membranes. Biochim Biophys Acta 1746:172–185
Veatch SL, Soubias O, Keller SL, Gawrisch K (2007) Critical fluctuations in domain-forming lipid mixtures. Proc Natl Acad Sci U S A 104:17650–17655
Verkleij AJ, Zwaal RF, Roelofsen B, Comfurius P, Kastelijn D, van Deenen LL (1973) The asymmetric distribution of phospholipids in the human red cell membrane. A combined study

using phospholipases and freeze-etch electron microscopy. Biochim Biophys Acta 323: 178–193

Viard M, Parolini I, Rawat SS, Fecchi K, Sargiacomo M, Puri A, Blumenthal R (2003) The role of glycosphingolipids in HIV signaling, entry and pathogenesis. Glycoconj J 20:213–222

Vieira CR, Munoz-Olaya JM, Sot J, Jimenez-Baranda S, Izquierdo-Useros N, Abad JL, Apellaniz B, Delgado R, Martinez-Picado J, Alonso A, Casas J, Nieva JL, Fabrias G, Manes S, Goni FM (2010) Dihydrosphingomyelin impairs HIV-1 infection by rigidifying liquid-ordered membrane domains. Chem Biol 17:766–775

Wan C, Kiessling V, Tamm LK (2008) Coupling of cholesterol-rich lipid phases in asymmetric bilayers. Biochemistry 47:2190–2198

Wartewig S, Neubert RHH (2007) Properties of ceramides and their impact on the stratum corneum structure: a review. Skin Pharmacol Physiol 20:220–229

Westerlund B, Grandell PM, Isaksson YJE, Slotte JP (2010) Ceramide acyl chain length markedly influences miscibility with palmitoyl sphingomyelin in bilayer membranes. Eur Biophys J 39: 1117–1128

Westerlund B, Slotte JP (2009) How the molecular features of glycosphingolipids affect domain formation in fluid membranes. Biochim Biophys Acta 1788:194–201

Williamson R, Usardi A, Hanger DP, Anderton BH (2008) Membrane-bound beta-amyloid oligomers are recruited into lipid rafts by a fyn-dependent mechanism. FASEB J 22:1552–1559

Young RM, Holowka D, Baird B (2003) A lipid raft environment enhances Lyn kinase activity by protecting the active site tyrosine from dephosphorylation. J Biol Chem 278:20746–20752

Young RM, Zheng X, Holowka D, Baird B (2005) Reconstitution of regulated phosphorylation of FcepsilonRI by a lipid raft-excluded protein-tyrosine phosphatase. J Biol Chem 280: 1230–1235

Zech T, Ejsing CS, Gaus K, de Wet B, Shevchenko A, Simons K, Harder T (2009) Accumulation of raft lipids in T-cell plasma membrane domains engaged in TCR signalling. EMBO J 28: 466–476

Zhang H, Abraham N, Khan LA, Hall DH, Fleming JT, Gobel V (2011) Apicobasal domain identities of expanding tubular membranes depend on glycosphingolipid biosynthesis. Nat Cell Biol 13:1189–1201

Zhao J, Wu J, Shao H, Kong F, Jain N, Hunt G, Feigenson G (2007) Phase studies of model biomembranes: macroscopic coexistence of Lalpha + Lbeta, with light-induced coexistence of Lalpha + Lo Phases. Biochim Biophys Acta 1768:2777–2786

Sphingolipid Metabolism and Neutral Sphingomyelinases

Michael V. Airola and Yusuf A. Hannun

Contents

Abstract Sphingolipids are an important class of lipid molecules that play fundamental roles in our cells and body. Beyond a structural role, it is now clearly established that sphingolipids serve as bioactive signaling molecules to regulate diverse processes including inflammatory signaling, cell death, proliferation, and pain sensing. Sphingolipid metabolites have been implicated in the onset and progression of various diseases including cancer, lung disease, diabetes, and lysosomal storage disorders. Here we review sphingolipid metabolism to introduce basic concepts as well as emerging complexities in sphingolipid function gained from modern technological advances and detailed cell and animal studies. Furthermore, we discuss the family of neutral sphingomyelinases (N-SMases), which generate ceramide through the hydrolysis of sphingomyelin and are key enzymes

M.V. Airola • Y.A. Hannun (✉)
Department of Medicine, Stony Brook University Cancer Center, Stony Brook University, Health Science Center, L-4, 178, Stony Brook, NY 11794, USA
e-mail: yusuf.hannun@stonybrookmedicine.edu

E. Gulbins and I. Petrache (eds.), *Sphingolipids: Basic Science and Drug Development*, Handbook of Experimental Pharmacology 215, DOI 10.1007/978-3-7091-1368-4_3,

in sphingolipid metabolism. Four mammalian N-SMase enzymes have now been identified. Most prominent is nSMase2 with established roles in bone mineralization, exosome formation, and cellular stress responses. Function for the other N-SMases has been more enigmatic and is an area of active investigation. The known properties and potential role(s) of each enzyme are discussed to help guide future studies.

Keywords Sphingolipid metabolism • Sphingomyelin • Ceramide • Sphingosine-1-phosphate • Neutral sphingomyelinase

1 Part One: Sphingolipid Metabolism

Sphingolipids encompass a broad range of lipid molecules that elicit a wide range of signaling properties and cellular functions (Hannun and Obeid 2008). To understand the diverse functionality of sphingolipids in health and disease, it is important to be familiar with the chemical structure of these lipid molecules, their cellular metabolism, and the sphingolipid-metabolizing enzymes. Here we present a brief review of these areas in an effort to introduce the main points and emerging concepts in the field.

1.1 Introduction

1.1.1 Sphingolipids

At the most basic level, sphingolipids can be defined as any lipid molecule that contains the sphingoid backbone, derived from the condensation of an amino acid (predominantly serine) and a fatty acid (predominantly palmitate) (Fig. 1b) (Merrill 2011). The presence or the absence of an acyl chain distinguishes ceramide (Cer) from sphingosine (Sph), while phosphorylation of the 1-hydroxy group generates ceramide-1-phosphate (C1P) or sphingosine-1-phosphate (S1P). Other common sphingolipids contain different headgroups at this position. Sphingomyelin (SM) contains a phosphorylcholine headgroup, and the basic glycosphingolipids, glucosylceramide (GluCer) and galactosylceramide (GalCer), contain a single sugar molecule linked to ceramide.

1.1.2 De Novo Biosynthesis

Sphingolipids can be synthesized de novo by a single biosynthetic pathway, with the end product being Cer (Bartke and Hannun 2009) (Fig. 1a). The first step is catalyzed by the serine palmitoyl transferase (SPT) complex and condenses the amino acid serine and the fatty acid palmitate to form 3-keto-dihydrosphingosine.

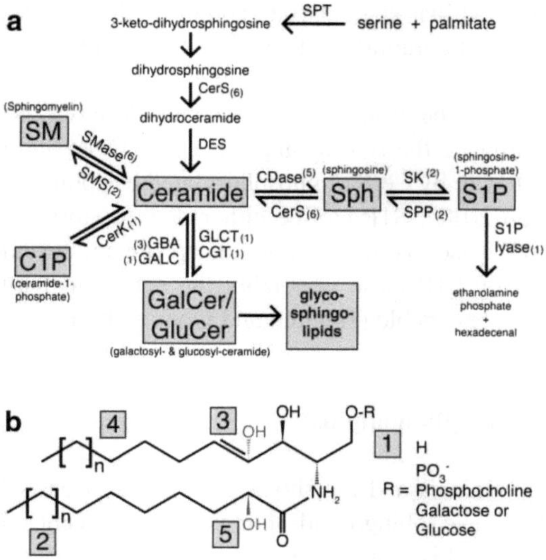

Fig. 1 Sphingolipid metabolism and chemical structure. (**a**) Diagram of sphingolipid metabolism showing the major lipid species in *grey* boxes and sphingolipid-metabolizing enzymes. The number of mammalian genes that catalyze each conversion are denoted by *brackets*. (**b**) Generic structure for sphingolipid molecules with modification points. The presence or the absence of the acyl chain (*lower chain*) distinguishes ceramide (shown) from sphingosine. Variation in the headgroup (1), attached to the terminal 1-oxygen, distinguishes each family of sphingolipids. The length of the sphingoid backbone (4) or acyl chain (2) generates subspecies within each family. Further modification at the 4,5 position on the sphingoid backbone (3) can occur with different saturation levels (e.g., dihydroCer vs. Cer) and hydroxylation. In addition, the acyl chain can be hydroxylated at the 2-position (5). SPT = serine phosphoryltransferase, CerS = (dihydro) ceramide synthase, DES = dihydroceramide desaturase, SMase = sphingomyelinase, SMS = sphingomyelin synthase, CerK = ceramide kinase, GBA = glucosylceramidase, GALC = galactosylceramidase, GCLT = ceramide glucosyltransferase, CGT = ceramide galactosyl-transferase, CDase = ceramidase, SK = sphingosine kinase, SPP = S1P-phosphatase

The carbonyl group of 3-keto-dihydrosphingosine is then reduced to form dihydrosphingosine, and the enzyme (dihydro)ceramide synthase (CerS) then adds a fatty acid chain (the acyl chain) by *N*-acylation to form dihydroceramide. Finally, desaturation of the 4,5 carbon–carbon bond on the sphingoid backbone generates Cer (Gault et al. 2010).

1.1.3 Sphingolipid Metabolism

From Cer as a central point, sphingolipid metabolism branches out in four main directions (Fig. 1a). Three of these reactions alter the headgroup of Cer. These include the phosphorylation of Cer by ceramide kinase (CerK) to produce C1P (Arana et al. 2010), the addition of phosphocholine by sphingomyelin synthase (SMS) to produce SM (via transfer of the phosphocholine headgroup from phosphatidylcholine) (Milhas et al. 2010a), and the addition of a sugar molecule by glucosyl- and galactosyl-ceramidesynthases to create GluCer or GalCer, respectively (Merrill 2011).

Alternatively, Cer can be broken down by ceramidase (CDase), which removes the acyl chain, to produce the lyso-sphingolipid Sph (Mao and Obeid 2008). Sph can be reconverted to Cer by CerS or phosphorylated by sphingosine kinase (SK) to produce S1P (Pitson 2011). S1P can be either dephosphorylated back to Sph or broken down by S1P lyase to ethanolamine phosphate and hexadecenal (Serra and Saba 2010). Action by S1P lyase is notably the sole exit point for sphingolipid breakdown and is not reversible (Hannun and Obeid 2008).

1.1.4 New Bioactive Sphingolipids

The advances in lipodomics and metabolomics provide a basis for the continuing discovery of new bioactive sphingolipids. For example, a recent study identified *N,N*-dimethylsphingosine (DMS) as a new bioactive molecule, inducing chronic pain in mice (Patti et al. 2012). The pathway and enzymes responsible for DMS synthesis have not been established but may simply involve the di-methylation of Sph.

1.2 Complexity in Sphingolipid Metabolism

1.2.1 Variations in Chemical Structure

Each lipid species described above comprises a family of molecules that share the same basic framework (i.e., same headgroup) but can differ in their chemical structure in specific ways (Fig. 1b). A common variation is the length of the acyl chain (Merrill 2011). This occurs through utilization of different length fatty acyl-CoAs by CerS enzymes (Pewzner-Jung et al. 2006). In addition, the length of the sphingoid base can vary if the SPT complex uses myristate or stearate as a substrate instead of palmitate (Han et al. 2009). Besides chain length, both the acyl chain and sphingoid base can be hydroxylated (Hama 2010) and the sphingoid base can be saturated or desaturated (Pruett et al. 2008). Overall, these variations create a family of related molecules that have the potential for distinct molecular and cellular functions.

1.2.2 Compartmentalization and Specificity

In mammalian sphingolipid metabolism, there is often more than one enzyme in the cell to catalyze each chemical reaction (Fig. 1a) (Bartke and Hannun 2009). From a cellular perspective, the functional reasons for enzyme multiplicity are twofold. First, isozymes can be localized in different cellular compartments. This allows organelle-specific activity and separate regulatory mechanisms. For example, there have been six identified SMase isoforms that localize to various compartments including the lysosome, inner or outer leaflet of the plasma membrane, ER, mitochondria, and nucleus (Jenkins et al. 2009; Wu et al. 2010a). Secondly, isozymes can have different substrate specificities to allow the formation or breakdown of particular molecular species. Representatively, the CerS isozymes appear to all localize to the ER but exhibit different preferences in fatty acid chain length (Mullen et al. 2012).

1.2.3 Interconnectivity of Metabolites

The metabolic network of sphingolipids is connected through a series of chemical reactions that transform one bioactive metabolite to another (Bartke and Hannun 2009) (Fig. 1a). Expanding beyond two-state equilibriums, we see a connectivity map that allows the production of distant metabolites from another; e.g., S1P from SM by the sequential action of SMases (SM to Cer), CDases (Cer to Sph), and SKs (Sph to S1P). Given the large excess of SM to S1P, it is a logical progression that a relatively small SM depletion has the ability to profoundly affect S1P production. Overall, this means the bioactive lipid generated by the activation of SMases or CDases, and the associated cellular response, does not have to directly correlate with the immediate reaction product (Hannun and Obeid 2008).

1.3 Emerging Concepts

1.3.1 Many Ceramides

A new paradigm for sphingolipid metabolism and function has emerged that integrates the significant advances described above into a working model. In the *many ceramides* hypothesis, each Cer molecular species, and related bioactive sphingolipid, has the potential to illicit a unique cellular response (Hannun and Obeid 2011). Unique functions may arise from differences in the bioactive molecule's structure, protein targets, and/or subcellular localization.

1.3.2 Molecular Specificity in Cellular Function

An elegant example of highly selective recognition of a lipid molecule was recently reported that embodies the essence of the *many ceramides* hypothesis (Patti et al. 2012). The authors found that the COPI machinery protein p24 specifically recognized and bound SM with 18 carbons. This recognition was required for efficient COPI-dependent transport. SM-18 was previously reported to be enriched in the Golgi, which suggests that this molecular specificity drives a compartment-specific cellular function (Patti et al. 2012).

Many of the other current examples include the chain length-dependent correlation of Cer molecules in various processes including autophagy, apoptosis, inflammation, and cancer (Grosch et al. 2012). In general, these processes have been associated with the action of CerS enzymes, which have different chain-length specificities. Based on their chain-length specificities, the involvement of individual CerS enzymes in these pathologies has been either inferred or tested experimentally (Mullen et al. 2011; Schiffmann et al. 2010; Mesicek et al. 2010; Ben-David and Futerman 2010).

As another example, the three-member family of alkaline ceramidases (AlkCDase1-3) has recently been characterized to have different specificities in the hydrolysis of Cer. AlkCDase1 hydrolyzes very-long-chain Cers (Mao and Obeid 2008), AlkCDase2 has broad substrate specificity (Sun et al. 2010), and AlkCDase3 hydrolyzes long-chain Cers (Hu et al. 2010) to regulate the levels of distinct Cer subclasses in the Golgi.

1.3.3 Mechanistic Implications

It is interesting to consider how this molecular specificity is functionally translated. Consider that all the distinguishing structural characteristics of different Cer molecules are embedded in the membrane and should only be recognizable by protein transmembrane domains. Proteins working only at the membrane interface should not be able to distinguish these molecular differences. This should limit the recognition of sphingolipid specificities to integral membrane proteins or proteins with significant hydrophobic pockets or membrane insertions.

1.4 Conclusions

Although new sphingolipids and their functions are still being identified, the basic outline for sphingolipid metabolism is well established. Involvement of sphingolipids and sphingolipid-metabolizing enzymes in new pathologies and signaling pathways continues to increase. Moving forward, the challenge is to understand the molecular mechanisms underlying sphingolipid functions using the advancing technology and our increased appreciation of sphingolipid complexities. Investigations of this nature will significantly advance our knowledge to benefit in the identification of novel targets and strategies for therapeutic intervention in health and disease.

2 Part Two: Neutral Sphingomyelinases

Neutral sphingomyelinases (N-SMases) are a family of related enzymes that catalyze the hydrolysis of SM to generate Cer and phosphorylcholine. Cer and sphingolipid metabolites are well-established regulators of many important cellular signaling pathways and are implicated in human health and disease (Wu et al. 2010b). In this section we discuss the mammalian N-SMase enzymes highlighting biochemical properties, localization, and roles in lipid metabolism, cellular signaling, and physiology.

2.1 Introduction

2.1.1 The Mammalian N-SMases

Since the original identification of N-SMase activity in 1967 (Schneider and Kennedy 1967) four mammalian N-SMase genes have been cloned or purified (Fig. 2a). These include nSMase1 (gene name = *SMPD2*), nSMase2 (*SMPD3*), nSMase3 (*SMPD4*), and MA-nSMase (mitochondrial-associated nSMase) (*SMPD5*). NSMase2 is currently the best-studied isoform with established roles in bone mineralization, cell growth arrest, exosome formation, and inflammatory response. The roles of the other N-SMases in mammalian physiology and biology are still ambiguous due to either an enigmatic function or an only recent identification.

2.1.2 Biochemical and Structural Features

Most N-SMases, from bacteria to mammals, share a DNase I-type catalytic core suggesting a common catalytic mechanism for SM hydrolysis (Clarke et al. 2006). NSMase3 is an exception and is discussed separately. N-SMases catalyze the hydrolysis of SM in a PLC-type manner to generate the reaction products: Cer and phosphorylcholine (Fig. 2c). The active site is defined by eight metal-binding residues, which together bind two Mg^{2+} ions (Ago et al. 2006). Other highly conserved residues are found near the active site and may be important for catalytic activity. Two examples are Asp and Lys residues found in a P-loop-like motif, which are both required for catalysis in the yeast N-SMase homolog Isc1 (Okamoto et al. 2003). The crystal structures of two N-SMase bacterial homologs have defined the general protein fold and position of these conserved residues (Ago et al. 2006; Openshaw et al. 2005).

All mammalian N-SMases contain an extra hydrophobic domain that tethers the catalytic domain to the membrane (Fig. 2a). The hydrophobic domain can also play other roles in phospholipid binding, subcellular localization, and enzyme activation to contribute to the activity and regulation of N-SMase enzymes in vivo.

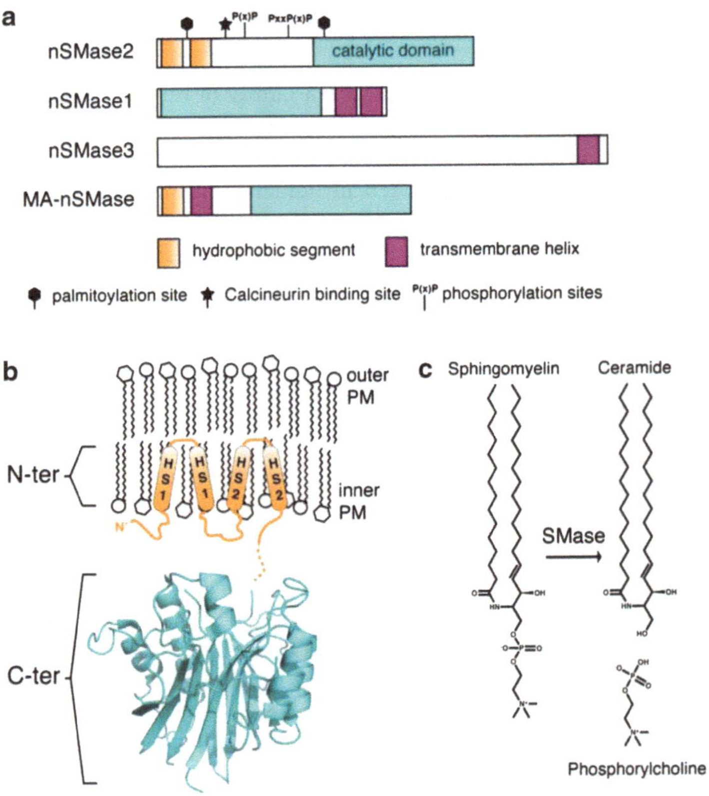

Fig. 2 Domain architecture and topology of N-SMase isoforms. (**a**) Domain architecture of N-SMase isoforms highlighting the catalytic domain, membrane-associated or transmembrane regions, and sites of protein binding or posttranslational modifications. (**b**) nSMase2 contains two domains: an N-terminal domain with two hydrophobic segments (HS1 and HS2) that associate with but do not span the membrane and a C-terminal catalytic domain (*blue*). For the catalytic domain, the structure of a bacterial homologue (bSMase, PDB: 2DDR) is shown with the active site towards the membrane. (**c**) Sphingomyelinases catalyze the hydrolysis of SM to generate Cer and phosphocholine

2.1.3 Subcellular Localization

In general, sphingolipids and sphingolipid metabolism are segregated into membrane compartments (Hannun and Obeid 2008). The localization of each N-SMase enzyme differs, suggesting that each enzyme is responsible for SM hydrolysis and ceramide formation in specific organelles. In addition, enzyme localization constrains the biologically relevant activation mechanisms, as the concentration of different lipid molecules is also organelle dependent.

2.2 *Neutral SMase2*

NSMase2 is the best-studied mammalian nSMase and has emerged as a key mediator of cellular stress-induced generation of Cer, as well as a somewhat surprising role in bone mineralization. Biochemical and physical characterization has identified a number of mechanisms for activation and regulation of nSMase2. These studies provide a template to investigate other N-SMase isoforms, as well as potential areas and modes for therapeutic intervention.

2.2.1 Biochemical Properties

Two domains have been identified in nSMase2: an N-terminal domain that is hydrophobic and tethers nSMase2 to the membrane and a C-terminal domain encompassing the catalytic domain (Clarke et al. 2011a) (Fig. 2b). Interestingly, the region separating these domains contains recently identified serine phosphorylation sites (Filosto et al. 2012) and a calcineurin-binding site (Filosto et al. 2010) suggesting that this may be a major area for activation and regulation of nSMase2.

NSMase2 exhibits low basal SMase activity in vitro and requires activation by anionic phospholipids (APLs) (Hofmann et al. 2000). The requirement of APLs is consistent with the location of nSMase2 at the plasma membrane (PM) (Milhas et al. 2010a, b), rich in the APL phosphatidylserine (PS) (Leventis and Grinstein 2010). Another potential biologically relevant APL is the minor lipid phosphatidic acid (PA) transiently produced at the PM in low concentrations (Stace and Ktistakis 2006). Given the different spatiotemporal dynamics of PS and PA at the PM, we speculate that constitutive function of nSMase2 may depend on PS, while PA generation may be one mechanism for acute activation.

The interaction of nSMase2 with the membrane has been studied in detail. The N-terminal domain contains two hydrophobic segments (Fig. 2), predicted to be helical, that associate with but do not span the membrane (Tani and Hannun 2007a) (Fig. 2b). Additionally, nSMase2 harbors two palmitoylation sites that contribute to nSMase2 membrane association (Tani and Hannun 2007b). Recently, the APL-binding domain was found to localize exclusively to the N-terminal domain (Wu et al. 2011). A binding motif consisting of three conserved Arg residues is necessary for APL binding, APL-mediated activation, and correct trafficking of nSMase2.

2.2.2 Localization

NSMase2 localizes to the plasma membrane, Golgi, and recycling compartments. TNF-α, PMA, H_2O_2, and cell confluence all induce nSMase2 translocation from the Golgi to the PM, and correlate with increased N-SMase activity. This suggests that the major site of nSMase2 action is the inner leaflet of the PM. Translocation may regulate activity by controlling access to substrate SM, activating APLs, or involve other mechanisms.

2.2.3 GW4869: An nSMase2-Specific Inhibitor

NSMase2 is the only N-SMase with a specific inhibitor, GW4869 (Luberto et al. 2002; Canals et al. 2011). GW4869 has been widely used as a tool to identify and confirm nSMase2-specific functions (Wu et al. 2010b). Mechanistically, GW4869 is thought to inhibit nSMase2 by interfering with APL activation (Luberto et al. 2002). We note that any inhibitory effect of GW4869 on the recently identified MA-nSMase has not been assessed and should be tested.

2.2.4 Cellular Signaling

NSMase2 has been implicated in the response to various cellular stresses and cytokines to affect a diverse set of signaling pathways including cancer pathogenesis (Kim et al. 2008; Ito et al. 2009), growth and development (Stoffel et al. 2005, 2007), and inflammatory responses (Nikolova-Karakashian et al. 2008; Devillard et al. 2010). It was originally cloned as a confluence arrest gene (Hayashi et al. 1997) and has a demonstrated role linking confluence to cell cycle arrest (Marchesini et al. 2004). Most recently, it has been linked as an upstream regulator of the mTOR/S6 kinase pathway (Clarke et al. 2011b) and shown that this pathway regulates the biosynthesis of the extracellular matrix component hyaluronan, which is a glycosaminoglycan produced at the PM (Qin et al. 2012).

Activation by TNF-α is currently the best-studied activation pathway and occurs through PKC-δ (Clarke et al. 2008) and p38 MAPK-dependent mechanisms (Clarke et al. 2007). Direct protein interaction with embryonic ectodermal development (EED) couples nSMase2 to TNF receptor 1 through FAN (factor associated with N-SMase activation) (Adam-Klages et al. 1996) and the scaffolding protein RACK1 (Philipp et al. 2010).

Phosphorylation has emerged as one mechanism regulating nSMase2 activity. Phosphorylation has been shown to modulate the activity of nSMase2, to be regulated by the phosphatase calcineurin (Filosto et al. 2010), and to occur through p38 MAPK- and PKC-dependent pathways (Clarke et al. 2007, 2008). Recently, two clusters of specific serine residues were identified as phosphorylation sites and found to affect protein stability (Filosto et al. 2012). The actual kinase responsible for directly phosphorylating nSMase2 is unknown.

A role for nSMase2 in Cer-mediated lung injury has emerged (Goldkorn and Filosto 2010). Oxidants in cigarette smoke and hydrogen peroxide increase nSMase2 activity and Cer levels leading to increased apoptosis in bronchial epithelial cells and the lung tissues of rodents (Levy et al. 2006, 2009; Filosto et al. 2011).

The effects of cigarette smoke can be blocked by the antioxidants glutathione and N-acetyl cysteine. In addition, the expression of nSMase2 was increased in lung tissues of smokers with emphysema (Filosto et al. 2011).

NSMase2 may play a role in tumorigenesis with mutations identified in the SMPD3 gene in several human leukemias (Kim et al. 2008). Two of these mutations affect function by either reducing protein stability or altering localization. Other studies suggest that nSMase2 may be important for cell death in transformed cells (Wu et al. 2010b). For example, NSMase2 mRNA and protein levels were induced by the anticancer drug daunorubicin, leading to increased N-SMase activity, Cer levels, and cell death (Ito et al. 2009).

Several studies have now demonstrated that ceramide and nSMase2 are key regulators of exosome formation and microRNA (miRNA) secretion (Chen et al. 2012; Aubin et al. 2005; Khavandgar et al. 2011). miRNAs are gene regulatory elements found both intra- and extracellularly (Chen et al. 2012). It is thought that they are encapsulated in exosomes, leading to their secretion outside the cell. Interestingly, overexpression of nSMase2 increased miRNA secretion, while inhibition by GW4869 or knockdown by siRNA of nSMase2 decreased miRNA secretion. This process was not dependent on the endosomal sorting complex required for transport (ESCRT) implicating a novel ceramide-dependent, ESCRT-independent secretion mechanism. In addition, these studies suggest that the Golgi may be a second site of action for nSMase2 SM hydrolysis.

2.2.5 Physiological Role in Skeletal Development

Animal studies of nSMase2-deficient mice have identified a role for nSMase2 in bone homeostasis. The nSMase2 knockout mice suffer from short stature (Aubin et al. 2005), while the nSMase2-inactivating *fro/fro* (for *fragilitas ossium*) mutation results in bone fragility (Stoffel et al. 2007). The observed skeletal abnormalities include short and bent limbs, as well as deformations in rib cages, long bones, and growth plate cartilage (Khavandgar et al. 2011).

Recently, it was shown that nSMase2 plays cell-specific roles in skeletal development (Khavandgar et al. 2011). Fro/fro mice are defective in both bone and cartilage mineralization and these two events are regulated by the different cell types: osteoblasts and chondrocytes, respectively. By selectively expressing nSMase2 in the osteoblasts of fro/fro mice, the authors were able to correct the osteoblast-specific bone defects but not affect the cartilage defects (Khavandgar et al. 2011).

The role of nSMase2 appears to involve a novel mechanism that does not involve the typical factors such as calcium, phosphate, and alkaline phosphates. This suggests a novel, nSMase2-dependent mechanism gating proper bone mineralization and may help in understanding cases of osteogenesis imperfecta in human patients that also share similar phenotypes and undetectable differences in typical mineralization parameters (Khavandgar et al. 2011; Glorieux et al. 2002). Overall, these discoveries are exciting and future work deciphering the molecular mechanisms of nSMase2 in bone mineralization is of great interest.

2.3 *Neutral SMase1*

2.3.1 Biochemical Properties

NSMase1 was the first identified and cloned mammalian N-SMase based on sequence homology to bacterial SMases (Tomiuk et al. 1998). The domain architecture of nSMase1 is identical to Isc1, the yeast homologue to N-SMases (Matmati and Hannun 2008), with a catalytic domain followed by two C-terminal transmembrane helices (Fig. 2a). Unlike other N-SMases, nSMase1 is not activated by phospholipids (Tomiuk et al. 1998).

2.3.2 Localization

The localization of nSMase1 appears to vary when comparing endogenous to overexpressed protein. Overexpressed nSMase1 mainly colocalizes with endoplasmic reticulum (ER) markers (Tomiuk et al. 2000). However, endogenous nSMase1, in contrast to overexpressed nSMase1, was reported to localize to the nuclear matrix (Mizutani et al. 2001).

2.3.3 A Role for nSMase1 in Lipid Metabolism?

Despite the activity of nSMase1 on SM in vitro, overexpression in cells does not affect SM metabolism (Tomiuk et al. 1998; Sawai et al. 1999). This ambiguity casts doubt on the role of nSMase1 in sphingolipid metabolism. Lyso-platelet-activating factor (lyso-PAF) may be a biologically relevant substrate, with nSMase1 displaying both PLC-lyso-PAF activity in vitro and in cells (Sawai et al. 1999). However, nSMase1 KO mice had no detectable changes in sphingolipid metabolism by high-performance thin-layer chromatography (HPTLC), a relatively insensitive method compared to mass spectrometry, including no alterations in SM or lyso-PAF metabolism.

Although a function for nSMase1 is not apparent, this does not preclude a role for nSMase1 in lipid metabolism. Given the significant advancements in lipid quantification by mass spectrometry and the emerging *many ceramides* model (Hannun and Obeid 2011), it would be interesting to measure lipid levels in nSMase1 KO mice to determine if discrete SM, Cer, or lyso-PAF species, undetectable by HPTLC, are altered. Quantification of ceramide levels in nSMase1 transiently transfected MCF7 cells did result in slight increases in some ceramide levels (Clarke et al. 2011c). Correct substrate identification may require in vivo measurements from KO animal tissues, as recently demonstrated for a different lipid-metabolizing enzyme (Long et al. 2011). With regard to nSMase1, this may be particularly relevant given the different subcellular localizations of endogenous and overexpressed proteins.

2.4 MA-nSMase (Mitochondrial-Associated Neutral SMase)

2.4.1 Biochemical Properties and Localization

MA-nSMase is the most recently identified mammalian N-SMase being discovered in 2010 (Wu et al. 2010a) by sequence homology with nSMase2 and a zebrafish mitochondrial N-SMase (Yabu et al. 2009). The subcellular localization of overexpressed MA-nSMase protein varied with cell type, showing strong or partial co-localization with mitochondrial markers, in addition to co-localization with ER markers (Wu et al. 2010a). The mouse MA-nSMase protein conserves a mitochondrial signal peptide with the zebrafish N-SMase, which may be responsible for its mitochondrial localization.

MA-nSMase has comparable domain architecture and biochemical properties with nSMase2. Catalytic activity requires Mg^{2+} or Mn^{2+} ions and is strongly increased by the presence of the phospholipids cardiolipin (CL), PS, or phosphatidylglycerol (PG) (Fig. 2). Both CL and PG are present in mitochondria providing a putative mechanism for activation and/or regulation of MA-nSMase activity in vivo (Wu et al. 2010a). The N-terminal hydrophobic domain of MA-nSMase is similar to nSMase2 by sequence homology, but the membrane topology and APL-binding motif has yet to be characterized.

2.4.2 Human vs. Murine MA-nSMase

The human homolog to murine MA-nSMase has yet to be cloned and characterized. However, an open reading frame for the human SMPD5 gene (XP_001714084) is present in the NCBI database. Surprisingly, the human and mouse proteins contain key differences in the N-terminal APL activation domain. Importantly, the putative mitochondrial signal peptide is conserved suggesting that the human protein will also localize to mitochondria. In the future it is important to clone the human gene and compare the amino acid sequence, biochemical properties, localization, and cellular functions.

2.4.3 Future Directions

At present, little is known about MA-nSMase beyond basic properties. However, the identification of a mammalian mitochondrial N-SMase presents another potential endogenous mechanism, in addition to the action of ceramide synthases, for mitochondrial ceramide generation. This is exciting considering the numerous studies linking ceramide to mitochondrial activation of apoptosis (Mullen and Obeid 2012).

Another potential role for MA-nSMase is in fertilization. Activation of an unidentified SMase during fertilization was inferred by a corresponding decrease in SM and increase in Cer levels (Petcoff et al. 2008). In support of this hypothesis, MA-nSMase gene expression in organ tissues is the highest in testis (Wu et al. 2010a) and the MA-nSMase gene appears to be bicistronic with a spermatogenesis-associated factor (NCBI database, unpublished observation). It is our hope that future studies taking advantage of gene silencing, antibodies to endogenous proteins, and knockout animals will better illuminate the role of MA-nSMase in mammalian physiology and biology.

2.5 Neutral SMase3

2.5.1 Biochemical Properties

Although human nSMase3 was identified in 2006 (Krut et al. 2006), relatively little work has been reported since regarding further biochemical and functional characterization. Identification was accomplished by sequence comparison using a peptide (KGLPYLEQLFR) from a previously purified N-SMase from bovine brain (Bernardo et al. 2000). The peptide sequence only matches 7 of the 11 residues in the identified human protein and bovine homolog. Given the short peptide sequence and low identity, this raises the question if the original purified bovine and identified human proteins correspond to the same protein.

NSMase3 shares no sequence homology with any N-SMases or any other characterized type of enzyme catalytic domain. The region comprising the catalytic domain is yet to be identified. A C-terminal transmembrane helix is predicted to embed nSMase3 in the membrane (Krut et al. 2006).

Two conflicting reports have characterized nSMase3 activity. In the original identification, nSMase3 activity is reported to occur at neutral pH and require Mg^{2+} or Mn^{2+} (Krut et al. 2006). The observed activity was slightly enhanced, approximately twofold, by the phospholipid PS. In a later study, MCF-7 cells transiently and stably overexpressing nSMase3 did not have significant N-SMase activity over vector controls (Clarke et al. 2011c). We hope that future studies will determine the underlying reasons behind this major discrepancy.

2.5.2 Localization

Subcellular localization studies found nSMase3 to display an ER distribution pattern (Krut et al. 2006). Later work confirmed this and found that localization was not dependent on a putative ER localization motif or the C-terminal transmembrane helix (Corcoran et al. 2008).

2.5.3 Cellular Signaling

It has been suggested that nSMase3 may play a role in TNF-α-mediated signaling (Krut et al. 2006; Corcoran et al. 2008). However, another report found that nSMase2 was the primary N-SMase activated by TNF-α in MCF-7 cells (Clarke et al. 2011c). In support of this, a comparison of mouse fibroblasts from normal and nSMase2-deficient *fro-/fro-* mice found that nSMase2 deficiency abrogated TNF-α-induced increases in N-SMase activity (Devillard et al. 2010) suggesting that nSMase3 is not required in these cells.

The other identified putative function of nSMase3 involves a role in tumorigenesis (Corcoran et al. 2008). In this study, nSMase3 expression was found to be both upregulated and downregulated to varying degrees (−70 to +70 %) in different tumor samples. In addition, nSMase3 mRNA levels were downregulated by the tumor suppressor p53.

Overall, a clear functional role for nSMase3 has yet to emerge. Further studies investigating the biological function and molecular mechanisms of nSMase3 are required to validate the current findings.

2.6 Conclusions

Four N-SMase isoforms have now been cloned and purified. However, beyond nSMase2 there is relatively little known about N-SMase physiology and cellular function. In addition, at a biochemical and mechanistic level, there is still much to learn about N-SMase activation and regulation. The roles of nSMase2 in bone homeostasis, exosome secretion of miRNAs, and cigarette-induced lung injury are essential and provide the potential for nSMase2-targeted therapies.

Acknowledgments This work was supported by the NIH grant R37GM43825 (Y.A.H.) and NIH NIGMS 1F32GM100679 (MVA). We thank Chris J. Clarke, Bill X. Wu, and David M. Perry for helpful review of this manuscript.

References

Adam-Klages S, Adam D, Wiegmann K, Struve S, Kolanus W, Schneider-Mergener J, Kronke M (1996) FAN, a novel WD-repeat protein, couples the p55 TNF-receptor to neutral sphingomyelinase. Cell 86(6):937–947, S0092-8674(00)80169-5 [pii]

Ago H, Oda M, Takahashi M, Tsuge H, Ochi S, Katunuma N, Miyano M, Sakurai J (2006) Structural basis of the sphingomyelin phosphodiesterase activity in neutral sphingomyelinase from Bacillus cereus. J Biol Chem 281(23):16157–16167. doi:10.1074/jbc.M601089200, M601089200 [pii]

Arana L, Gangoiti P, Ouro A, Trueba M, Gómez-Muñoz A (2010) Ceramide and ceramide 1-phosphate in health and disease. Lipids Health Dis 9:15

Aubin I, Adams CP, Opsahl S, Septier D, Bishop CE, Auge N, Salvayre R, Negre-Salvayre A, Goldberg M, Guenet JL, Poirier C (2005) A deletion in the gene encoding sphingomyelin phosphodiesterase 3 (Smpd3) results in osteogenesis and dentinogenesis imperfecta in the mouse. Nat Genet 37(8):803–805. doi:10.1038/ng1603, ng1603 [pii]

Bartke N, Hannun YA (2009) Bioactive sphingolipids: metabolism and function. J Lipid Res 50(Suppl):S91–S96. doi:10.1194/jlr.R800080-JLR200, R800080-JLR200 [pii]

Ben-David O, Futerman AH (2010) The role of the ceramide acyl chain length in neurodegeneration: involvement of ceramide synthases. Neuromolecular Med 12(4):341–350

Bernardo K, Krut O, Wiegmann K, Kreder D, Micheli M, Schafer R, Sickman A, Schmidt WE, Schroder JM, Meyer HE, Sandhoff K, Kronke M (2000) Purification and characterization of a magnesium-dependent neutral sphingomyelinase from bovine brain. J Biol Chem 275 (11):7641–7647

Canals D, Perry DM, Jenkins RW, Hannun YA (2011) Drug targeting of sphingolipid metabolism: sphingomyelinases and ceramidases. Br J Pharmacol 163(4):694–712. doi:10.1111/j.1476-5381.2011.01279.x

Chen X, Liang H, Zhang J, Zen K, Zhang CY (2012) Secreted microRNAs: a new form of intercellular communication. Trends Cell Biol 22:125–132

Clarke CJ, Snook CF, Tani M, Matmati N, Marchesini N, Hannun YA (2006) The extended family of neutral sphingomyelinases. Biochemistry 45(38):11247–11256. doi:10.1021/bi061307z

Clarke CJ, Truong TG, Hannun YA (2007) Role for neutral sphingomyelinase-2 in tumor necrosis factor alpha-stimulated expression of vascular cell adhesion molecule-1 (VCAM) and intercellular adhesion molecule-1 (ICAM) in lung epithelial cells: p38 MAPK is an upstream regulator of nSMase2. J Biol Chem 282(2):1384–1396. doi:10.1074/jbc.M609216200, M609216200 [pii]

Clarke CJ, Guthrie JM, Hannun YA (2008) Regulation of neutral sphingomyelinase-2 (nSMase2) by tumor necrosis factor-alpha involves protein kinase C-delta in lung epithelial cells. Mol Pharmacol 74(4):1022–1032. doi:10.1124/mol.108.046250, mol.108.046250 [pii]

Clarke CJ, Wu BX, Hannun YA (2011a) The neutral sphingomyelinase family: identifying biochemical connections. Adv Enzyme Regul 51(1):51–58. doi:10.1016/j.advenzreg.2010.09.016, S0065-2571(10)00075-0 [pii]

Clarke CJ, Mediwala K, Jenkins RW, Sutton CA, Tholanikunnel BG, Hannun YA (2011b) Neutral sphingomyelinase-2 mediates growth arrest by retinoic acid through modulation of ribosomal S6 kinase. J Biol Chem 286(24):21565–21576. doi:10.1074/jbc.M110.193375, M110.193375 [pii]

Clarke CJ, Cloessner EA, Roddy PL, Hannun YA (2011c) Neutral sphingomyelinase 2 (nSMase2) is the primary neutral sphingomyelinase isoform activated by tumour necrosis factor-alpha in MCF-7 cells. Biochem J 435(2):381–390. doi:10.1042/BJ20101752, BJ20101752 [pii]

Corcoran CA, He Q, Ponnusamy S, Ogretmen B, Huang Y, Sheikh MS (2008) Neutral sphingomyelinase-3 is a DNA damage and nongenotoxic stress-regulated gene that is deregulated in human malignancies. Mol Cancer Res 6(5):795–807. doi:10.1158/1541-7786. MCR-07-2097, 6/5/795 [pii]

Devillard R, Galvani S, Thiers JC, Guenet JL, Hannun Y, Bielawski J, Negre-Salvayre A, Salvayre R, Auge N (2010) Stress-induced sphingolipid signaling: role of type-2 neutral sphingomyelinase in murine cell apoptosis and proliferation. PLoS One 5(3):e9826. doi:10.1371/journal.pone.0009826

Filosto S, Fry W, Knowlton AA, Goldkorn T (2010) Neutral sphingomyelinase 2 (nSMase2) is a phosphoprotein regulated by calcineurin (PP2B). J Biol Chem 285(14):10213–10222. doi:10.1074/jbc.M109.069963, M109.069963 [pii]

Filosto S, Castillo S, Danielson A, Franzi L, Khan E, Kenyon N, Last J, Pinkerton K, Tuder R, Goldkorn T (2011) Neutral sphingomyelinase 2: a novel target in cigarette smoke-induced apoptosis and lung injury. Am J Respir Cell Mol Biol 44(3):350–360. doi:10.1165/rcmb.2009-0422OC, 2009-0422OC [pii]

Filosto S, Ashfaq M, Chung S, Fry W, Goldkorn T (2012) Neutral sphingomyelinase 2 activity and protein stability are modulated by phosphorylation of five conserved serines. J Biol Chem 287:514–522. doi:10.1074/jbc.M111.315481, M111.315481 [pii]

Gault CR, Obeid LM, Hannun YA (2010) An overview of sphingolipid metabolism: from synthesis to breakdown. Adv Exp Med Biol 688:1–23

Glorieux FH, Ward LM, Rauch F, Lalic L, Roughley PJ, Travers R (2002) Osteogenesis imperfecta type VI: a form of brittle bone disease with a mineralization defect. J Bone Miner Res 17(1):30–38. doi:10.1359/jbmr.2002.17.1.30

Goldkorn T, Filosto S (2010) Lung injury and cancer: mechanistic insights into ceramide and EGFR signaling under cigarette smoke. Am J Respir Cell Mol Biol 43(3):259

Grosch S, Schiffmann S, Geisslinger G (2012) Chain length-specific properties of ceramides. Prog Lipid Res 51(1):50–62. doi:10.1016/j.plipres.2011.11.001, S0163-7827(11)00042-7 [pii]

Hama H (2010) Fatty acid 2-hydroxylation in mammalian sphingolipid biology. Biochim Biophys Acta 1801(4):405–414

Han G, Gupta SD, Gable K, Niranjanakumari S, Moitra P, Eichler F, Brown RH, Harmon JM, Dunn TM (2009) Identification of small subunits of mammalian serine palmitoyltransferase that confer distinct acyl-CoA substrate specificities. Proc Natl Acad Sci 106(20):8186

Hannun YA, Obeid LM (2008) Principles of bioactive lipid signalling: lessons from sphingolipids. Nat Rev Mol Cell Biol 9(2):139–150. doi:10.1038/nrm2329, nrm2329 [pii]

Hannun YA, Obeid LM (2011) Many ceramides. J Biol Chem 286(32):27855–27862. doi:10.1074/jbc.R111.254359, R111.254359 [pii]

Hayashi Y, Kiyono T, Fujita M, Ishibashi M (1997) cca1 is required for formation of growth-arrested confluent monolayer of rat 3Y1 cells. J Biol Chem 272(29):18082–18086

Hofmann K, Tomiuk S, Wolff G, Stoffel W (2000) Cloning and characterization of the mammalian brain-specific, Mg2 + -dependent neutral sphingomyelinase. Proc Natl Acad Sci U S A 97 (11):5895–5900, 97/11/5895 [pii]

Hu W, Xu R, Sun W, Szulc ZM, Bielawski J, Obeid LM, Mao C (2010) Alkaline ceramidase 3 (ACER3) hydrolyzes unsaturated long-chain ceramides, and its down-regulation inhibits both cell proliferation and apoptosis. J Biol Chem 285(11):7964–7976. doi:10.1074/jbc.M109.063586, M109.063586 [pii]

Ito H, Murakami M, Furuhata A, Gao S, Yoshida K, Sobue S, Hagiwara K, Takagi A, Kojima T, Suzuki M, Banno Y, Tanaka K, Tamiya-Koizumi K, Kyogashima M, Nozawa Y, Murate T (2009) Transcriptional regulation of neutral sphingomyelinase 2 gene expression of a human breast cancer cell line, MCF-7, induced by the anti-cancer drug, daunorubicin. Biochim Biophys Acta 1789(11–12):681–690. doi:10.1016/j.bbagrm.2009.08.006, S1874-9399(09)00100-X [pii]

Jenkins RW, Canals D, Hannun YA (2009) Roles and regulation of secretory and lysosomal acid sphingomyelinase. Cell Signal 21(6):836–846

Khavandgar Z, Poirier C, Clarke CJ, Li J, Wang N, McKee MD, Hannun YA, Murshed M (2011) A cell-autonomous requirement for neutral sphingomyelinase 2 in bone mineralization. J Cell Biol 194(2):277–289. doi:10.1083/jcb.201102051, jcb.201102051 [pii]

Kim WJ, Okimoto RA, Purton LE, Goodwin M, Haserlat SM, Dayyani F, Sweetser DA, McClatchey AI, Bernard OA, Look AT, Bell DW, Scadden DT, Haber DA (2008) Mutations in the neutral sphingomyelinase gene SMPD3 implicate the ceramide pathway in human leukemias. Blood 111(9):4716–4722. doi:10.1182/blood-2007-10-113068, blood-2007-10-113068 [pii]

Krut O, Wiegmann K, Kashkar H, Yazdanpanah B, Kronke M (2006) Novel tumor necrosis factor-responsive mammalian neutral sphingomyelinase-3 is a C-tail-anchored protein. J Biol Chem 281(19):13784–13793. doi:10.1074/jbc.M511306200, M511306200 [pii]

Leventis PA, Grinstein S (2010) The distribution and function of phosphatidylserine in cellular membranes. Annu Rev Biophys 39:407–427. doi:10.1146/annurev.biophys.093008.131234

Levy M, Castillo SS, Goldkorn T (2006) nSMase2 activation and trafficking are modulated by oxidative stress to induce apoptosis. Biochem Biophys Res Commun 344(3):900–905. doi:10.1016/j.bbrc.2006.04.013, S0006-291X(06)00788-1 [pii]

Levy M, Khan E, Careaga M, Goldkorn T (2009) Neutral sphingomyelinase 2 is activated by cigarette smoke to augment ceramide-induced apoptosis in lung cell death. Am J Physiol Lung Cell Mol Physiol 297(1):L125–L133. doi:10.1152/ajplung.00031.2009, 00031.2009 [pii]

Long JZ, Cisar JS, Milliken D, Niessen S, Wang C, Trauger SA, Siuzdak G, Cravatt BF (2011) Metabolomics annotates ABHD3 as a physiologic regulator of medium-chain phospholipids. Nat Chem Biol 7(11):763–765. doi:10.1038/nchembio.659, nchembio.659 [pii]

Luberto C, Hassler DF, Signorelli P, Okamoto Y, Sawai H, Boros E, Hazen-Martin DJ, Obeid LM, Hannun YA, Smith GK (2002) Inhibition of tumor necrosis factor-induced cell death in MCF7 by a novel inhibitor of neutral sphingomyelinase. J Biol Chem 277(43):41128–41139. doi:10.1074/jbc.M206747200, M206747200 [pii]

Mao C, Obeid LM (2008) Ceramidases: regulators of cellular responses mediated by ceramide, sphingosine, and sphingosine-1-phosphate. Biochim Biophys Acta 1781(9):424–434

Marchesini N, Osta W, Bielawski J, Luberto C, Obeid LM, Hannun YA (2004) Role for mammalian neutral sphingomyelinase 2 in confluence-induced growth arrest of MCF7 cells. J Biol Chem 279(24):25101–25111. doi:10.1074/jbc.M313662200, M313662200 [pii]

Matmati N, Hannun YA (2008) Thematic review series: sphingolipids. ISC1 (inositol phosphosphingolipid-phospholipase C), the yeast homologue of neutral sphingomyelinases. J Lipid Res 49(5):922–928. doi:10.1194/jlr.R800004-JLR200, R800004-JLR200 [pii]

Merrill AH Jr (2011) Sphingolipid and glycosphingolipid metabolic pathways in the era of sphingolipidomics. Chem Rev 111(10):6387

Mesicek J, Lee H, Feldman T, Jiang X, Skobeleva A, Berdyshev EV, Haimovitz-Friedman A, Fuks Z, Kolesnick R (2010) Ceramide synthases 2, 5, and 6 confer distinct roles in radiation-induced apoptosis in HeLa cells. Cell Signal 22(9):1300–1307

Milhas D, Clarke CJ, Hannun YA (2010a) Sphingomyelin metabolism at the plasma membrane: implications for bioactive sphingolipids. FEBS Lett 584(9):1887–1894. doi:10.1016/j.febslet.2009.10.058, S0014-5793(09)00846-1 [pii]

Milhas D, Clarke CJ, Idkowiak-Baldys J, Canals D, Hannun YA (2010b) Anterograde and retrograde transport of neutral sphingomyelinase-2 between the Golgi and the plasma membrane. Biochim Biophys Acta 1801(12):1361–1374. doi:10.1016/j.bbalip. 2010.08.001, S1388-1981(10)00179-4 [pii]

Mizutani Y, Tamiya-Koizumi K, Nakamura N, Kobayashi M, Hirabayashi Y, Yoshida S (2001) Nuclear localization of neutral sphingomyelinase 1: biochemical and immunocytochemical analyses. J Cell Sci 114(Pt 20):3727–3736

Mullen TD, Obeid LM (2012) Ceramide and apoptosis: exploring the enigmatic connections between sphingolipid metabolism and programmed cell death. Anticancer Agents Med Chem 12:340–363, BSP/ACAMC/E-Pub/00187 [pii]

Mullen TD, Jenkins RW, Clarke CJ, Bielawski J, Hannun YA, Obeid LM (2011) Ceramide synthase-dependent ceramide generation and programmed cell death. J Biol Chem 286(18):15929

Mullen TD, Hannun YA, Obeid LM (2012) Ceramide synthases at the centre of sphingolipid metabolism and biology. Biochem J 441(3):789–802. doi:10.1042/BJ20111626, BJ20111626 [pii]

Nikolova-Karakashian M, Karakashian A, Rutkute K (2008) Role of neutral sphingomyelinases in aging and inflammation. Subcell Biochem 49:469–486. doi:10.1007/978-1-4020-8831-5_18

Okamoto Y, Vaena de Avalos S, Hannun YA (2003) Functional analysis of ISC1 by site-directed mutagenesis. Biochemistry 42(25):7855–7862. doi:10.1021/bi0341354

Openshaw AE, Race PR, Monzo HJ, Vazquez-Boland JA, Banfield MJ (2005) Crystal structure of SmcL, a bacterial neutral sphingomyelinase C from Listeria. J Biol Chem 280 (41):35011–35017. doi:10.1074/jbc.M506800200, M506800200 [pii]

Patti GJ, Yanes O, Shriver LP, Courade JP, Tautenhahn R, Manchester M, Siuzdak G (2012) Metabolomics implicates altered sphingolipids in chronic pain of neuropathic origin. Nat Chem Biol 8(3):232–234. doi:10.1038/nchembio.767, nchembio.767 [pii]

Petcoff DW, Holland WL, Stith BJ (2008) Lipid levels in sperm, eggs, and during fertilization in Xenopus laevis. J Lipid Res 49(11):2365–2378. doi:10.1194/jlr.M800159-JLR200, M800159-JLR200 [pii]

Pewzner-Jung Y, Ben-Dor S, Futerman AH (2006) When do Lasses (longevity assurance genes) become CerS (ceramide synthases)?: insights into the regulation of ceramide synthesis. J Biol Chem 281(35):25001–25005. doi:10.1074/jbc.R600010200, R600010200 [pii]

Philipp S, Puchert M, Adam-Klages S, Tchikov V, Winoto-Morbach S, Mathieu S, Deerberg A, Kolker L, Marchesini N, Kabelitz D, Hannun YA, Schutze S, Adam D (2010) The Polycomb group protein EED couples TNF receptor 1 to neutral sphingomyelinase. Proc Natl Acad Sci U S A 107(3):1112–1117. doi:10.1073/pnas.0908486107, 0908486107 [pii]

Pitson SM (2011) Regulation of sphingosine kinase and sphingolipid signaling. Trends Biochem Sci 36(2):97–107

Pruett ST, Bushnev A, Hagedorn K, Adiga M, Haynes CA, Sullards MC, Liotta DC, Merrill AH Jr (2008) Biodiversity of sphingoid bases ("sphingosines") and related amino alcohols. J Lipid Res 49(8):1621–1639. doi:10.1194/jlr.R800012-JLR200, R800012-JLR200 [pii]

Qin J, Berdyshev E, Poirier C, Schwartz NB, Dawson G (2012) Neutral sphingomyelinase 2 deficiency increases hyaluronan synthesis by up-regulation of hyaluronan Synthase 2 through decreased ceramide production and activation of Akt. J Biol Chem 287:13620–13632. doi:10.1074/jbc.M111.304857, M111.304857 [pii]

Sawai H, Domae N, Nagan N, Hannun YA (1999) Function of the cloned putative neutral sphingomyelinase as lyso-platelet activating factor-phospholipase C. J Biol Chem 274 (53):38131–38139

Schiffmann S, Ziebell S, Sandner J, Birod K, Deckmann K, Hartmann D, Rode S, Schmidt H, Angioni C, Geisslinger G (2010) Activation of ceramide synthase 6 by celecoxib leads to a selective induction of C16: 0-ceramide. Biochem Pharmacol 80(11):1632–1640

Schneider PB, Kennedy EP (1967) Sphingomyelinase in normal human spleens and in spleens from subjects with Niemann-Pick disease. J Lipid Res 8(3):202–209

Serra M, Saba JD (2010) Sphingosine 1-phosphate lyase, a key regulator of sphingosine 1-phosphate signaling and function. Adv Enzyme Regul 50(1):349

Stace CL, Ktistakis NT (2006) Phosphatidic acid- and phosphatidylserine-binding proteins. Biochim Biophys Acta 1761(8):913–926. doi:10.1016/j.bbalip. 2006.03.006, S1388-1981 (06)00067-9 [pii]

Stoffel W, Jenke B, Block B, Zumbansen M, Koebke J (2005) Neutral sphingomyelinase 2 (smpd3) in the control of postnatal growth and development. Proc Natl Acad Sci U S A 102(12):4554–4559. doi:10.1073/pnas.0406380102, 0406380102 [pii]

Stoffel W, Jenke B, Holz B, Binczek E, Gunter RH, Knifka J, Koebke J, Niehoff A (2007) Neutral sphingomyelinase (SMPD3) deficiency causes a novel form of chondrodysplasia and dwarfism that is rescued by Col2A1-driven smpd3 transgene expression. Am J Pathol 171(1):153–161. doi:10.2353/ajpath.2007.061285, S0002-9440(10)61951-7 [pii]

Sun W, Jin J, Xu R, Hu W, Szulc ZM, Bielawski J, Obeid LM, Mao C (2010) Substrate specificity, membrane topology, and activity regulation of human alkaline ceramidase 2 (ACER2). J Biol Chem 285(12):8995–9007. doi:10.1074/jbc.M109.069203, M109.069203 [pii]

Tani M, Hannun YA (2007a) Analysis of membrane topology of neutral sphingomyelinase 2. FEBS Lett 581(7):1323–1328. doi:10.1016/j.febslet.2007.02.046, S0014-5793(07)00214-1 [pii]

Tani M, Hannun YA (2007b) Neutral sphingomyelinase 2 is palmitoylated on multiple cysteine residues. Role of palmitoylation in subcellular localization. J Biol Chem 282(13):10047–10056. doi:10.1074/jbc.M611249200, M611249200 [pii]

Tomiuk S, Hofmann K, Nix M, Zumbansen M, Stoffel W (1998) Cloned mammalian neutral sphingomyelinase: functions in sphingolipid signaling? Proc Natl Acad Sci U S A 95(7):3638–3643

Tomiuk S, Zumbansen M, Stoffel W (2000) Characterization and subcellular localization of murine and human magnesium-dependent neutral sphingomyelinase. J Biol Chem 275 (8):5710–5717

Wu BX, Rajagopalan V, Roddy PL, Clarke CJ, Hannun YA (2010a) Identification and characterization of murine mitochondria-associated neutral sphingomyelinase (MA-nSMase), the mammalian sphingomyelin phosphodiesterase 5. J Biol Chem 285(23):17993–18002. doi:10.1074/jbc.M110.102988, M110.102988 [pii]

Wu BX, Clarke CJ, Hannun YA (2010b) Mammalian neutral sphingomyelinases: regulation and roles in cell signaling responses. Neuromolecular Med 12(4):320–330. doi:10.1007/s12017-010-8120-z

Wu BX, Clarke CJ, Matmati N, Montefusco D, Bartke N, Hannun YA (2011) Identification of novel anionic phospholipid binding domains in neutral sphingomyelinase 2 with selective binding preference. J Biol Chem 286(25):22362–22371. doi:10.1074/jbc.M110.156471, M110.156471 [pii]

Yabu T, Shimuzu A, Yamashita M (2009) A novel mitochondrial sphingomyelinase in zebrafish cells. J Biol Chem 284(30):20349–20363. doi:10.1074/jbc.M109.004580, M109.004580 [pii]

Acid Sphingomyelinase

Brian Henry, Regan Ziobro, Katrin Anne Becker, Richard Kolesnick, and Erich Gulbins

Contents

Abstract The enzyme acid sphingomyelinase catalyzes the hydrolysis of sphingomyelin to ceramide. The importance of the enzyme for cell functions was first recognized in Niemann–Pick disease type A and B, the genetic disorders with a massive accumulation of sphingomyelin in many organs. Studies in the last years demonstrated that the enzyme also has an important role in cell signalling. Thus, the acid sphingomyelinase has a central function for the re-organization of molecules within the cell upon stimulation and thereby for the response of cells to stress

B. Henry • E. Gulbins (✉)
Department of Molecular Biology, University of Duisburg-Essen, Hufelandstrasse 55, 45122 Essen, Germany

Department of Surgery, University of Cincinnati, 231 Albert Sabin Way, Cincinnati, OH 45267, USA
e-mail: erich.gulbins@uni-due.de

R. Ziobro • K.A. Becker
Department of Molecular Biology, University of Duisburg-Essen, Hufelandstrasse 55, 45122 Essen, Germany

R. Kolesnick
Memorial Sloan Kettering Cancer Center, 1275 York Avenue, New York, NY 10065, USA

E. Gulbins and I. Petrache (eds.), *Sphingolipids: Basic Science and Drug Development*, Handbook of Experimental Pharmacology 215, DOI 10.1007/978-3-7091-1368-4_4,

and the induction of cell death but also proliferation and differentiation. Here, we discuss the current state of the art of the structure, regulation, and function of the acid sphingomyelinase.

Keywords Acid sphingomyelinase • Ceramide • Membrane domains • Cell stress

1 Introduction

Ceramide is generated by hydrolysis of sphingomyelin by sphingomyelinases, via the de novo synthesis pathway, by the conversion of sphingosine via the retrograde activity of ceramidases, or by the hydrolysis of complex glycosylated lipids or ceramide 1-phosphate (for recent review see Hannun and Obeid 2011). Here, we focus on the acid sphingomyelinase (Enzyme Commission Classification number 3.1.4.12). Mechanistically, sphingomyelinases function as hydrolases of phosphodiester bonds. Peak activity of sphingomyelinases depends on pH and, at present, acid, neutral, and alkaline sphingomyelinases have been identified and classified in regard to respective pH optimum of activity. Deficiency of acid sphingomyelinase results in the lysosomal storage disease, Niemann–Pick-type A and B (Niemann 1914; Brady et al. 1966; Schuchman et al. 1992).

In vitro, the optimum pH for acid sphingomyelinase activity is 4.5–5.0. Originally, this indicated that the enzyme was solely lysosomal (Fowler 1969). However, more recent data indicate that the lipid composition of the membrane alters the Km of the enzyme, thus permitting acid sphingomyelinase activity at higher pH (Schissel et al. 1998a, b). Further, recent studies demonstrate that acidic microenvironments exist not only in lysosomes but also in domains on the outer leaflet of plasma membrane, where acid sphingomyelinase and NADPH-oxidases localize after certain stimuli (Xu et al. 2012). Collectively, these data indicate that activity of acid sphingomyelinase is not restricted to lysosomes.

2 Structure of the Acid Sphingomyelinase

The acid sphingomyelinase gene, referred to via the symbol *SMPD1* for the human and *Smpd1* for the murine gene, is 5–6 kb long, contains six exons, and localizes to the p15.1–p15.4 region of chromosome 11 (da Veiga Pereira et al. 1991). Human and murine acid sphingomyelinases are encoded by one conserved gene. Human cDNA codes for a protein of 629 amino acids, with approximately 82 % amino acid similarity to the murine version. Partially purified acid sphingomyelinase has a Km value for sphingomyelin of 2 μM (Gulbins and Kolesnick 2000). V_{max} values were reported as 4.3 μmol/mg protein/h, which increases after stimulation of the enzyme to 13.5 μmol/mg protein/h (Gulbins and Kolesnick 2000).

Seven different acid sphingomyelinase isoforms have been identified to date (Quintern et al. 1989; Schuchman et al. 1991; Rhein et al. 2012), although only acid sphingomyelinase-1 has been reported to be catalytically active (GenBank Accession Number NM_000543.4). The active transcript contains exons 1–6 with a total coding length of 1,896 bp (Quintern et al. 1989; Schuchman et al. 1991; Rhein et al. 2012). Alternatively spliced acid sphingomyelinase type 2 transcript contains a 40 bp insertion derived from intron 2 and lacks exon 3. Alternatively spliced acid sphingomyelinase transcript type 3 lacks exon 3, resulting in a frameshift and premature termination during translation (Quintern et al. 1989; Schuchman et al. 1991; Rhein et al. 2012). Alternatively spliced transcript 4 lacks 652 bp of exon 2, whereas alternatively spliced acid sphingomyelinase transcripts 5, 6, and 7 lack portions of the catalytic- and/or carboxy-terminal domains and are also inactive, but may function as dominant-negative proteins in vivo (Quintern et al. 1989; Schuchman et al. 1991; Rhein et al. 2012).

Acid sphingomyelinase protein is processed by partial cleavage of a precursor form to the mature protein (Hurwitz et al. 1994; Jenkins et al. 2011). Cleavage depends on the subcellular location of the protein, which is present in classic and secretory lysosomes, and is also secreted (Schissel et al. 1998b). Lysosomal acid sphingomyelinase is cleaved at Gly66, whereas secretory acid sphingomyelinase initiates at His60 (Schissel et al. 1998b). In addition to initial cleavage, the enyzme can be further processed to a protein with a size of 57 kDa (Hurwitz et al. 1994), although details of this cleavage process are unknown.

Acid sphingomyelinase is glycosylated at six glycosylation sites, and enzyme glycosylation is required for activity (Newrzella and Stoffel 1996; Ferlinz et al. 1994). While lysosomal acid sphingomyelinase is glycosylated by N-linked oligomannose groups, the secretory form exhibits complex glycosylation. Glycosylation pattern also determines trafficking of the enzyme, as only mannose-6 phosphate glycosylation permits binding to mannose-6-phosphate receptors, which target the enzyme to lysosomes (Schissel et al. 1998b). Alternatively, complex-glycosylated acid sphingomyelinase is secreted.

Many stimuli induce translocation of acid sphingomyelinase onto the extracellular leaflet of the cell membrane (Abdel Shakor et al. 2004; Avota et al. 2011; Bao et al. 2010; Cremesti et al. 2001; Dumitru and Gulbins 2006; Esen et al. 2001; Grassmé et al. 2001a, b; 2002; 2003a, b; 2005; Hauck et al. 2000; Jia et al. 2008; Jin et al. 2008a, b; Rotolo et al. 2005). Acid sphingomyelinase reaches the surface upon fusion of secretory lysosomes with the plasma membrane resulting in exposure of acid sphingomyelinase to the cell membrane extracellular leaflet as detailed below. However, it is also likely that in addition secreted acid sphingomyelinase binds to and acts at the cell surface.

Acid sphingomyelinase is a metalloenzyme, i.e., it contains several Zn^{2+}-binding motifs and binding of Zn^{2+} is required for activity (Schissel et al. 1998b). While the lysosomal form binds Zn^{2+} tightly, secretory acid sphingomyelinase remains segregated from Zn^{2+} throughout the Golgi secretory pathway and requires addition of Zn^{2+} for activation.

3 Activation of Acid Sphingomyelinase

The molecular mechanisms regarding regulation of acid sphingomyelinase are only partially characterized. In vitro studies show that the enzyme is activated directly by oxidation (Zhang and Li 2010). These studies employed purified acid sphingomyelinase and demonstrated direct oxidation of acid sphingomyelinase at cysteine residue 629, resulting in enzyme dimerization and activation (Qiu et al. 2003). To date, however, acid sphingomyelinase dimerization after cellular stimulation has not been observed in vivo. Several other studies demonstrate that reactive oxygen radical scavengers prevent acid sphingomyelinase activation by stimuli such as DR5 or Cu^{2+} (Dumitru et al. 2007; Lang et al. 2007). These studies did not investigate whether reactive oxygen species directly regulate enzyme activity or whether this is accomplished by unknown intermediates. Thus, it is still not certain whether acid sphingomyelinase is regulated by redox mechanisms in vivo at baseline or during signal transduction.

The role of proteases, in particular caspases, to regulate acid sphingomyelinase has also been implied (Brenner et al. 1998; Edelmann et al. 2011). Initial studies demonstrated that caspase inhibitors prevent acid sphingomyelinase activation by CD95 (Brenner et al. 1998). Further studies demonstrated that TNF-α triggers the interaction of the TNF-receptor with caspase-7, resulting in receptor internalization and rapid formation of endosomes (Edelmann et al. 2011). Fusion of these endosomes with lysosomes results in multivesicular bodies that bring intra-lysosomal acid sphingomyelinase into contact with caspase-7. Caspase-7 then cleaves and activates the enzyme. Whether this mechanism of acid sphingomyelinase activation also applies to other stimuli than TNF-α is presently unknown.

A few studies described PKC-mediated phosphorylation of acid sphingomyelinase at serine 508 (Zeidan and Hannun 2007). Similar to the interaction of acid sphingomyelinase with caspase-7, formation of multivesicular bodies and digestion of membranes within these vesicles may bring PKC and acid sphingomyelinase together, thus permitting transregulation.

In contrast to the TNF-receptor, other receptors, such as CD95 or DR5, or infections with *P. aeruginosa* or some viruses, trigger translocation of acid sphingomyelinase onto the extracellular leaflet (Avota et al. 2011; Dumitru and Gulbins 2006; Grassmé et al. 2001a, b, 2002, 2003a, b, 2005). Recent studies by Clementi et al. demonstrated that acid sphingomyelinase resides in secretory lysosomes that are mobilized upon stimulation to fuse with the cell membrane (Perrotta et al. 2010). This exocytic pathway required the t-SNARE protein syntaxin-4. Interestingly, this process can be regulated by nitric oxide, which activates G-kinase that in turn phosphorylates syntaxin-4 on S78. This phosphorylation induces rapid proteasomal degradation of syntaxin-4, thereby inhibiting acid sphingomyelinase translocation and activation. Further studies indicate a requirement of dysferlin for translocation of these vesicles (Han et al. 2012).

Acid sphingomyelinase is inhibited by inositol-phosphates (Kolzer et al. 2003; Roth et al. 2009; Preuss et al. 2012), although the physiological significance of this finding remains to be determined. The regulation of the acid sphingomyelinase by inositol-phosphates may represent a novel mechanism for pharmacological inhibition of the enzyme, for instance in lung (Preuss et al. 2012).

4 Acid Sphingomyelinase-Mediated Signalling

4.1 General Principles

The cell membrane predominantly contains sphingolipids, cholesterol, and (glycero) phospholipids. Ceramide is the backbone of sphingolipids and consists of D-erythro-sphingosine connected via an amide ester bond to a fatty acid containing 2–32 carbon atoms in the acyl chain. Long-chain ceramides are very hydrophobic molecules. Attachment of a hydrophilic headgroup to ceramide results in complete sphingolipids, for instance sphingomyelin with phosphorylcholine as headgroup. Sphingolipids interact with each other and with cholesterol molecules. These interactions are mediated by hydrophilic forces between sphingolipid headgroups and the hydroxy group of cholesterol on one side, and via hydrophobic van der Waal interactions between ceramide moieties and the sterol ring system on the other (Simons and Ikonen 1997; Brown and London 1998; Kolesnick et al. 2000). These interactions are relatively tight, resulting in a liquid-ordered membrane phase separated from other plasma membrane phospholipids and consequential formation of small distinct domains—also named “rafts” (Simons and Ikonen 1997).

Generation of ceramide dramatically alters biophysical properties of biological membranes both within and outside of rafts. As mentioned, hydrophobic ceramide molecules separate from other lipids in membranes, and self-associate into small ceramide-enriched membrane rafts. These microdomains have the tendency to spontaneously fuse into large ceramide-enriched membrane domains, termed platforms, that can be easily detected by light/fluorescence microscopy in living or fixed cells (Simons and Ikonen 1997; Brown and London 1998; Kolesnick et al. 2000). The formation of ceramide-enriched membrane platforms was shown to occur in cells after stimulation with a variety of receptors or non-receptor-stimuli including CD95, CD40, DR5, FcγRII, and PAF-receptor; infection with *P. aeruginosa*, *Neisseriae gonorrhoeae*, Rhinovirus, and measles virus; application of stress stimuli such as UV-light, gamma-irradiation, cisplatin, or Cu^{2+}-treatment, to name a few (Abdel Shakor et al. 2004; Avota et al. 2011; Dumitru and Gulbins 2006; Gassert et al. 2009; Göggel et al. 2004; Grassmé et al. 1997, 2001a, b, 2002, 2003a, b, 2005; Lacour et al. 2004; Lang et al. 2007; Rotolo et al. 2005, 2012). Since these diverse stimuli act differently on cells, ceramide-enriched membrane platforms appear to act at least in part as a general mechanism to reorganize potential signalling molecules within membranes. This is in contrast to a classic second messenger such as DAG that acts specifically to stoichiometrically activate target molecules.

4.2 Receptor-Mediated Signals

We originally demonstrated that ceramide-enriched membrane domains serve to trap and cluster receptor molecules and intracellular signalling molecules in the cell membrane (Grassmé et al. 2003b). This reorganization of a given "signalosome" facilitates transmission of the sequestered molecules' cognate signal across the bilayer into the cell. Trapping and clustering of receptor molecules in ceramide-enriched membrane domains are likely mediated by the unique biophysical properties created by the domain's microenvironment. However, at present molecular details regarding receptor clustering and preferential sorting of molecules into ceramide-enriched membrane domains are unknown. Trapping and clustering of receptor molecules may be mediated by direct interaction of proteins with ceramide molecules or with ceramide-enriched membrane domains acting as a hydrophobic platform. In addition, the length and composition of the transmembranous domain may determine whether a protein preferentially traps in a ceramide-enriched membrane domain or whether it is excluded. This principle was shown for the CD40 receptor, a protein, which clusters in ceramide-enriched membrane domains (Bock and Gulbins 2003). Replacement of its transmembrane domain by that of CD45, which does not cluster after stimulation, abrogated CD40 clustering in ceramide-enriched domains, indicating that the composition and structure of the transmembranous domain are critically involved in trapping of this receptor (Bock and Gulbins 2003). Thus, alteration of the transmembrane domain upon binding of ligand to its receptor may also serve to cluster a receptor in ceramide-enriched membrane domains and provide specificity for protein sorting.

Clustering of receptors and associated signalling molecules in ceramide-enriched membrane domains results in a high receptor density, spatial association of activated receptors with intracellular molecules, exclusion of inhibitory enzymes, and/or transactivation of intracellular enzymes. Thus, clustering may also coordinate the spatial and temporal organization of the signalosome generated by a specific receptor. This reorganization of a given signalosome enables the receptor to generate a strong localized signal. This concept was first shown with CD95 by demonstrating that clustered CD95 amplifies initial signalling, i.e., the early activation of caspase-8 approximately 100-fold (Grassmé et al. 2003b).

In addition to alteration of biophysical properties of plasma membrane, ceramide also directly interacts and regulates cathepsin D (Heinrich et al. 1999), phospholipase A_2 (Huwiler et al. 2001), kinase suppressor of Ras (Zhang et al. 1997), ceramide-activated protein serine–threonine phosphatases (CAPP) (Dobrowsky and Hannun 1993), and protein kinase C isoforms (Huwiler et al. 1998; Müller et al. 1995). Moreover, ceramide regulates the activity of the potassium channel Kv1.3 and calcium release-activated calcium (CRAC) channels (Gulbins et al. 1997; Lepple-Wienhues et al. 1999; Samapati et al. 2012). Finally, studies by Siskind and Columbini demonstrated that ceramide molecules form channels in the outer mitochondrial membrane, which might contribute to the induction of apoptosis (Siskind and Colombini 2000).

4.3 Non-receptor-Mediated Signalling

The studies above describe how the acid sphingomyelinase–ceramide system enables receptor-initiated signalling. However, the acid sphingomyelinase is also critically involved in the mediation of non-receptor-initiated signals. Irradiation is one of the best studied receptor-independent stimuli activating the acid sphingomyelinase and may serve as a paradigm for the activation of this enzyme by physical stimuli.

For decades DNA was considered the only relevant target for ionizing radiation with unrepaired or misrepaired double-strand breaks representing the lethal lesions. In 1994, Haimovitz-Friedman et al. published that ionizing radiation induces sphingomyelinase activation in intact bovine aortic endothelial cells within seconds to minutes, and in membranes devoid of nuclei. Subsequently, using acid sphingomyelinase knockout mice, it was shown that endothelial cells lacking acid sphingomyelinase were entirely resistant to apoptotic cell death (Santana et al. 1996). Similarly, lymphoblastoid cells derived from patients with Niemann–Pick disease were resistant to radiation-induced apoptosis. Interestingly these same patient cells were made vulnerable to radiation-induced apoptosis by genetic restoration of the enzyme via retroviral transduction. These studies provided definitive evidence that at least one form of radiation-induced death required membrane damage.

These reports served as the basis for a group of studies that attempted to address an important question in the field of cancer therapy: Did it matter whether tumor blood vessels underwent cell death for cancer therapy to succeed? Utilizing a mouse model to measure the tumor response upon the radiation gastrointestinal (GI) syndrome (Withers and Elkind 1970), studies uncovered that acid sphingomyelinase knockout mice failed to display endothelial apoptosis and were resistant to the radiation GI syndrome (Paris et al. 2001). Next, tumors were implanted in mice exposed to single, high-intensity radiation doses. In this case the tumor vasculature, recruited in response to VEGF and other growth factors, was either apoptosis competent (in tumors in acid sphingomyelinase wild-type hosts) or apoptosis resistant (in tumor of acid sphingomyelinase-deficient hosts) (Garcia-Barros et al. 2003). As in the GI tract, if tumor vasculature was apoptosis resistant, as supplied by an acid sphingomyelinase-deficient host, the tumor itself became radiation resistant. These studies provided the first solid evidence that a target other than the tumor cell controlled the fate of a tumor after irradiation.

At present, it is unknown how physical stimuli such as UV- or gamma-irradiation activate the acid sphingomyelinase. Using a series of genetic and pharmacologic manipulations in Jurkat cells, it was determined that there are two distinct mechanisms for acid sphingomyelinase activation. One required caspase, as in the case of CD95, whereas UV-C and ionizing radiation induced caspase-independent acid sphingomyelinase activation (Rotolo et al. 2005).

Within seconds of cytotoxic irradiation, ceramide-enriched platforms were detected, indicating that membrane platforms may be responsible for irradiation-induced cell death. Using an antibody that binds ceramide generated on the endothelial surface prevented radiation-induced platform formation, thereby protecting mice from radiation GI syndrome lethality (Rotolo et al. 2012). UV-C radiation, like ionizing radiation and CD95, also induced rapid extracellular membrane ceramide generation and cell death in Jurkat T cells, events inhibitable by anti-ceramide antibody (Rotolo et al. 2005).

Thus, as described above for receptor molecules, it seems very likely that the signal generated by activation of the acid sphingomyelinase via both forms of radiation is also transmitted into the cell via ceramide-enriched membrane platforms. At present, the molecules sorted and regulated by ceramide-enriched membrane platforms upon UV- or gamma-irradiation still require definition.

5 Perspective

A plethora of studies employing genetically modified mouse models, mouse disease models, and human tissues provide solid evidence that acid sphingomyelinase and ceramide play an important role in mediation of cell stress; receptor-, gamma-irradiation-, or UV-light-induced cell death; bacterial and viral infection; and receptor-initiated signalling. However, the molecular mechanisms mediating acid sphingomyelinase activation in vivo still require a great deal of definition. Ceramide alters the membrane microenvironment and subsequently forms ceramide-enriched membrane platforms that serve to trap and cluster receptor and signalling molecules. The molecular mechanisms mediating clustering of proteins remain to be determined. Ceramide itself is also thought to directly interact with some signalling molecules, although the motifs that mediate binding of ceramide to these molecules in many cases remain to be defined.

Acknowledgements Work in this review was supported by the DFG-SPP1267 program.

References

Abdel Shakor AB, Kwiatkowska K, Sobota A (2004) Cell surface ceramide generation precedes and controls FcgammaRII clustering and phosphorylation in rafts. J Biol Chem 279:36778–36787

Avota E, Gulbins E, Schneider-Schaulies S (2011) Ceramide generation is essential for enhancement of viral uptake in dendritic cells. PLoS Pathog 7:e1001290

Bao JX, Xia M, Poklis JL, Han WQ, Brimson C, Li PL (2010) Triggering role of acid sphingomyelinase in endothelial lysosome-membrane fusion and dysfunction in coronary arteries. Am J Physiol Heart Circ Physiol 298:H992–H1002

Bock J, Gulbins E (2003) The transmembranous domain of CD40 determines CD40 partitioning into lipid rafts. FEBS Lett 534:169–174

Brady RO, Kanfer JN, Mock MB, Fredrickson DS (1966) The metabolism of sphingomyelin. II. Evidence of an enzymatic deficiency in Niemann-Pick diseae. Proc Natl Acad Sci U S A 55:366–369

Brenner B, Ferlinz K, Grassmé H, Weller M, Koppenhoefer U, Dichgans J, Sandhoff K, Lang F, Gulbins E (1998) Fas/CD95/Apo-I activates the acidic sphingomyelinase via caspases. Cell Death Differ 5:29–37

Brown DA, London E (1998) Functions of lipid rafts in biological membranes. Annu Rev Cell Dev Biol 14:111–136

Cremesti A, Paris F, Grassmé H, Holler N, Tschopp J, Fuks Z, Gulbins E, Kolesnick R (2001) Ceramide enables fas to cap and kill. J Biol Chem 276:23954–23961

da Veiga Pereira L, Desnick RJ, Adler DA, Disteche CM, Schuchman EH (1991) Regional assignment of the human acid sphingomyelinase gene (SMPD1) by PCR analysis of somatic cell hybrids and in situ hybridization to 11p15.1-p15.4. Genomics 9:229–234

Dobrowsky RT, Hannun YA (1993) Ceramide-activated protein phosphatase: partial purification and relationship to protein phosphatase 2A. Adv Lipid Res 25:91–104

Dumitru CA, Gulbins E (2006) TRAIL activates acid sphingomyelinase via a redox mechanism and releases ceramide to trigger apoptosis. Oncogene 25:5612–5625

Dumitru CA, Carpinteiro A, Trarbach T, Hengge UR, Gulbins E (2007) Doxorubicin enhances TRAIL-induced cell death via ceramide-enriched membrane platforms. Apoptosis 12:1533–1541

Edelmann B, Bertsch U, Tchikov V, Winoto-Morbach S, Perrotta C, Jakob M, Adam-Klages S, Kabelitz D, Schutze S (2011) Caspase-8 and caspase-7 sequentially mediate proteolytic activation of acid sphingomyelinase in TNF-R1 receptosomes. EMBO J 30:379–394

Esen M, Schreiner B, Jendrossek V, Lang F, Fassbender K, Grassmé H, Gulbins E (2001) Mechanisms of Staphylococcus aureus induced apoptosis of human endothelial cells. Apoptosis 6:431–439

Ferlinz K, Hurwitz R, Vielhaber G, Suzuki K, Sandhoff K (1994) Occurrence of two molecular forms of human acid sphingomyelinase. Biochem J 301(Pt 3):855–862

Fowler S (1969) Lysosomal localization of sphingomyelinase in rat liver. Biochim Biophys Acta 191:481–484

Garcia-Barros M, Paris F, Cordon-Cardo C, Lyden D, Rafii S, Haimovitz-Friedman A, Fuks Z, Kolesnick R (2003) Tumor response to radiotherapy regulated by endothelial cell apoptosis. Science 300:1155–1159

Gassert E, Avota E, Harms H, Krohne G, Gulbins E, Schneider-Schaulies S (2009) Induction of membrane ceramides: a novel strategy to interfere with T lymphocyte cytoskeletal reorganisation in viral immunosuppression. PLoS Pathog 5:e1000623

Göggel R, Winoto-Morbach S, Vielhaber G, Imai Y, Lindner K, Brade L, Brade H, Ehlers S, Slutsky AS, Schütze S, Gulbins E, Uhlig S (2004) PAF-mediated pulmonary edema: a new role for acid sphingomyelinase and ceramide. Nat Med 10:155–160

Grassmé H, Gulbins E, Brenner B, Ferlinz K, Sandhoff K, Harzer K, Lang F, Meyer TF (1997) Acidic sphingomyelinase mediates entry of *N. gonorrhoeae* into nonphagocytic cells. Cell 91:605–615

Grassmé H, Jekle A, Riehle A, Schwarz H, Berger J, Sandhoff K, Kolesnick R, Gulbins E (2001a) CD95 signaling via ceramide-rich membrane rafts. J Biol Chem 276:20589–20596

Grassmé H, Schwarz H, Gulbins E (2001b) Molecular mechanisms of ceramide-mediated CD95 clustering. Biochem Biophys Res Commun 284:1016–1030

Grassmé H, Jendrossek V, Bock J, Riehle A, Gulbins E (2002) Ceramide-rich membrane rafts mediate CD40 clustering. J Immunol 168:298–307

Grassmé H, Jendrossek V, Riehle A, von Kürthy G, Berger J, Schwarz H, Weller M, Kolesnick R, Gulbins E (2003a) Host defense against *Pseudomonas aeruginosa* requires ceramide-rich membrane rafts. Nat Med 9:322–330

Grassmé H, Cremesti A, Kolesnick R, Gulbins E (2003b) Ceramide-mediated clustering is required for CD95-DISC formation. Oncogene 22:5457–5470

Grassmé H, Riehle A, Wilker B, Gulbins E (2005) Rhinoviruses infect human epithelial cells via ceramide-enriched membrane platforms. J Biol Chem 280:26256–26262

Gulbins E, Szabo I, Baltzer K, Lang F (1997) Ceramide-induced inhibition of T lymphocyte voltage-gated potassium channel is mediated by tyrosine kinases. Proc Natl Acad Sci U S A 94:7661–7666

Gulbins E, Kolesnick R (2000) Measurement of sphingomyelinase activity. Methods Enzymol 322:382–388

Haimovitz-Friedman A, Kan CC, Ehleiter D, Persaud RS, McLoughlin M, Fuks Z, Kolesnick RN (1994) Ionizing radiation acts on cellular membranes to generate ceramide and initiate apoptosis. J Exp Med 180:525–535

Han WQ, Xia M, Xu M, Boini KM, Ritter JK, Li NJ, Li PL (2012) Lysosome fusion to the cell membrane is mediated by the dysferlin C2A domain in coronary arterial endothelial cells. J Cell Sci 125:1225–1234

Hannun YA, Obeid LM (2011) Many ceramides. J Biol Chem 286:27855–27862

Hauck CR, Grassmé H, Bock J, Jendrossek V, Ferlinz K, Meyer TF, Gulbins E (2000) Acid sphingomyelinase is involved in CEACAM receptor-mediated phagocytosis of Neisseria gonorrhoeae. FEBS Lett 478:260–266

Heinrich M, Wickel M, Schneider-Brachert W, Sandberg C, Gahr J, Schwandner R, Weber T, Saftig P, Peters C, Brunner J, Krönke M, Schütze S (1999) Cathepsin D targeted by acid sphingomyelinase-derived ceramide. EMBO J 18:5252–5263

Hurwitz R, Ferlinz K, Vielhaber G, Moczall H, Sandhoff K (1994) Processing of human acid sphingomyelinase in normal and I-cell fibroblasts. J Biol Chem 269:5440–5445

Huwiler A, Fabbro D, Pfeilschifter J (1998) Selective ceramide binding to protein kinase C-alpha and -delta isoenzymes in renal mesangial cells. Biochemistry 37:14556–14562

Huwiler A, Johansen B, Skarstad A, Pfeilschifter J (2001) Ceramide binds to the CaLB domain of cytosolic phospholipase A_2 and facilitates its membrane docking and arachidonic acid release. FASEB J 15:7–9

Jia SJ, Jin S, Zhang F, Yi F, Dewey WL, Li PL (2008) Formation and function of ceramide-enriched membrane platforms with CD38 during M1-receptor stimulation in bovine coronary arterial myocytes. Am J Physiol Heart Circ Physiol 295:H1743–H1752

Jenkins RW, Idkowiak-Baldys J, Simbari F, Canals D, Roddy P, Riner CD, Clarke CJ, Hannun YA (2011) A novel mechanism of lysosomal acid sphingomyelinase maturation: requirement for carboxyl-terminal proteolytic processing. J Biol Chem 286:3777–3788

Jin S, Yi F, Zhang F, Poklis JL, Li PL (2008a) Lysosomal targeting and trafficking of acid sphingomyelinase to lipid raft platforms in coronary endothelial cells. Arterioscler Thromb Vasc Biol 28:2056–2062

Jin S, Zhang Y, Yi F, Li PL (2008b) Critical role of lipid raft redox signaling platforms in endostatin-induced coronary endothelial dysfunction. Arterioscler Thromb Vasc Biol 28:485–490

Kolesnick RN, Goni FM, Alonso A (2000) Compartmentalization of ceramide signaling: physical foundations and biological effects. J Cell Physiol 184:285–300

Kolzer M, Arenz C, Ferlinz K, Werth N, Schulze H, Klingenstein R, Sandhoff K (2003) Phosphatidylinositol-3,5-Bisphosphate is a potent and selective inhibitor of acid sphingomyelinase. Biol Chem 384:1293–1298

Lacour S, Hammann A, Grazide S, Lagadic-Gossmann D, Athias A, Sergent O, Laurent G, Gambert P, Solary E, Dimanche-Boitrel MT (2004) Cisplatin-induced CD95 redistribution into membrane lipid rafts of HT29 human colon cancer cells. Cancer Res 64:3593–3598

Lang PA, Schenck M, Nicolay JP, Becker JU, Kempe DS, Lupescu A, Koka S, Eisele K, Klarl BA, Rubben H, Schmid KW, Mann K, Hildenbrand S, Hefter H, Huber SM, Wieder T, Erhardt A, Häussinger D, Gulbins E, Lang F (2007) Liver cell death and anemia in Wilson disease involve acid sphingomyelinase and ceramide. Nat Med 13:164–170

Lepple-Wienhues A, Belka C, Laun T, Jekle A, Walter B, Wieland U, Welz M, Heil L, Kun J, Busch G, Weller M, Bamberg M, Gulbins E, Lang F (1999) Stimulation of CD95 (Fas) blocks T lymphocyte calcium channels through sphingomyelinase and sphingolipids. Proc Natl Acad Sci U S A 96:13795–13800

Müller G, Ayoub M, Storz P, Rennecke J, Fabbro D, Pfizenmaier K (1995) PKC zeta is a molecular switch in signal transduction of TNF-alpha, bifunctionally regulated by ceramide and arachidonic acid. EMBO J 14:1961–1969

Newrzella D, Stoffel W (1996) Functional analysis of the glycosylation of murine acid sphingomyelinase. J Biol Chem 271:32089–32095

Niemann A (1914) Ein unbekanntes Krankheitsbild. Jahrb Kinderheilkd 79:1–10

Paris F, Fuks Z, Kang A, Capodieci P, Juan G, Ehleiter D, Haimovitz-Friedman A, Cordon-Cardo C, Kolesnick R (2001) Endothelial apoptosis as the primary lesion initiating intestinal radiation damage in mice. Science 293:293–297

Perrotta C, Bizzozero L, Cazzato D, Morlacchi S, Assi E, Simbari F, Zhang Y, Gulbins E, Bassi MT, Rosa P, Clementi E (2010) Syntaxin 4 is required for acid sphingomyelinase activity and apoptotic function. J Biol Chem 285:40240–40251

Preuss S, Omam FD, Scheiermann J, Stadelmann S, Winoto-Morbach S, von Bismarck P, Adam-Klages S, Knerlich-Lukoschus F, Lex D, Wesch D, Held-Feindt J, Uhlig S, Schütze S, Krause MF (2012) Topical application of phosphatidyl-inositol-3,5-bisphosphate for acute lung injury in neonatal swine. J Cell Mol Med 16:2813–2826

Qiu H, Edmunds T, Baker-Malcolm J, Karey KP, Estes S, Schwarz C, Hughes H, Van Patten SM (2003) Activation of human acid sphingomyelinase through modification or deletion of C-terminal cysteine. J Biol Chem 278:32744–32752

Quintern LE, Schuchman EH, Levran O, Suchi M, Ferlinz K, Reinke H, Sandhoff K, Desnick RJ (1989) Isolation of cDNA clones encoding human acid sphingomyelinase: occurrence of alternatively processed transcripts. EMBO J 8:2469–2473

Rhein C, Tripal P, Seebahn A, Konrad A, Kramer M, Nagel C, Kemper J, Bode J, Mühle C, Gulbins E, Reichel M, Becker CM, Kornhuber J (2012) Functional implications of novel human acid sphingomyelinase splice variants. PLoS One 7:e35467

Roth AG, Redmer S, Arenz C (2009) Potent inhibition of acid sphingomyelinase by phosphoinositide analogues. Chembiochem 10:2367–2374

Rotolo JA, Zhang J, Donepudi M, Lee H, Fuks Z, Kolesnick R (2005) Caspase-dependent and -independent activation of acid sphingomyelinase signaling. J Biol Chem 280:26425–26434

Rotolo J, Stancevic B, Zhang J, Hua G, Fuller J, Yin X, Haimovitz-Friedman A, Kim K, Qian M, Cardo-Vila M, Fuks Z, Pasqualini R, Arap W, Kolesnick R (2012) Anti-ceramide antibody prevents the radiation gastrointestinal syndrome in mice. J Clin Invest 122:1786–1790

Samapati R, Yang Y, Yin J, Stoerger C, Arenz C, Dietrich A, Gudermann T, Adam D, Wu S, Freichel M, Flockerzi V, Uhlig S, Kuebler WM (2012) Lung endothelial Ca^{2+} and permeability response to platelet-activating factor is mediated by acid sphingomyelinase and transient receptor potential classical 6. Am J Respir Crit Care Med 185:160–170

Santana P, Pena LA, Haimovitz-Friedman A, Martin S, Green D, McLoughlin M, Cordon-Cardo C, Schuchman EH, Fuks Z, Kolesnick R (1996) Acid sphingomyelinase-deficient human lymphoblasts and mice are defective in radiation-induced apoptosis. Cell 86:189–199

Schissel SL, Jiang X, Tweedie-Hardman J, Jeong T, Camejo EH, Najib J, Rapp JH, Williams KJ, Tabas I (1998a) Secretory sphingomyelinase, a product of the acid sphingomyelinase gene, can hydrolyze atherogenic lipoproteins at neutral pH. Implications for atherosclerotic lesion development. J Biol Chem 273:2738–2746

Schissel SL, Keesler GA, Schuchman EH, Williams KJ, Tabas I (1998b) The cellular trafficking and zinc dependence of secretory and lysosomal sphingomyelinase, two products of the acid sphingomyelinase gene. J Biol Chem 273:18250–18259

Schuchman EH, Suchi M, Takahashi T, Sandhoff K, Desnick RJ (1991) Human acid sphingomyelinase. Isolation, nucleotide sequence and expression of the full-length and alternatively spliced cDNAs. J Biol Chem 266:8531–8539

Schuchman EH, Levran O, Pereira LV, Desnick RJ (1992) Structural organization and complete nucleotide sequence of the gene encoding human acid sphingomyelinase (SMPD1). Genomics 12:197–205

Simons K, Ikonen E (1997) Functional rafts in cell membranes. Nature 387:569–572

Siskind LJ, Colombini M (2000) The lipids C2- and C16-ceramide form large stable channels. Implications for apoptosis. J Biol Chem 275:38640–38644

Withers HR, Elkind MM (1970) Microcolony survival assay for cells of mouse intestinal mucosa exposed to radiation. Int J Radiat Biol Relat Stud Phys Chem Med 17:261–267

Xu M, Xia M, Li XX, Han WQ, Boini KM, Zhang F, Zhang Y, Ritter JK, Li PL (2012) Requirement of translocated lysosomal V1 H(+)-ATPase for activation of membrane acid sphingomyelinase and raft clustering in coronary endothelial cells. Mol Biol Cell 23:1546–1557

Zeidan YH, Hannun YA (2007) Activation of acid sphingomyelinase by protein kinase Cdelta-mediated phosphorylation. J Biol Chem 282:11549–11561

Zhang C, Li PL (2010) Membrane raft redox signalosomes in endothelial cells. Free Radic Res 44:831–842

Zhang Y, Yao B, Delikat S, Bayoumy S, Lin XH, Basu S, McGinley M, Chan-Hui PY, Lichenstein H, Kolesnick R (1997) Kinase suppressor of Ras is ceramide-activated protein kinase. Cell 89:63–72

Ceramide Synthases: Reexamining Longevity

Joo-Won Park and Yael Pewzner-Jung

Contents

Abstract The ceramide synthase (CerS) enzymes catalyze the formation of (dihydro) ceramide, and thereby provide critical complexity to all sphingolipids (SLs) with respect to their acyl chain length. This review summarizes the progress in the field of CerS from the time of their discovery more than a decade ago as *L*ongevity *ass*urance (*Lass*) genes in yeast, until the recent development of CerS-deficient mouse models. Human hereditary CerS disorders are yet to be discovered. However, the recent findings in CerS mutant animals highlight the important physiological role of these enzymes. The fundamental findings with respect to CerS structure, function, localization, and regulation are discussed, as well as CerS roles in maintaining longevity in vivo.

Keywords Ceramide synthase (CerS) deficiency • Mice • Fatty acyl-CoA

J.-W. Park
Department of Biochemistry, School of Medicine, Ewha Womans University, Seoul 158-710, South Korea

Y. Pewzner-Jung (✉)
Department of Biological Chemistry, Weizmann Institute of Science, 234 Herzl St., Rehovot 76100, Israel
e-mail: yael.pewzner-jung@weizmann.ac.il

E. Gulbins and I. Petrache (eds.), *Sphingolipids: Basic Science and Drug Development*, Handbook of Experimental Pharmacology 215, DOI 10.1007/978-3-7091-1368-4_5,

1 Introduction

Sphingolipids (SLs) are membrane structural components and exist in all forms of life, including some species of viruses (Bidle and Vardi 2011), bacteria (Heung et al. 2006), algae (Pagarete et al. 2009), as well as yeast (Obeid et al. 2002), nematodes (Deng et al. 2008), insects (Bauer et al. 2009), plants (Sperling and Heinz 2003), and other higher organisms (Levy and Futerman 2010). In the past decade it was established that specific types of SLs act as bioactive signaling molecules (Hannun and Obeid 2008). Recently, moreover, it was also shown that individual SL species can interact directly and specifically with a protein's transmembrane (TM) domain and act as a cofactor to actively modulate the protein's function (Contreras et al. 2012). This regulatory capacity and the known "second messenger" properties of some SLs may explain the large variety and intricacy of this class of lipids. Complex SLs consist of a hydrophobic ceramide backbone generated de novo in the endoplasmic reticulum (ER) and a hydrophilic moiety made up of phosphorylcholine, sugars, and sialic acids that are added onto the ceramide backbone either in the ER or in the Golgi apparatus (Futerman and Hannun 2004). Complex SLs that are localized to the plasma membrane (PM) can be recycled and hydrolyzed by specific "sphingolipases" (i.e., sphingomyelinases, gluco- and galactosidases, hexosaminidases, etc.), regenerating ceramide in the PM or lysosome. Ceramide can be further hydrolyzed to form sphingosine, which can be metabolized to regenerate ceramide via the salvage pathway (Kitatani et al. 2008).

Ceramide consists of two hydrocarbon chains: a sphingoid long-chain base (LCB) and a fatty acid that is attached at the C-2 position of the LCB via N-acylation. The length, saturation, and hydroxylation status of these two hydrophobic chains varies among ceramide molecules (Zheng et al. 2006). Thus, ceramides are a family of related molecules that can be further modified to generate even more complex SLs that are localized to the PM and other cell organelles (Hannun and Obeid 2011). It is conceivable that membrane proteins select a specific set of SLs to optimize their functional environment, and indeed this notion finds support in the literature (Iwabuchi et al. 2008; Contreras et al. 2012). Furthermore, upon protein activation, ceramide, "hidden" in the structure of the complex SL, is released in situ adjacent to the activated protein by "sphingolipases," and acts as a second messenger. This may generate a synchronized change of the protein and the biophysical properties of its lipid environment that allow information to be relayed into the cell. The discrepancies found in the literature regarding the role of specific ceramide molecules in physiological processes, such as C16-ceramide involvement in apoptosis, as well as in cell growth (Osawa et al. 2005; Koyanagi et al. 2003), can therefore be attributed to the type of protein with which these specific SLs or ceramides are associated.

Although many enzymes are obviously involved in the generation of this SL complexity, the most critical ones are the ceramide synthases (CerS), which contribute to the variability of the fatty acid length (Pewzner-Jung et al. 2006). Originally, CerS were found while screening for genes that are important for yeast longevity, and thus named *L*ongevity *Ass*urance (*Lass*) genes (D'Mello et al. 1994; Jiang et al. 1998). The name was replaced with "CerS" to better define the actual enzymatic activity (Pewzner-Jung et al. 2006), but recent data (see below) reaffirm the possible role of all CerS orthologs in maintaining physiological processes that influence longevity.

CerS orthologs are found in many species and are evolutionarily conserved in eukaryotes (Pewzner-Jung et al. 2006). Some species have a single CerS, whereas others have several paralogs. However, the complexity of the CerS family does not necessarily correlate with evolutionary development: for example, *Caenorhabditis elegans* harbors three CerS paralogs, whereas *Drosophila melanogaster* has only one. Mammals have six CerS (named CerS1-6); each can acylate an LCB with a fatty acyl-CoA of distinct length to generate (dihydro) ceramide (using sphinganine—in the de novo biosynthesis pathway) or ceramide (using sphingosine—in the recycling salvage pathway of SL). Thus, the specificity of CerS is limited to fatty acyl chain length, whereas the saturation and hydroxylation states of the fatty acids, as well as the nature of the LCB substrate, are not restricted (Levy and Futerman 2010). Hence, CerS1 uses mostly C18-fatty acyl CoA (Venkataraman et al. 2002), and CerS2 can utilize a wider range of very-long-chain (VLC) fatty acyl CoAs (C20 to C26) (Laviad et al. 2008). CerS3 incorporates ultra-long-chain (ULC) fatty acyl CoAs (C26 to C32) (Mizutani et al. 2006; Rabionet et al. 2008), and CerS4 uses C18- and C20-fatty acyl CoAs (Riebeling et al. 2003). Interestingly, CerS5 has specificity only for C16-fatty acyl CoA (Lahiri and Futerman 2005), whereas CerS6 can use both C14- and C16-fatty acyl CoAs (Mizutani et al. 2005). Notably, these specificity assignments are largely based on results obtained from in vitro CerS activity assays. However, it was shown that in complementation experiments involving *Saccharomyces cerevisiae* mutants lacking CerS activity, the introduction of mammalian CerS1 allowed utilization of the naturally abundant C26-fatty acyl CoA to generate C26-ceramide in vivo (Guillas et al. 2003). This may suggest that the availability of specific fatty acids and the particular CerS activity in a given cell type determine the composition of specific SLs containing distinct fatty acids.

In addition to CerS, de novo synthesis of ceramide depends on other ER enzymes such as serine palmitoyl transferase (SPT), the rate-limiting enzyme in SL biosynthesis, which condensates mainly L-serine and palmitoyl CoA to form 3-ketosphinganine. The latter is reduced by 3-ketosphinganine reductase to sphinganine, which upon N-acylation by CerS forms dihydro-ceramide that is desaturated by dihydro-ceramide desaturase to ceramide (Levy and Futerman 2010). The interaction and potential feedback regulation between these enzymes are yet to be explored, although it is conceivable that all of them closely cooperate in a united and well-regulated "Ceramidosome."

2 Ceramide Synthase Localization and Structure

CerS are integral membrane ER proteins and ceramide was shown to be synthesized at the cytosolic surface of the ER membrane (Hirschberg et al. 1993). CerS activity has also been attributed to the mitochondria (Shimeno et al. 1998; Bionda et al. 2004). As mitochondria play an important role in ceramide-induced apoptosis, this may imply the direct involvement of mitochondrial ceramide formation in the cell death process. However, mitochondria obtained by subcellular fractionation may contain ER membranes (mitochondria-associated membrane—MAM) (Wang et al. 1991), and thus, contaminations of ER-derived CerS activity cannot be excluded. Regulated re-localization of CerS1 from the ER to the Golgi apparatus following stress stimuli was recently demonstrated, and could suggest a more general phenomenon of CerS compartmentalization (Sridevi et al. 2009).

CerS orthologs harbor common conserved domains (Fig. 1a). The Tram-Lag-CLN8 (TLC) domain, consisting of ~200 AA residues, is also found in other proteins (with yet unknown function) (Winter and Ponting 2002). The Lag1p motif within the TLC domain contains an evolutionarily conserved 52 AA stretch shown to have an important role in CerS activity (Venkataraman et al. 2002; Spassieva et al. 2006; Kageyama-Yahara and Riezman 2006). Finally, CerS2–CerS6, but not CerS1 and its orthologs, harbor a Homeobox (Hox)-like domain in their N-terminal region. This domain lacks the DNA-binding motif of standard homeobox domains, and thus is not considered to be involved in the usual function of homeobox-containing proteins. The fact that CerS1 lacks this domain reduces the possibility that the Hox-like domain serves as one of the substrate-binding domains in all other CerS. However, the terminal 12 AA residues of the Hox domain (black box, Fig. 1a) were shown to play a pivotal role in CerS5 and CerS6 activity (Mesika et al. 2007). Therefore, the Hox-like domain is most likely important in regulatory aspects of CerS containing this motif. Recent data suggested a critical role for the end of the second putative TM domain, and the loop between the fifth and sixth putative TM domains (Fig. 1a—dashed boxes) of CerS5 activity (Tidhar et al. 2012). As the six mammalian CerS exhibit different specificities for fatty acyl-CoAs, the binding domain for the acyl group is expected to differ accordingly, but could be similar among CerS with overlapping specificities. Unexpectedly, the 150 residues (open box, Fig. 1a) that determine CerS5 specificity are highly conserved among all CerS (Tidhar et al. 2012). CerS enzymes have similar K_m values for sphinganine (Lahiri et al. 2007), and thus—in contrast to the fatty acyl-CoA-binding domains—sphinganine-binding domains are expected to be similar in all CerS. Yet the exact mechanism that determines substrate specificity remains to be elucidated.

So far CerS topology has been predicted using only bioinformatics prediction software, and as of today no crystal structures of CerS are available. Eight putative transmembrane domains were suggested to exist in the *S. cerevisiae* CerS Lag1p and Lac1p, and the conserved Lag motif was suggested to be embedded in the membrane (Kageyama-Yahara and Riezman 2006). In another study it was

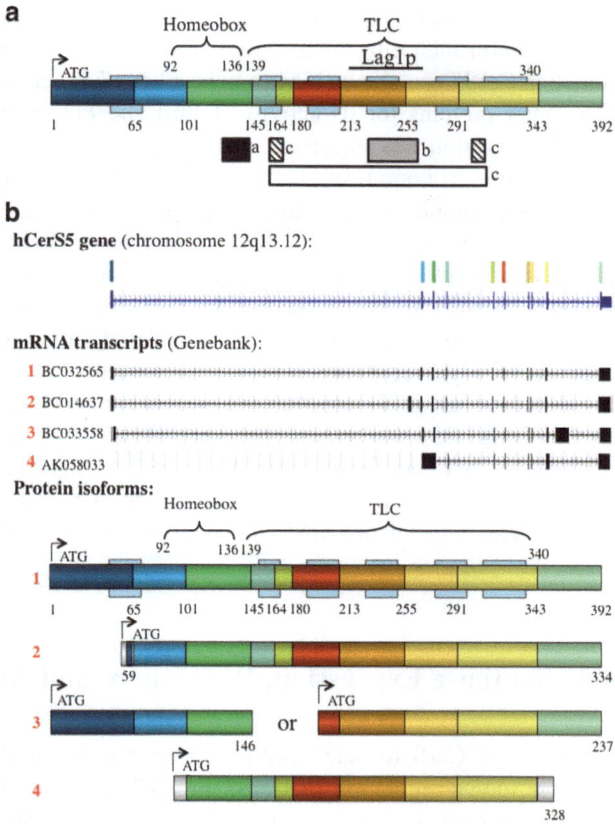

Fig. 1 Structural and functional domains in human CerS5. (**a**) Protein domains in human CerS5 protein. The full-length hCerS5 protein is encoded by ten exons. Each color indicates a stretch of amino acids (AA) encoded by a specific exon marked with an identical color. The numbers *beneath* the bar represent the positions of the amino acids (AA) in the protein. The numbers *above* the bar mark the various known domains of the protein (see text). The putative transmembrane domains are highlighted as *blue boxes* behind the bar. Important residues shown to be involved in the enzymatic activity are marked as *small filled rectangles*. The *open rectangle* represents the minimal region involved in specificity. The letters next to the rectangles signify the following references: (a) Mesika et al. (2007). (b) Spassieva et al. (2006). (c) Tidhar et al. (2012). (**b**) Protein isoforms derived from splice variants of human CerS5 transcripts. Protein isoforms are presented by the same color code as in (**a**). The numbers *beneath* the bar represent AA positions. Transcript number 1 represents the full-length protein; transcripts 2–4 represent four possible isoforms (transcript 3 could encode two different putative proteins). The transcripts shown are representative of many others encoding the same protein isoform. All the information was retrieved from the UCSC genome browser database (http://genome.ucsc.edu—under human CerS5)

proposed that CerS6 has five putative transmembrane domains. In addition, it was considered that the N-terminal end of CerS6 might face the luminal side of the ER membrane, the C-terminal region stretching out to the cytosol (Mizutani et al. 2005). However, the precise topology of CerS proteins in the ER membrane remains to be defined.

In yeast, Lag1p and Lac1p depend on an additional single-transmembrane domain protein (Lip1) for their enzymatic activity (Vallee and Riezman 2005). In contrast, mammalian CerS5 was shown in vitro to be a *bona fide* ceramide synthase, and required no other proteins for its activity (Lahiri and Futerman 2005). The concept of CerS dimerization was suggested (Lahiri et al. 2009), and recently, the formation of heterodimeric complexes between CerS2, CerS5, and CerS6 was demonstrated, using co-immunoprecipitation methods (Mesicek et al. 2010; Laviad et al. 2012). In addition, co-expression of CerS2 with CerS5 or CerS6 significantly increased CerS activity in cells, as well as the treatment of curcumin that promoted CerS dimerization (Laviad et al. 2012). Interestingly, in a mouse model for CerS3 deficiency, where ULC-ceramides were absent from the epidermis, there was also a significant reduction in VLC-ceramides, which are the product of CerS2 enzyme activity (highly abundant in skin along with CerS3) (Jennemann et al. 2011). These results suggest that CerS3 may catalyze, in addition to ULC-ceramides, the formation of VLC-ceramides, but could also imply that CerS2 collaborates with CerS3 for its maximum activity. Taken together, it remains to be explored whether, as in yeast, mammalian CerS (other than CerS5) are regulated by other CerS or other proteins.

3 Ceramide Synthase Expression, Regulation, and Activity

Selective knockdown of CerS revealed complex inter-regulation (Mullen et al. 2011). For example, down-regulation of CerS2 in MCF-7 human breast cancer cells increased the mRNA levels of CerS4, CerS5, and CerS6, as well as their products (Mullen et al. 2011). Similar results were obtained in other cell types (Spassieva et al. 2009), as well as in a mouse model of CerS2 deficiency, where CerS5 mRNA expression was upregulated along with C16-ceramide (Pewzner-Jung et al. 2010b). This may be attributed to the activity of specific miRNAs to CerS transcripts, or of some unknown transcription factors (TFs). Recently, miR-221/222 was shown to bind directly to the 3'-untranslated region of CerS2, thus reducing both mRNA and protein levels of CerS2 (Yu et al. 2012). In addition, miR-574-5p was found to be involved in the degradation of a CerS1 splice variant (Meyers-Needham et al. 2011). At the same time, it was shown that CerS1 transcription was repressed by the inability of Sp1 TF to bind to its promoter due to histone deacetylase 1 (HDAC1)-dependent inhibition of Sp1. In addition, other proteins, apart from TFs, were also shown to be involved in the expression of CerS. For example, CerS6 transcription was downregulated by the expression of the dominant-negative ER stress kinase (protein kinase R-like ER kinase) PERK (Yacoub et al. 2010). These data imply the involvement of many factors in the regulation of CerS mRNA expression.

All CerS have putative splice variants (Pewzner-Jung et al. 2006), CerS5 being the champion (Fig. 1b). CerS2 has a short isoform called tumor metastasis suppressor gene-1 (TMSG-1), which lacks the N-terminal Hox-like domain and part of the

TLC domain (Ma et al. 2003). Transcription factors KLF6 and Sp1 were recently reported to be involved in TMSG-1 expression (Gong et al. 2011). TMSG-1 was suggested to function as a tumor suppressor gene, as it promotes apoptosis and is absent in metastatic cancer (Ma et al. 2003; Fei et al. 2004). In contrast, CerS1 isoforms were shown to be highly expressed in cancer cells (Meyers-Needham et al. 2011). With the exception of CerS3, all human CerS have similar splice variants lacking their N-terminal regions, including the Hox-like domains (for CerS5 see Fig. 1b). This repeating molecular pattern suggests biologically significant posttranslational regulation of CerS proteins, which may possibly yield a critical dominant negative effect if CerS indeed form heterocomplexes. Interestingly, human CerS5 also has a short isoform lacking the C-terminal portion of the protein (Fig. 1b), which may be involved in other regulatory aspects of this enzyme. Evidence for the in vivo expression of some of these isoforms has been reported (van Hall et al. 2006), highlighting the potential function of CerS variants in cell physiology.

CerS mRNA expression and protein activity were shown to be correlated in some cell types and organs but not in all. Thus, despite the high mRNA expression of CerS2 in the kidney, C16-ceramide is the most abundant ceramide in this organ, which is consistent with the relatively low CerS2 activity found in mouse kidney homogenates (Laviad, EL. and Futerman, AH. unpublished data). These findings support the notion of possible posttranscriptional regulation of CerS, and highlight the importance of studying CerS enzymatic activity in specific cell types. Currently, only few antibodies for CerS proteins are commercially available, and most CerS expression data are based on profiling transcript expression of whole organs or specific cell types (summarized in Table 1). However, these data sets should be taken with caution owing to the possible discrepancy between the expression and activity of CerS proteins.

CerS activity can be directly regulated by their binding to specific proteins. The pro-apoptotic molecule Bak was suggested to be in a stoichiometric complex with CerS producing long-chain (LC) ceramides. Furthermore, Bak expression was shown to be required for the production of LC-ceramide during UV irradiation-induced apoptosis (Siskind et al. 2010). The formation of pores in the outer mitochondria leads to apoptosis and has been associated with both Bax/Bak translocation to the mitochondria and to the formation of ceramide-rich domains in these organelles (Lee et al. 2011). Bax/Bak thus potentially promotes apoptosis not only by direct pore formation but also by increasing CerS activity, which can lead to an increase in LC-ceramide in the outer mitochondrial membrane. Interestingly, not only proteins bind CerS but also sphingosine-1 phosphate (S1P), which has been proposed to be a noncompetitive inhibitor of CerS2, directly interacting via an S1P receptor-like motif found uniquely in this enzyme (Laviad et al. 2008).

CerS proteins were also shown to undergo posttranslational modifications such as glycosylation and phosphorylation. CerS2, CerS5, and CerS6 were found to be N-glycosylated at the Asn residue in the N-terminal portion of the protein. However, these modifications did not appear to affect CerS activity (Mizutani et al. 2005). CerS1 was shown to be phosphorylated by protein kinase C (PKC) at one or

Table 1 CerS mRNA distribution in tissue and specific cell types

Ceramide synthase	Tissue	Cell	Reference
CerS1	Brain[a]	Neurons[a]	Becker et al. (2008); Laviad et al. (2008)
		Purkinje neurons	Zhao et al. (2011)
	Skeletal muscle[a]		Laviad et al. (2008)
CerS2	Liver[a]	Hepatocytes[a]	Laviad et al. (2008); Pewzner-Jung et al. (2010a); Imgrund et al. (2009); Mizutani et al. (2005)
	Brain	Oligodendrocytes[a]	Becker et al. (2008); Ben-David et al. (2011)
	Spinal cord	Schwann cells[a]	
	Kidney[a]	HEK293[a]	Imgrund et al. (2009); Becker et al. (2008)
	Spleen[a]		Laviad et al. (2008); Mizutani et al. (2005)
	Prostate[a]		Laviad et al. (2008)
	Intestine[a]		Laviad et al. (2008)
	Bone marrow[a]		Laviad et al. (2008)
	Thymus[a]		Laviad et al. (2008)
	Lymph nodes[a]		Laviad et al. (2008)
	Lung[a]		Laviad et al. (2008); Mizutani et al. (2005); Pewzner-Jung, Y. and Futerman, AH., unpublished data
	Adrenal gland[a]		Park, WJ. and Futerman AH., unpublished data
	White adipose tissue[a]		Bonzon-Kulichenko et al. (2009); Park et al. (2012)
		Mouse embryonic stem cells[a]	Park et al. (2010)
		Mouse embryoid bodies[a]	Park et al. (2010)
		Erythroleukemic cell line	http://www.genecards.org/ http://www.immgen.org/
		Macrophages (raw cells)[a]	Lingaraju, MH. and Futerman AH., unpublished data
CerS3	Skin	Keratinocytes	Laviad et al. (2008); Mizutani et al. (2006); Jennemann et al. (2011)
	Testis[a]		
CerS4	Heart		Laviad et al. (2008)
	Skin[a]		Laviad et al. (2008)
	Skeletal muscle		Laviad et al. (2008)
	White adipose tissue		Bonzon-Kulichenko et al. (2009)
	Liver		Laviad et al. (2008)
	Lung		Pewzner-Jung, Y. and Futerman, AH., unpublished data
		Leukocytes[a]	Laviad et al. (2008)
CerS5	Skeletal muscle		Laviad et al. (2008)
	Testis		Laviad et al. (2008); Mizutani et al. (2005)

(continued)

Table 1 (continued)

Ceramide synthase	Tissue	Cell	Reference
	Kidney		Laviad et al. (2008); Mizutani et al. (2005)
	Liver		Laviad et al. (2008)
	Lung	Lung epithelial cells[a]	Mizutani et al. (2005); Xu et al. (2005)
	Brain		Mizutani et al. (2005)
	Prostate		Laviad et al. (2008)
CerS6	Brain		Mizutani et al. (2005)
	Kidney		Mizutani et al. (2006); Laviad et al. (2008)
	Intestine		Laviad et al. (2008)
	Liver		Mizutani et al. (2006); Laviad et al. (2008)
	White adipose tissue		Bonzon-Kulichenko et al. (2009)
		T cell lines[a]	http://www.genecards.org/
		Neutrophiles[a]	http://www.immgen.org/

[a]The major CerS in this tissue or cells

more of five putative serine and threonine phosphorylation sites that play a role in the translocation of CerS1 from the ER to the Golgi apparatus (Sridevi et al. 2009, 2010). In addition, the results of a large-scale phosphorylation analysis suggested phosphorylation of CerS2 and CerS5 (Villen et al. 2007). Interestingly, CerS1 was found to be degraded by ubiquitination, implying a relatively short half-life for CerS1 (Sridevi et al. 2009). Further analysis of these modifications is required to firmly establish their role in CerS regulation.

4 In Vivo Models of Ceramide Synthase Deficiency

The yeast *S. cerevisiae* was the first genetic model in which ceramide synthase activity was abolished (Guillas et al. 2001; Schorling et al. 2001). When the two CerS homologs Lag1p and Lac1p were deleted, the cells, lacking the ability to generate C26-ceramide, accumulated 20 times more free LCB, and exhibited impaired transport of glycosylphosphatidylinositol (GPI)-anchored proteins from the ER to the Golgi (Guillas et al. 2001). However, the cells were viable and their phenotype could be partially corrected by overexpression of ceramidases (*YPC1* or *YDC1*) due to the enzymes' reverse activity. Quadruple mutant cells (*lag1*Δ*lac1*Δ*ypc1*Δ*ydc1*Δ) did not make any SLs, but were still viable, probably due to production of novel lipids that compensated to some extent for SL function. However, these strains were found to multiply much more slowly, exhibited compromised colony formation from single cells, and did not survive at higher temperatures (Schorling et al. 2001). Interestingly, Guillas et al. later reported that *lag1*Δ*lac1*Δ double mutants were viable or lethal, depending on the cell-type background (W303 cells vs. YPK9, respectively) (Guillas et al. 2003). This suggests that other genetic factors may determine viability in the absence of CerS, especially under stress or environmental changes.

C. elegans has three CerS enzymes: *hyl-1*, *hyl-2*, and *lagr-1*. Generation of C20–C22 ceramides requires HYL-2, whereas that of C24–26 ceramides depends mainly on HYL-1 (Menuz et al. 2009). The *hyl-2* promoter was found to be active from the larval to the adult stage, with strong expression in the gut, the posterior bulb of the pharynx, the hypoderm, as well as in unidentified cells in the head and tail. Mutation of the double-histidine motif in the Lag domain (conserved in all CerS) of *hyl-2* or complete depletion of the gene resulted in hypoxia sensitivity, which was CED-3 (caspase) independent. In addition, *hyl-2* mutants were also more sensitive to heat shock. In contrast, lack of *hyl-1* led to resistance to hypoxia, but introduction of *hyl-1* into worms lacking HYL-2 did not fully restore hypoxia resistance. This suggests that C20–C22-SLs, but not C24–C26-SLs, play a role in the resistance to hypoxia. Of note, in both *hyl-1* and *hyl-2* mutants, iso-sphingoid bases were accumulated, but experiments showed that this did not influence hypoxia sensitivity (Menuz et al. 2009). Interestingly, simultaneous deletion of both *hyl-1* and *hyl-2* was reported to be lethal (Menuz et al. 2009). In a different set of experiments, screening for mutants resistant to radiation-induced germ-cell apoptosis revealed cell death arrest in *hyl-1* and *lagr-1* but not in *hyl-2* mutants. In contrast, *hyl-1* and *lagr-1* mutations did not affect somatic cell death, nor did the *hyl-2* mutant. Germline injection of C16-ceramide restored apoptosis in *hyl-1* and *lagr-1* deletion mutants (Deng et al. 2008). These studies suggest that both *hyl-1* and *lagr-1* are crucial for radiation-induced apoptosis of germline cells (Deng et al. 2008). Thus, each of the *C. elegans* CerS paralogs has distinct nonredundant functions, which, under specific stress conditions, cannot be compensated for.

The *drosophila* Lass gene 1 homolog (*Dlag1*) was recently renamed *Schlank* (slim) to emphasize the main phenotype described for the *Dlag1* mutant (Bauer et al. 2009). Three mutants were characterized; two of them $schlank^{G0061}$ and $schlank^{G0349}$ went through embryogenesis and hatched from eggs to become larvae, whereas in the complete null mutant no eggs were obtained. The two mutants showed arrest at different larval development stages, in correlation with reduced total ceramide levels and fat storage (triacylglycerols—TG), whereas overexpression of *schlank* led to increased levels of ceramide and fat. These results were confirmed in RNAi-mediated knockdown and genetic rescue experiments in *drosophila*. Immunohistochemical analyses indicate that *schlank* is strongly expressed in fat cells. Moreover, *schlank* positively regulates fatty acid synthesis by promoting the expression of sterol-responsive element-binding protein SREBP and its target genes. This study suggested that *schlank* acts as a regulator of the balance between lipogenesis and lipolysis during *drosophila* larval growth (Bauer et al. 2009). Taken together, complete ablation of CerS activity in the fruit fly causes lethality, whereas residual *schlank* activity allows normal embryonic development, albeit limiting larval development.

Recently, CerS1-, 2-, and 3-deficient mice were generated in an attempt to elucidate the role and function of specific CerS in mammalian organisms.

Two mouse strains harboring spontaneous CerS1 mutations were discovered (Zhao et al. 2011). The flincher (*fln*) mutation lacks a single nucleotide in exon 5, resulting in a frameshift mutation significantly reducing CerS1 transcript levels. These mice developed Purkinje cell degeneration and ataxia starting at around 3 weeks of age. The Purkinje neurons of this mutant exhibited shorter dendritic arbors with lower order branches and complexity. Surprisingly, a similar phenotype was reported for toppler (*to*) mutant mice that harbor a missense mutation in CerS1 exon 5, adjacent to the *fln* mutation (Zhao et al. 2011). In both mouse strains CerS1 enzyme activity was nearly absent. As a result, total ceramide in the brain was reduced by half due to the drop in C18:0-ceramide (the major ceramide in brain), albeit a twofold increase in C16-ceramide. In addition, sphinganine and sphinganine 1-phosphate were significantly elevated, which could account for the reduction in total ceramide or reflect a feedback mechanism sensing the lack of neuronal CerS1 activity. In their neurons, CerS1-deficient mice also accumulated lipofuscin—an undigested storage material containing lipids and proteins usually found in normal aging brains and in some pathological conditions (Jung et al. 2007). It has been suggested that lipofuscin production results from impaired autophagy, especially mitophagy, and chronic oxidative stress (Sulzer et al. 2008). Thus, a change in the overall SL profile, which leads to altered membrane-biophysical properties, could harm the lysosomal hydrolytic capacity. Interestingly, only Purkinje neurons were shown to be affected in CerS1-deficient mice, despite the fact that it was reported to be expressed in most neurons (Becker et al. 2008). However, since the data are derived from mRNA expression levels, it may well be that CerS1 activity is different in various types of neurons due to posttranscriptional or -translational regulation. So far, information on the effect of CerS1 deficiency on other organs, such as the skeletal muscle, where CerS1 mRNA is highly expressed (Laviad et al. 2008) is not available. A human genetic deficiency in the CerS1 gene remains to be identified, but the fact that *fln* mice that have almost no CerS1 activity are viable and survive to adulthood suggests that individuals with genetic CerS1 deficiencies exist in human populations and may contribute to one of the inherited acute ataxia disorders of unknown origin.

Owing to the abundance of its mRNA in most organs, CerS2 was for a while considered a "housekeeping" gene (Laviad et al. 2008). However, analysis of mice harboring a lacZ reporter gene under the control of the endogenous CerS2 promoter revealed a pronounced cell type-specific CerS2 expression pattern in different organs (Pewzner-Jung et al. 2010a; Imgrund et al. 2009). CerS2-deficient mice exhibited chronic alterations in liver and in brain physiology, but not in kidney homeostasis where CerS2 activity was—despite high mRNA expression levels—low (Pewzner-Jung et al. 2010a). Phenotypic analysis of "unchallenged" CerS2-deficient mice is expected to reveal the pathology of cells in which the enzyme is highly active. Accordingly, hepatocytes that showed high CerS2 activity and high levels of VLC-SLs (Park, JW., and Futerman, AH., unpublished data) exhibited increased apoptosis in CerS2-deficient mice already at 1 month of age (Pewzner-Jung et al. 2010a). As the liver has a high regeneration capacity, hepatocyte death was associated with massive proliferation. Moreover, the high hepatocyte turnover eventually led to the generation of adenomas and hepatocellular carcinomas in CerS2-deficient mice above 8 months of age (Pewzner-Jung et al. 2010a; Imgrund et al. 2009). The primary mechanism causing hepatocyte death in the absence of CerS2 is not fully understood. CerS2-deficient livers exhibit altered SL profiles due to the almost complete lack of VLC-SLs, with an elevation in C16-fatty acyl chains containing SLs, and a dramatic increase in LCB, in particular sphinganine. However, the total SL levels in CerS2-deficient livers were found to be similar to those in wild-type mice. Therefore, the SL profile alterations are unlikely to be the only explanation accounting for the increased hepatocyte turnover, because in the kidneys of "unchallenged" CerS2-deficient mice, there was a similarly altered SL profile but no apparent pathology. Rather, hepatocyte-specific factor/s are likely altered in CerS2-deficient mice as a result of the change in the biophysical properties of the hepatocyte membranes (Pewzner-Jung et al. 2010b), resulting in hepatocyte death already at an early age.

The CerS2 promoter is highly active in the periportal region of the liver, rather than in the centrilobular area, as detected by X-gal staining of heterozygous liver. As gluconeogenesis is attributed mainly to this liver zone (Jungermann and Katz 1989), that might explain, at least in part, the hypoglycemia observed in CerS2-deficient mice (Pewzner-Jung et al. 2010a). However, it should be stressed, hepatocyte death is not restricted to the periportal zone, but found scattered in all the liver zones. This issue may be resolved in the near future with the development of a good antibody for mouse CerS2 and improved resolution of SL MALDI imaging mass spectrometry techniques (Chen et al. 2010), which will allow detection of the specific products of each CerS in situ. Interestingly, CerS2-deficient mice also exhibit reduced body weight and fat accumulation (Pewzner-Jung et al. 2010a), reminiscent of the phenotype found in drosophila—*schlank* mutants, where TG are dramatically reduced (Bauer et al. 2009). This suggests that CerS2 and *schlank* are involved in energy metabolism, and maybe even in the function of mitochondria, which are known to play a role in energy metabolism (see note added in proof).

In the brain, the CerS2 promoter was found to be highly active in all the white matter tracts, including the corpus callosum, striatum, white matter of the cerebellum, and brain stem (Ben-David et al. 2011; Imgrund et al. 2009). This expression pattern is in accordance with CerS2 expression in mature myelinating oligodendrocytes, as well as in Schwann cells of the peripheral nervous system (Becker et al. 2008). CerS2-deficient mice exhibited chronic and progressive myelin degeneration and detachment, starting as early as 2 months of age. However, myelin development was normal (Ben-David et al. 2011; Imgrund et al. 2009). Demyelination was accompanied by massive gliosis of both microglia and astrocytes, albeit with no effect on neuronal survival (Ben-David et al. 2011). The demyelination was most likely a result of the dramatic reduction in VLC-galactosylceramide (GalCer) and sulfatide (Ben-David et al. 2011; Imgrund et al. 2009), which are major myelin sheath components. Interestingly, mutant mice lacking the enzyme UDP-galactose:ceramide galactosyltransferase (CGT), and thus depleted of GalCer and sulfatide, show a similar deficiency in myelin maintenance (Coetzee et al. 1996). CerS2-deficient brain exhibited an increase in C18-ceramide. However, total GalCer was dramatically reduced, suggesting that CGT cannot use the abundant C18-ceramide, most likely due to the different subcellular localization of CGT and CerS1 (responsible for C18-ceramide generation) (Ben-David et al. 2011). CerS2-deficient mice also exhibited convulsions, pronounced in older mice mainly upon audiogenic stress, although the absence of cortical EEG abnormality suggested a subcortical (basal ganglia or brain stem) dysfunction. Moreover, undefined membranous storage material was detected in lysosomes of glial cells in the brain as early as at 1.5 months of age.

Cumulatively, it is intriguing to speculate that CerS2 deficiency may be attributed to one of the etiologically unknown human lysosomal storage disorders, which in CerS2 null mice prominently feature both liver and brain metabolic dysfunction.

CerS3 mRNA was found to be highly expressed in skin (Laviad et al. 2008). Jennemann et al. further localized CerS3 expression in skin to keratinocytes of the upper stratum spinosum and granulosum (Jennemann et al. 2011). Recently, homozygous CerS3-deficient mice were generated by homologous recombination deleting exon 7 of the CerS3 gene (Jennemann et al. 2011). Such mice die shortly after birth owing to the impaired water permeability of their skin. SL analysis of epidermal extracts of newborns showed that all SL with C26:0 and ULC-fatty acid (FA) were absent, and that VLC-SLs in particular with C24:0 FA were unexpectedly reduced Jennemann et al. reported an overall 90 % reduction in all SLs in the skin of CerS3-deficient mice. This dramatic change led to an epidermal maturation arrest at the embryonic pre-barrier stage (E18.5) and impaired cornification. Moreover, cultured neonatal mutant skin inoculated with *Candida albicans* exhibited high susceptibility to the fungal invasion, stressing the importance of ULC/VLC-SLs and sphingosine (also found to be absent in CerS3-deficient skin) for maintenance of the cutaneous barrier against pathogens. For further details, see Chap. 18 in Vol 216 by Kleuser and Japtok "Sphingolipids and Inflammatory Diseases of the Skin".

CerS3 was also found to be highly expressed in the testis (Mizutani et al. 2006) and was reported to dramatically increase during postnatal testis development along with ULC-glycosphingolipids (Rabionet et al. 2008). Thus, CerS3 may be linked to a subset of etiologically unknown human ichtyosis, which could be associated with male infertility (Sandhoff 2010). A current combined search for "ichtyosis," "male infertility," and "chromosome 15q26.3" (the genomic location of CerS3) in the Online Mendelian Inheritance in Man (OMIM) database does not define a human disease. However, it is likely that complete deletion of a region within the CerS3 gene leads—as in CerS3-deficient mice—to human lethality, whereas a point mutation, leading to some residual CerS3 activity, will allow survival of patients, albeit with skin impairment and/or spermatopathy. An inducible or conditional approach to generate CerS3-deficient mice is required to further explore the role of CerS3 in adult testis physiology.

It is intriguing to note that despite the huge evolutionary distance between *S. cerevisiae* and *Mus musculus*, the overall findings in the mutant organisms exhibit striking similarities. For example, both yeast mutants and CerS-deficient mice display a pronounced accumulation of LCB, especially sphinganine (Guillas et al. 2001; Zhao et al. 2011; Pewzner-Jung et al. 2010b; Ben-David et al. 2011).

Moreover, in both species CerS paralogs are dispensable for viability or embryonic development.[1] To date, the generation of CerS4-, CerS5-, and CerS6-deficient mice has not been reported. However, efforts are under way, which should result in a better understanding of the regulation and the exact role of all CerS in vivo. Cumulatively, CerS enzymes exhibit specific cell type expression, and generate particular ceramides that serve as a basis for complex SLs, which have precise functions in the maintenance of many important processes.

5 Conclusions

Complete depletion of all SLs results in lethality (Hojjati et al. 2005; Guillas et al. 2003; Menuz et al. 2009; Bauer et al. 2009). However, so far it may be concluded that in organisms in which only one CerS paralog is depleted, viability is preserved. This is most likely due to compensation by other SLs that maintain a basic homeostasis. Yet, upon challenge or stress stimuli, this fragile compensation by other SLs seems to be insufficient, and viability is at risk. A major future challenge for the field will be to identify hereditary human CerS disorders. However, altered human and mouse genotypes do not necessarily always correlate phenotypically and, therefore, specific CerS mutations could still be lethal in humans. On the other hand, we predict the existence of gene variants in the human population resulting from point mutations that could lead to phenotypes milder than the ones observed in the mouse models, as residual enzymatic activity would be preserved. High-throughput sequencing of DNA samples from etiologically unknown human diseases with some similarities to the CerS-deficient animal models should help to resolve this issue. Clearly, understanding the biochemical properties of CerS and the elements responsible for their activity and regulation is of highest priority in order to reveal the fundamental role of SLs in cell physiology.

Note added in proof: A recent paper by Zigdon H. et al., (2013) doi: 10.1074/jbc.M112.402719 shows that CerS2 deficiency lead to liver mitochondrial dysfunction, which results in chronic oxidative stress.

Acknowledgments We would like to thank Drs. Anthony H. Futerman, Elad Lavee Laviad, Shifra Ben-Dor, Steffen Jung, and Alexandra Mahler, as well as Oshrit Ben-David and Ruth Seiden, for their constructive comments.

[1] CerS3 is unique in this respect, as it was found to be important for the last stages of proper skin development, and thus, these mice do not survive after birth (see above).

References

Bauer R, Voelzmann A, Breiden B, Schepers U, Farwanah H, Hahn I, Eckardt F, Sandhoff K, Hoch M (2009) Schlank, a member of the ceramide synthase family controls growth and body fat in Drosophila. EMBO J 28(23):3706–3716

Becker I, Wang-Eckhardt L, Yaghootfam A, Gieselmann V, Eckhardt M (2008) Differential expression of (dihydro)ceramide synthases in mouse brain: oligodendrocyte-specific expression of CerS2/Lass2. Histochem Cell Biol 129(2):233–241

Ben-David O, Pewzner-Jung Y, Brenner O, Laviad EL, Kogot-Levin A, Weissberg I, Biton IE, Pienik R, Wang E, Kelly S, Alroy J, Raas-Rothschild A, Friedman A, Brugger B, Merrill AH Jr, Futerman AH (2011) Encephalopathy caused by ablation of very long acyl chain ceramide synthesis may be largely due to reduced galactosylceramide levels. J Biol Chem 286 (34):30022–30033

Bidle KD, Vardi A (2011) A chemical arms race at sea mediates algal host-virus interactions. Curr Opin Microbiol 14(4):449–457

Bionda C, Portoukalian J, Schmitt D, Rodriguez-Lafrasse C, Ardail D (2004) Subcellular compartmentalization of ceramide metabolism: MAM (mitochondria-associated membrane) and/or mitochondria? Biochem J 382(Pt 2):527–533

Bonzon-Kulichenko E, Schwudke D, Gallardo N, Molto E, Fernandez-Agullo T, Shevchenko A, Andres A (2009) Central leptin regulates total ceramide content and sterol regulatory element binding protein-1C proteolytic maturation in rat white adipose tissue. Endocrinology 150(1): 169–178

Chen Y, Liu Y, Allegood J, Wang E, Cachon-Gonzalez B, Cox TM, Merrill AH Jr, Sullards MC (2010) Imaging MALDI mass spectrometry of sphingolipids using an oscillating capillary nebulizer matrix application system. Methods Mol Biol 656:131–146

Coetzee T, Fujita N, Dupree J, Shi R, Blight A, Suzuki K, Popko B (1996) Myelination in the absence of galactocerebroside and sulfatide: normal structure with abnormal function and regional instability. Cell 86(2):209–219

Contreras FX, Ernst AM, Haberkant P, Bjorkholm P, Lindahl E, Gonen B, Tischer C, Elofsson A, von Heijne G, Thiele C, Pepperkok R, Wieland F, Brugger B (2012) Molecular recognition of a single sphingolipid species by a protein's transmembrane domain. Nature 481(7382):525–529

D'Mello NP, Childress AM, Franklin DS, Kale SP, Pinswasdi C, Jazwinski SM (1994) Cloning and characterization of LAG1, a longevity-assurance gene in yeast. J Biol Chem 269 (22):15451–15459

Deng X, Yin X, Allan R, Lu DD, Maurer CW, Haimovitz-Friedman A, Fuks Z, Shaham S, Kolesnick R (2008) Ceramide biogenesis is required for radiation-induced apoptosis in the germ line of C. elegans. Science 322(5898):110–115

Fei P, Junyu N, Jiangfeng Y, Jingpin Y, Yuping W, Zhihui H, Jieliang W, Xianglin C, Shaomin Y, Jie Z (2004) Monoclonal antibodies against human tumor metastasis suppressor gene-1 (TMSG-1): preparation, characterization, and application. Hybrid Hybridomics 23(5):318–325

Futerman AH, Hannun YA (2004) The complex life of simple sphingolipids. EMBO Rep 5(8):777–782

Gong MZ, You JF, Pei F, Cui XL, Li G, Zheng J (2011) Transcriptional activation of TMSG-1 by complex of KLF6 and Sp1. Zhonghua Bing Li Xue Za Zhi 40(8):542–548

Guillas I, Jiang JC, Vionnet C, Roubaty C, Uldry D, Chuard R, Wang J, Jazwinski SM, Conzelmann A (2003) Human homologues of LAG1 reconstitute Acyl-CoA-dependent ceramide synthesis in yeast. J Biol Chem 278(39):37083–37091

Guillas I, Kirchman PA, Chuard R, Pfefferli M, Jiang JC, Jazwinski SM, Conzelmann A (2001) C26-CoA-dependent ceramide synthesis of Saccharomyces cerevisiae is operated by Lag1p and Lac1p. EMBO J 20(11):2655–2665

Hannun YA, Obeid LM (2008) Principles of bioactive lipid signalling: lessons from sphingolipids. Nat Rev Mol Cell Biol 9(2):139–150

Hannun YA, Obeid LM (2011) Many ceramides. J Biol Chem 286(32):27855–27862

Heung LJ, Luberto C, Del Poeta M (2006) Role of sphingolipids in microbial pathogenesis. Infect Immun 74(1):28–39

Hirschberg K, Rodger J, Futerman AH (1993) The long-chain sphingoid base of sphingolipids is acylated at the cytosolic surface of the endoplasmic reticulum in rat liver. Biochem J 290(Pt 3): 751–757

Hojjati MR, Li Z, Jiang XC (2005) Serine palmitoyl-CoA transferase (SPT) deficiency and sphingolipid levels in mice. Biochim Biophys Acta 1737(1):44–51

Imgrund S, Hartmann D, Farwanah H, Eckhardt M, Sandhoff R, Degen J, Gieselmann V, Sandhoff K, Willecke K (2009) Adult ceramide synthase 2 (CERS2)-deficient mice exhibit myelin sheath defects, cerebellar degeneration, and hepatocarcinomas. J Biol Chem 284(48):33549–33560

Iwabuchi K, Prinetti A, Sonnino S, Mauri L, Kobayashi T, Ishii K, Kaga N, Murayama K, Kurihara H, Nakayama H, Yoshizaki F, Takamori K, Ogawa H, Nagaoka I (2008) Involvement of very long fatty acid-containing lactosylceramide in lactosylceramide-mediated superoxide generation and migration in neutrophils. Glycoconj J 25(4):357–374

Jennemann R, Rabionet M, Gorgas K, Epstein S, Dalpke A, Rothermel U, Bayerle A, van der Hoeven F, Imgrund S, Kirsch J, Nickel W, Willecke K, Riezman H, Grone HJ, Sandhoff R (2011) Loss of ceramide synthase 3 causes lethal skin barrier disruption. Hum Mol Genet 21(3):586–608

Jiang JC, Kirchman PA, Zagulski M, Hunt J, Jazwinski SM (1998) Homologs of the yeast longevity gene LAG1 in Caenorhabditis elegans and human. Genome Res 8(12):1259–1272

Jung T, Bader N, Grune T (2007) Lipofuscin: formation, distribution, and metabolic consequences. Ann N Y Acad Sci 1119:97–111

Jungermann K, Katz N (1989) Functional specialization of different hepatocyte populations. Physiol Rev 69(3):708–764

Kageyama-Yahara N, Riezman H (2006) Transmembrane topology of ceramide synthase in yeast. Biochem J 398(3):585–593

Kitatani K, Idkowiak-Baldys J, Hannun YA (2008) The sphingolipid salvage pathway in ceramide metabolism and signaling. Cell Signal 20(6):1010–1018

Koyanagi S, Kuga M, Soeda S, Hosoda Y, Yokomatsu T, Takechi H, Akiyama T, Shibuya S, Shimeno H (2003) Elevation of de novo ceramide synthesis in tumor masses and the role of microsomal dihydroceramide synthase. Int J Cancer 105(1):1–6

Lahiri S, Futerman AH (2005) LASS5 is a bona fide dihydroceramide synthase that selectively utilizes palmitoyl-CoA as acyl donor. J Biol Chem 280(40):33735–33738

Lahiri S, Lee H, Mesicek J, Fuks Z, Haimovitz-Friedman A, Kolesnick RN, Futerman AH (2007) Kinetic characterization of mammalian ceramide synthases: determination of K(m) values towards sphinganine. FEBS Lett 581(27):5289–5294

Lahiri S, Park H, Laviad EL, Lu X, Bittman R, Futerman AH (2009) Ceramide synthesis is modulated by the sphingosine analog FTY720 via a mixture of uncompetitive and noncompetitive inhibition in an Acyl-CoA chain length-dependent manner. J Biol Chem 284(24): 16090–16098

Laviad EL, Albee L, Pankova-Kholmyansky I, Epstein S, Park H, Merrill AH Jr, Futerman AH (2008) Characterization of ceramide synthase 2: tissue distribution, substrate specificity, and inhibition by sphingosine 1-phosphate. J Biol Chem 283(9):5677–5684

Laviad EL, Kelly S, Merrill AH Jr, Futerman AH (2012) Modulation of ceramide synthase activity via dimerization. J Biol Chem 287:21025–21033

Lee H, Rotolo JA, Mesicek J, Penate-Medina T, Rimner A, Liao WC, Yin X, Ragupathi G, Ehleiter D, Gulbins E, Zhai D, Reed JC, Haimovitz-Friedman A, Fuks Z, Kolesnick R (2011) Mitochondrial ceramide-rich macrodomains functionalize Bax upon irradiation. PLoS One 6(6):e19783

Levy M, Futerman AH (2010) Mammalian ceramide synthases. IUBMB Life 62(5):347–356

Ma C, Ning J, You J, Liu L, Wang J, Cui X, Wu B, Zheng J (2003) Primary functional identification of gene TMSG-1. Sci China C Life Sci 46(6):641–650

Menuz V, Howell KS, Gentina S, Epstein S, Riezman I, Fornallaz-Mulhauser M, Hengartner MO, Gomez M, Riezman H, Martinou JC (2009) Protection of C. elegans from anoxia by HYL-2 ceramide synthase. Science 324(5925):381–384

Mesicek J, Lee H, Feldman T, Jiang X, Skobeleva A, Berdyshev EV, Haimovitz-Friedman A, Fuks Z, Kolesnick R (2010) Ceramide synthases 2, 5, and 6 confer distinct roles in radiation-induced apoptosis in HeLa cells. Cell Signal 22(9):1300–1307

Mesika A, Ben-Dor S, Laviad EL, Futerman AH (2007) A new functional motif in Hox domain-containing ceramide synthases: identification of a novel region flanking the Hox and TLC domains essential for activity. J Biol Chem 282(37):27366–27373

Meyers-Needham M, Ponnusamy S, Gencer S, Jiang W, Thomas RJ, Senkal CE, Ogretmen B (2011) Concerted functions of HDAC1 and microRNA-574-5p repress alternatively spliced ceramide synthase 1 expression in human cancer cells. EMBO Mol Med 4(2):78–92

Mizutani Y, Kihara A, Igarashi Y (2005) Mammalian Lass6 and its related family members regulate synthesis of specific ceramides. Biochem J 390(Pt 1):263–271

Mizutani Y, Kihara A, Igarashi Y (2006) LASS3 (longevity assurance homologue 3) is a mainly testis-specific (dihydro)ceramide synthase with relatively broad substrate specificity. Biochem J 398(3):531–538

Mullen TD, Jenkins RW, Clarke CJ, Bielawski J, Hannun YA, Obeid LM (2011) Ceramide synthase-dependent ceramide generation and programmed cell death: involvement of salvage pathway in regulating postmitochondrial events. J Biol Chem 286(18):15929–15942

Obeid LM, Okamoto Y, Mao C (2002) Yeast sphingolipids: metabolism and biology. Biochim Biophys Acta 1585(2–3):163–171

Osawa Y, Uchinami H, Bielawski J, Schwabe RF, Hannun YA, Brenner DA (2005) Roles for C16-ceramide and sphingosine 1-phosphate in regulating hepatocyte apoptosis in response to tumor necrosis factor-alpha. J Biol Chem 280(30):27879–27887

Pagarete A, Allen MJ, Wilson WH, Kimmance SA, de Vargas C (2009) Host-virus shift of the sphingolipid pathway along an Emiliania huxleyi bloom: survival of the fattest. Environ Microbiol 11(11):2840–2848

Park H, Haynes CA, Nairn AV, Kulik M, Dalton S, Moremen K, Merrill AH Jr (2010) Transcript profiling and lipidomic analysis of ceramide subspecies in mouse embryonic stem cells and embryoid bodies. J Lipid Res 51(3):480–489

Park JW, Park WJ, Kuperman Y, Boura-Halfon S, Pewzner-Jung Y, Futerman AH (2012) Ablation of very long acyl chain sphingolipids causes hepatic insulin resistance in mice due to altered getergent-resistant membranes. Hepatology. Aug 22. doi:10.1002/hep.26015

Pewzner-Jung Y, Ben-Dor S, Futerman AH (2006) When do Lasses (longevity assurance genes) become CerS (ceramide synthases)?: Insights into the regulation of ceramide synthesis. J Biol Chem 281(35):25001–25005

Pewzner-Jung Y, Brenner O, Braun S, Laviad EL, Ben-Dor S, Feldmesser E, Horn-Saban S, Amann-Zalcenstein D, Raanan C, Berkutzki T, Erez-Roman R, Ben-David O, Levy M, Holzman D, Park H, Nyska A, Merrill AH Jr, Futerman AH (2010a) A critical role for ceramide synthase 2 in liver homeostasis: II. Insights into molecular changes leading to hepatopathy. J Biol Chem 285(14):10911–10923

Pewzner-Jung Y, Park H, Laviad EL, Silva LC, Lahiri S, Stiban J, Erez-Roman R, Brugger B, Sachsenheimer T, Wieland F, Prieto M, Merrill AH Jr, Futerman AH (2010b) A critical role for ceramide synthase 2 in liver homeostasis: I. Alterations in lipid metabolic pathways. J Biol Chem 285(14):10902–10910

Rabionet M, van der Spoel AC, Chuang CC, von Tumpling-Radosta B, Litjens M, Bouwmeester D, Hellbusch CC, Korner C, Wiegandt H, Gorgas K, Platt FM, Grone HJ, Sandhoff R (2008) Male germ cells require polyenoic sphingolipids with complex glycosylation for completion of meiosis: a link to ceramide synthase-3. J Biol Chem 283(19):13357–13369

Riebeling C, Allegood JC, Wang E, Merrill AH Jr, Futerman AH (2003) Two mammalian longevity assurance gene (LAG1) family members, trh1 and trh4, regulate dihydroceramide synthesis using different fatty acyl-CoA donors. J Biol Chem 278(44):43452–43459

Sandhoff R (2010) Very long chain sphingolipids: tissue expression, function and synthesis. FEBS Lett 584(9):1907–1913
Schorling S, Vallee B, Barz WP, Riezman H, Oesterhelt D (2001) Lag1p and Lac1p are essential for the Acyl-CoA-dependent ceramide synthase reaction in Saccharomyces cerevisae. Mol Biol Cell 12(11):3417–3427
Shimeno H, Soeda S, Sakamoto M, Kouchi T, Kowakame T, Kihara T (1998) Partial purification and characterization of sphingosine N-acyltransferase (ceramide synthase) from bovine liver mitochondrion-rich fraction. Lipids 33(6):601–605
Siskind LJ, Mullen TD, Romero Rosales K, Clarke CJ, Hernandez-Corbacho MJ, Edinger AL, Obeid LM (2010) The BCL-2 protein BAK is required for long-chain ceramide generation during apoptosis. J Biol Chem 285(16):11818–11826
Spassieva S, Seo JG, Jiang JC, Bielawski J, Alvarez-Vasquez F, Jazwinski SM, Hannun YA, Obeid LM (2006) Necessary role for the Lag1p motif in (dihydro)ceramide synthase activity. J Biol Chem 281(45):33931–33938
Spassieva SD, Mullen TD, Townsend DM, Obeid LM (2009) Disruption of ceramide synthesis by CerS2 down-regulation leads to autophagy and the unfolded protein response. Biochem J 424(2):273–283
Sperling P, Heinz E (2003) Plant sphingolipids: structural diversity, biosynthesis, first genes and functions. Biochim Biophys Acta 1632(1–3):1–15
Sridevi P, Alexander H, Laviad EL, Min J, Mesika A, Hannink M, Futerman AH, Alexander S (2010) Stress-induced ER to Golgi translocation of ceramide synthase 1 is dependent on proteasomal processing. Exp Cell Res 316(1):78–91
Sridevi P, Alexander H, Laviad EL, Pewzner-Jung Y, Hannink M, Futerman AH, Alexander S (2009) Ceramide synthase 1 is regulated by proteasomal mediated turnover. Biochim Biophys Acta 1793(7):1218–1227
Sulzer D, Mosharov E, Talloczy Z, Zucca FA, Simon JD, Zecca L (2008) Neuronal pigmented autophagic vacuoles: lipofuscin, neuromelanin, and ceroid as macroautophagic responses during aging and disease. J Neurochem 106(1):24–36
Tidhar R, Ben-Dor S, Wang E, Kelly S, Merrill AH Jr, Futerman AH (2012) Acyl chain specificity of ceramide synthases is determined within a region of 150 residues in the Tram-Lag-CLN8 (TLC) domain. J Biol Chem 287(5):3197–3206
Vallee B, Riezman H (2005) Lip1p: a novel subunit of acyl-CoA ceramide synthase. EMBO J 24 (4):730–741
van Hall T, Wolpert EZ, van Veelen P, Laban S, van der Veer M, Roseboom M, Bres S, Grufman P, de Ru A, Meiring H, de Jong A, Franken K, Teixeira A, Valentijn R, Drijfhout JW, Koning F, Camps M, Ossendorp F, Karre K, Ljunggren HG, Melief CJ, Offringa R (2006) Selective cytotoxic T-lymphocyte targeting of tumor immune escape variants. Nat Med 12(4):417–424
Venkataraman K, Riebeling C, Bodennec J, Riezman H, Allegood JC, Sullards MC, Merrill AH Jr, Futerman AH (2002) Upstream of growth and differentiation factor 1 (uog1), a mammalian homolog of the yeast longevity assurance gene 1 (LAG1), regulates N-stearoyl-sphinganine (C18-(dihydro)ceramide) synthesis in a fumonisin B1-independent manner in mammalian cells. J Biol Chem 277(38):35642–35649
Villen J, Beausoleil SA, Gerber SA, Gygi SP (2007) Large-scale phosphorylation analysis of mouse liver. Proc Natl Acad Sci U S A 104(5):1488–1493
Wang E, Norred WP, Bacon CW, Riley RT, Merrill AH Jr (1991) Inhibition of sphingolipid biosynthesis by fumonisins. Implications for diseases associated with Fusarium moniliforme. J Biol Chem 266(22):14486–14490
Winter E, Ponting CP (2002) TRAM, LAG1 and CLN8: members of a novel family of lipid-sensing domains? Trends Biochem Sci 27(8):381–383
Xu Z, Zhou J, McCoy DM, Mallampalli RK (2005) LASS5 is the predominant ceramide synthase isoform involved in de novo sphingolipid synthesis in lung epithelia. J Lipid Res 46(6):1229–1238

Yacoub A, Hamed HA, Allegood J, Mitchell C, Spiegel S, Lesniak MS, Ogretmen B, Dash R, Sarkar D, Broaddus WC, Grant S, Curiel DT, Fisher PB, Dent P (2010) PERK-dependent regulation of ceramide synthase 6 and thioredoxin play a key role in mda-7/IL-24-induced killing of primary human glioblastoma multiforme cells. Cancer Res 70(3):1120–1129

Yu B, Zhou S, Wang Y, Qian T, Ding G, Ding F, Gu X (2012) miR-221/222 promote Schwann cell proliferation and migration by targeting LASS2 following sciatic nerve injury. J Cell Sci 125:2675–2683

Zhao L, Spassieva SD, Jucius TJ, Shultz LD, Shick HE, Macklin WB, Hannun YA, Obeid LM, Ackerman SL (2011) A deficiency of ceramide biosynthesis causes cerebellar purkinje cell neurodegeneration and lipofuscin accumulation. PLoS Genet 7(5):e1002063

Zheng W, Kollmeyer J, Symolon H, Momin A, Munter E, Wang E, Kelly S, Allegood JC, Liu Y, Peng Q, Ramaraju H, Sullards MC, Cabot M, Merrill AH Jr (2006) Ceramides and other bioactive sphingolipid backbones in health and disease: lipidomic analysis, metabolism and roles in membrane structure, dynamics, signaling and autophagy. Biochim Biophys Acta 1758(12):1864–1884

Membrane Channels Formed by Ceramide

Marco Colombini

Contents

Abstract The sphingolipid, ceramide, forms channels in the mitochondrial outer membrane and in lipid membranes composed of only phospholipid/cholesterol, using lipids typically found in the natural membrane. These channels are large, allowing proteins to cross membranes. Experimental results are consistent with ceramide forming barrel-stave channels that are rigid and highly organized. Bcl-2 family proteins control these channels in a manner expected from their physiological function: anti-apoptotic proteins destabilize the channels whereas pro-apoptotic proteins act synergistically with ceramide to increase membrane permeability. The use of ceramide analogs has allowed one to gain insight into the features of the molecule that are most important for channel formation. These analogs have also been useful in identifying the sites of interaction between ceramide and both Bax and Bcl-xL. The pores formed in phospholipid membranes by ceramide were visualized by electron microscopy. The most common pore size

M. Colombini (✉)
Department of Biology, University of Maryland, College Park, MD 20742, USA
e-mail: colombini@umd.edu

E. Gulbins and I. Petrache (eds.), *Sphingolipids: Basic Science and Drug Development*, Handbook of Experimental Pharmacology 215, DOI 10.1007/978-3-7091-1368-4_6,

was 10 nm in diameter, consistent with results obtained from electrophysiological recordings. All indications point to a role for ceramide channels in the release of proteins from mitochondria, a key decision-making step in the apoptotic process.

Keywords Ceramide • Channel • Apoptosis • Bax • Bcl-xL

1 Introduction: Ceramide Permeabilizes Membranes

The ability of ceramide to permeabilize the mitochondrial outer membrane (MOM) has been reported by a number of different research groups (Ghafourifar et al. 1999; Di Paola et al. 2000, 2004; Birbes et al. 2001; Siskind et al. 2002, 2006; Colombini 2010) using different approaches. It has also been demonstrated that ceramide can permeabilize phospholipid membranes in the absence of any proteins (Siskind and Colombini 2000; Siskind et al. 2003; Montes et al. 2002; Pajewski et al. 2005; Stiban et al. 2006). There is agreement that these permeability pathways are large (Siskind et al. 2002, 2003; Montes et al. 2002; Samanta et al. 2011) and thus require the cooperation of many ceramide molecules. When supra-physiologically high mole fractions of ceramide are used, there is compelling evidence to conclude that, under these conditions, ceramide can participate in the formation of separate lipid phases (Montes et al. 2002). These investigators suggested that the interface between these lipid phases, formed under these conditions, may contain nonspecific pathways by which polar molecules may cross the membrane. If these exist, these pathways are not the topic of this review. In contrast, when physiologically relevant mole fractions of ceramide are present in phospholipid membranes, there is ample evidence for the existence of highly organized cylindrical channels large enough to allow proteins to cross membranes (Siskind et al. 2002, 2006; Samanta et al. 2011). A number of features of ceramide channels, including stereospecificity (Perera et al. 2012a), provide compelling evidence for highly organized structures, which are the focus of this chapter. These ceramide channels can be regulated by proteins in ways that are consistent with these channels playing an important role in the early, decision-making stages of apoptosis (Siskind et al. 2008; Ganesan et al. 2010).

2 The Working Model of the Ceramide Channel

It is useful to be able to put experimental results into a framework and this is embodied in the working model of the ceramide channel. This model was first proposed in 2000 (Siskind and Colombini 2000) when electrophysiological measurements were made on conductances produced by the addition of both C_2- and C_{16}-ceramide to planar phospholipid membranes. It was modified in 2003 by arranging the ceramide columns in an antiparallel fashion (Siskind et al. 2003). It was tested by performing molecular dynamic (MD) simulations (Anishkin et al. 2006) and

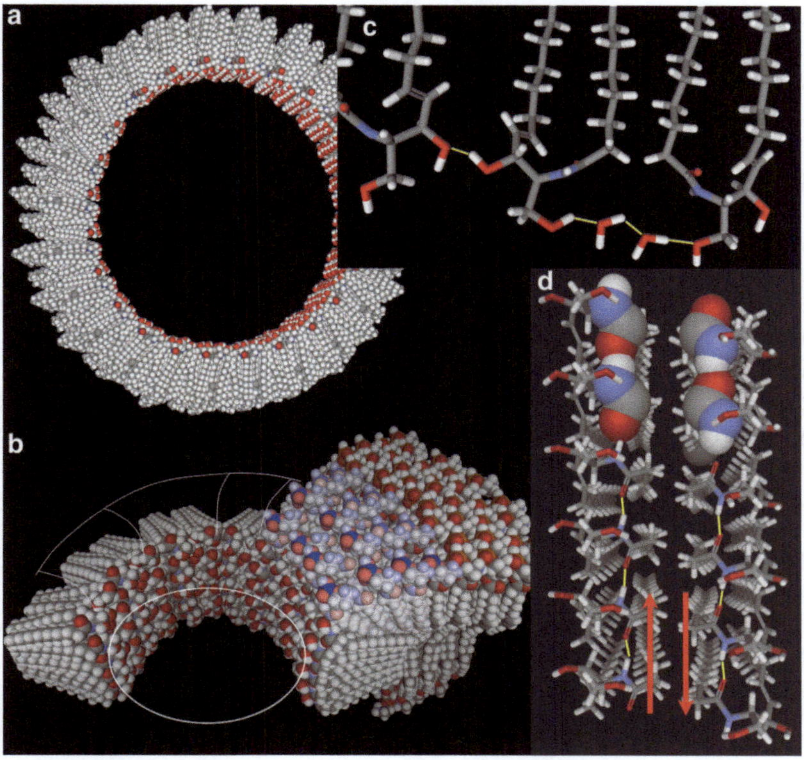

Fig. 1 Model of a ceramide channel. (**a**) Top view of a 48-column channel forming a 10 nm pore. (**b**) Cross section showing how the channel would interface with the phospholipid bilayer. The *lighter colored* phospholipids are shown as curving to meet the channel and cover the aliphatic chains of ceramide. The channel takes on a somewhat hourglass shape to meet the phospholipids. The result is a continuous polar surface interfacing with the aqueous phase. (**c**) Hydrogen bonding (*yellow*) at the ceramide–water interface would take place among the hydroxyl groups and bridging water molecules. (**d**) Two ceramide columns are illustrated. Ceramide columns are formed by six ceramide molecules held together by hydrogen bonding (*yellow*) between adjacent amide linkages (shown as full-size atoms for the *top* four ceramide molecules). Columns are arranged in an antiparallel fashion allowing favorable dipole–dipole interactions (*red arrows*) between adjacent ceramide columns

found to be stable during the lifetime of the simulation. The simulations helped to address the issue of how the channel might interface with the phospholipid bilayer.

As illustrated in Fig. 1, ceramide molecules are proposed to form columns that span the hydrophobic inner core of the membrane. The column is proposed to consist of six ceramides, although four ceramide-columns are also possible as indicated by MD simulations. The monomers are held together by hydrogen bonding through the amide linkage that connects the sphingoid base to the acyl chain (Fig. 1c). Thus the hydrogen bonding of the amide linkage, responsible for the secondary structure of proteins, is proposed to be the primary organizing interaction in the ceramide channel. The alignment of these amide linkages in the column would produce a strong dipole similar to that formed in an alpha helix. This recognition and results obtained from electrophysiological recordings (Sect. 3) caused us to propose (Siskind et al. 2003) that the columns are organized in an antiparallel fashion. The number of columns, arranged as staves of a barrel, determines the physical size of the ceramide channel. Adjacent columns are proposed to be held together by hydrogen bonding of the twin hydroxyl groups either directly or via a water bridge (Fig. 1c).

Experimental observations indicate that ceramide channels are in dynamic equilibrium with ceramide monomers or ceramide aggregates in the membrane. There is evidence (Siskind et al. 2003) that groups of columns can leave the channel as a unit, presumably ending up as an aggregate, or perhaps an organized structure, in a monolayer of the membrane. Certainly the high degree of insolubility of the medium (C_{16}-, C_{18}-ceramide) and long-chain ceramides (C_{24}-ceramide) makes it very unlikely that ceramide monomers would leave the membrane for the aqueous phase. The short-chain ceramide, C_2-ceramide, is more water soluble and there is evidence that it can leave the membrane, binding to proteins in solution (Siskind et al. 2002).

To form a continuous polar surface to interface with the aqueous phase requires a distortion at the boundary of both the ceramide channel and the phospholipid bilayer (Fig. 1b). Molecular dynamic simulations (Anishkin et al. 2006) indicate a channel with a somewhat hourglass shape. The distortion of the phosphoglycerolipids is illustrated by a lipid annulus with a somewhat different coloration.

3 Electrophysiological Evidence for the Structure of Ceramide Channels

Discrete changes in conductance, i.e., sudden changes in the membrane's ability to allow ions to cross a membrane, are a defining characteristic of the formation of a channel. The nearly instantaneous change in the ability of ions to flow through the membrane by 10^5 and higher ions per second cannot be explained by anything save a continuous water-filled pathway that allows for the constant flow of ions. While this observation distinguishes a channel from an ion carrier, it says nothing about the structure that forms the flowing pathway by which ions cross membranes. The first publication on ceramide channels (Siskind and Colombini 2000) interpreted the discrete conductance changes as individual channels, the simplest hypothesis. A subsequent publication demonstrated that the conductance increments and decrements represented changes in the permeability (most probably size) of one channel-forming structure (Siskind et al. 2003). In a typical electrophysiological experiment (e.g., Fig. 7A in Samanta et al. 2011), during which a ceramide channel is formed in a planar phospholipid membrane, one observes a variable lag time prior to an increase in membrane permeability. The membrane conductance increases in a wide variety of incremental steps and these are the growth of a single channel. The conductance reaches a steady value that varies from experiment to experiment because the size of the final channel that is formed varies. Among the experimental evidence for the formation of one large channel were observations of multiple conductance increments followed by a sudden total loss of conductance. This demonstrated that the apparently individual permeability pathways being formed in the membrane were actually one pathway because it disassembled in one event. The alternative interpretation of some form of incredible cooperativity was countered by other observations such as selectivity measurements. The overall selectivity of the membrane containing the ceramide-induced ion flow declined as the conductance increased, consistent with a single channel growing in size and therefore losing selectivity because the permeating ions would have little or no interaction with the wall of the channel. However, the most compelling evidence was the demonstration of stochastic behavior. Channel disassembly following the addition of La^{+3} did not have the properties expected of a population of channels (exponential decay) but rather showed a variable delay followed by rapid disassembly (Siskind et al. 2003). The delay varied from experiment to experiment in a way expected for a single energy barrier experienced by a single channel. Pooling many experiments together resulted in the generation of an artificial population of channels and this pool showed the expected exponential decay with a time constant of 17 s.

Thus, these results are consistent with ceramide addition resulting in the formation of a single large aqueous pathway in the phospholipid membrane under study. This membrane has a large surface area of approximately 8,000 μm^2. Why would ceramide addition not result in the formation of multiple channels? The answer may come from the observation that an increase in conductance following the addition of ceramide to the aqueous solution bathing the membrane was usually delayed, sometimes for hours. Following the first conductance increment, the conductance usually increased rapidly before stabilizing at a certain level. It appears that the formation of the first "protochannnel" is an unfavorable process, similar to the formation of the first microtubule disk or a seed crystal. Once the seed is formed, growth is rapid. Thus the generation of a new protochannel is less likely to occur than the growth of the existing channel.

Strong evidence for the barrel-stave model of the ceramide channel came from a careful analysis of the conductance decrements, either the spontaneous decrements or decrements induced by the addition of La^{3+} (Siskind et al. 2003). Whereas the small decrements showed the expected decrease in frequency with increase in decrement size, the large decrements showed a remarkable pattern. Beginning at 12 nS, there was a strong preference for decrements that were a multiple of 4 nS. This pattern could only be explained by the loss of the circumference of a channel by multiples of a fundamental structure. Calculations of the size of the segment of a circle that would explain the results were made based on the standard equation:

$$G = \frac{(\kappa_{sp}\pi \cdot r^2)}{(L + 0.5\pi \cdot r)},$$

where G is the conductance of the channel, κ_{sp} is the specific conductivity of the solution (0.112 S cm^{-1} for 1.0 M KCl), L is the length of the channel, and r is the radius of the aqueous pore. Thus the loss of a segment of the circumference would result in a change in conductance and from this change the arc of the circumference can be calculated. For very large channels $L << 0.5\pi r$ and so:

$$\text{Segment of circumference} = \frac{\pi}{\kappa_{sp}}(\Delta G).$$

This works out to 1.12 nm for a change in conductance of 4.0 nS. For the typical 10 nm channel, calculations without the approximation yield 1.3 nm. This corresponds to a pair of ceramide columns in the 48-column model illustrated in Fig. 1a. The loss of an even number of columns could be explained if the columns were arranged in an antiparallel fashion, held together by dipole–dipole interactions. Like the alpha helix, the hydrogen bonding between adjacent ceramides in each column results in the combination of electric dipoles forming one large overall dipole on one side of the ceramide column. This would have a favorable interaction with the opposite dipole of the adjacent column oriented in an opposite direction. Simulated annealing studies performed by Andriy Anishkin (personal communication) confirmed that the antiparallel orientation is favored.

Molecular dynamic simulations (Anishkin et al. 2006) showed that the proposed structure of the ceramide channel is stable. These also showed how the channel could interface with the lipid bilayer to form a continuous polar surface interfacing with the aqueous phase.

4 Regulation of Ceramide Channels by Bcl-2 Family Proteins

Bcl-2 family proteins influence the structure/stability of ceramide channels in a way that matches the role of these proteins in controlling the onset of apoptosis. The anti-apoptotic proteins tested (Bcl-xL and CED-9) were found (Siskind et al. 2008) to inhibit and reverse the permeabilization of the MOM by ceramide (Fig. 2a). This was interpreted as favoring the disassembly of ceramide channels. In sharp contrast, the pro-apoptotic protein, Bax, was found to act synergistically with ceramide to increase the permeabilization of the MOM (Ganesan et al. 2010), Fig. 2b. Note that CED-9 is an anti-apoptotic protein found in *Caenorhabditis elegans* whereas Bcl-xL is found in mammals including humans. It seems highly unlikely, but not impossible, that this harmony between the actions of these proteins on ceramide channels and their established action on influencing protein release from mitochondria in living cells is just coincidental. Control experiments and the details of the regulation support the conclusion that this regulation is specific and likely related to the apoptotic function of these proteins.

The permeabilization of the outer membrane of isolated mammalian mitochondria by the addition of ceramide could be the result of ceramide channel

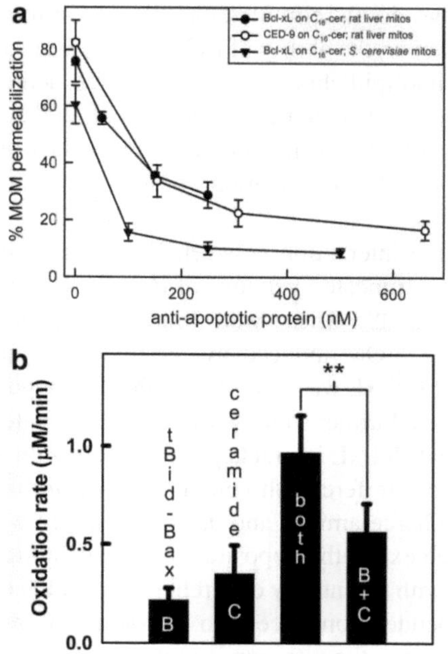

Fig. 2 Bcl-2 family proteins influence the ability of ceramide to permeabilize the mitochondrial outer membranes (MOM) of rat liver and yeast mitochondria. (**a**) Anti-apoptotic proteins destabilize ceramide channels. The ceramide-induced permeabilization of the outer membrane of mitochondria (rat liver, *circles*; *S. cerevisiae*, *triangles*) is decreased in a dose-dependent manner by the addition of Bcl-xL (*solid symbols*) and CED-9 (*open symbol*). In each set of experiments, the permeability induced by C_{16}-ceramide alone was different due to variations in sensitivity of different batches of mitochondria to added ceramide. The data were taken from Siskind et al. (2008). (**b**) Bax and ceramide act synergistically to permeabilize the MOM. The first two *bars* report the permeabilization achieved with Bax or C_{16}-ceramide alone, the third is the result from combined additions, the fourth is the summation of the individual additions. To observe the synergy the Bax needed to be activated with t-Bid. Without activation Bax had no effect. Statistics: *double asterisk*, 99 % confidence. The data were taken from Ganesan et al. (2010)

formation or ceramide acting on other outer membrane components. Similarly, the addition of Bcl-xL to mitochondria could interfere with the permeabilization in an indirect manner. Isolated rat liver mitochondria contain other pro-apoptotic agents such as BAK, and Bcl-xL could instead be reducing the perceived ceramide-induced permeabilization by binding to BAK. However, essentially the same results were obtained (Siskind et al. 2008) using mitochondria isolated from the yeast, *Saccharomyces cerevisiae* (Fig. 2a). Since this yeast lacks proteins homologous to those of the Bcl-2 family, the results are most consistent with the notion that Bcl-xL is interfering with ceramide's ability to permeabilize the MOM in a direct manner, rather than an indirect effect through the neutralization of other pro-apoptotic factors known to interact with Bcl-xL. Furthermore, the expression of the protein Bcl-2 in yeast resulted in isolated yeast mitochondria containing Bcl-2. These mitochondria were resistant to ceramide permeabilization. These results with yeast mitochondria reduce but do not eliminate the likelihood that the action of Bcl-xL might be indirect. However, the ability of Bcl-xL to disassemble ceramide channels formed in pure phospholipid/cholesterol membranes demonstrated that Bcl-xL must act directly on ceramide channels (Siskind et al. 2008). Furthermore, CED-9 had a similar direct effect. Although each individual experiment could be interpreted in other ways, the combination of all these results makes a compelling case.

The specificity of the interaction between Bcl-xL and ceramide channels was demonstrated by testing truncated versions of Bcl-xL shown to lack anti-apoptotic activity (Siskind et al. 2008). These were found to also be unable to reduce the permeabilization of the MOM induced by ceramide, thus providing more confidence that the action of Bcl-xL was indeed specific. The opposite effect of Bax adds further to this confidence. Further studies with ceramide analogs (Perera et al. 2011, 2012b) demonstrate that Bcl-xL interacts preferentially with the aliphatic chains of ceramide. Its ability to interfere with ceramide permeabilization of the MOM is optimal for C_{16}- and C_{18}-ceramides and declines sharply with longer and shorter acyl chains. Bax, which exerts the opposite effect on ceramide channels as Bcl-xL, seems to also interact with an entirely different region of the ceramide channel. The action of Bax is dependent on access to the amide nitrogen of ceramide, an interaction that is lost when the nitrogen is methylated.

Both octyl glucoside-activated Bax and t-Bid-activated Bax permeabilize the MOM and phospholipid membranes to proteins without the need for the addition of ceramide (although ceramide was naturally present in all published experiments on mitochondria of which the author is aware). Similarly, ceramide addition permeabilizes these membranes without the requirement for the presence of either Bax or BAK. Nevertheless, the combination of both treatments (Fig. 2b), at low permeability levels, resulted in a level of permeabilization that exceeded that observed with either agent alone (Ganesan et al. 2010). Thus it was concluded that activated Bax and ceramide act synergistically, although it is possible that just one of these agents somehow enhances the channel-forming ability of the other. This synergy requires that both agents produce a relatively low level of outer membrane permeabilization. In its absence, no synergy was observed. Lee and coworkers (2011) also reported compelling evidence that ceramide and Bax act synergistically to permeabilize the MOM. They showed the formation of ceramide-rich microarrays into which Bax oligomerizes resulting in outer membrane permeabilization. Pastorino et al. (1999) reported synergy but interpreted their results in terms of the mitochondrial permeability transition.

Insight into the mechanism by which Bax might act on ceramide channels was obtained by noting a change in the apparent affinity between Bax and the ceramide channel at different levels of MOM permeabilization. The concentration of Bax that produced half-maximal enhancement of ceramide permeabilization decreased as the level of ceramide-induced permeabilization without Bax addition was increased. Thus less Bax was needed to enhance the ceramide permeabilization of the MOM when that permeabilization was already elevated. However, the maximal percent enhancement by Bax was also reduced. If the permeabilization of the MOM by ceramide alone was too great, Bax addition had no effect. That is not because the permeabilization was maximal because a further addition of ceramide could easily increase the level of permeabilization even further as could hypotonic shock. Rather it appeared that Bax leads to a saturating level of permeabilization. If permeabilization was equated to channel size, perhaps Bax favors a specific channel size. It was speculated that activated Bax acts as a molecular scaffold, driving the ceramide channel to a specific radius of curvature. This would explain both the apparent maximal Bax enhancement of ceramide permeabilization of MOM and the apparent increase in affinity as the degree of ceramide-induced permeabilization was increased. If the enhancement of MOM permeabilization was related to the formation of more channels rather than larger channels, it is difficult to see why Bax enhancement would be limited. Thus the former hypothesis seems more likely.

5 Features of Ceramide That Are Important for Channel Formation

The use of analogs provided important insights into the aspects of the ceramide molecule that are important for channel formation. Ceramide has several structural features that could contribute to the stability of a channel. The hydrocarbon chains not only anchor ceramide to the membrane by hydrophobic interactions, but the largely saturated chains could also be important in the packing and orientation of each ceramide molecule making up the channel. The *trans* double bonds should add to the ordering of the hydrocarbon chains. The location of the double bond relative to the hydroxyl on carbon 3 makes the latter an allylic hydroxyl, changing its hydrogen-bonding characteristics. The two hydroxyl groups and the amide linkage must result in a variety of hydrogen bonds and these could provide both mechanical strength to the channel structure and a degree of rigidity that would allow the transmission of mechanical stress. The latter could either stabilize or destabilize the channel resulting in either channel growth or disassembly. Ceramide has two chiral centers and a planar amide linkage, all of which limit possible conformational structures and intra- or intermolecular interactions. Each of these features was altered and the ability of the resulting structures to form channels assessed (Perera et al. 2012a).

The ability of ceramide to form channels was not significantly altered by shortening either one of the two hydrocarbon chains. Any changes in the ability to permeabilize the MOM could be attributed to differences in the ability to deliver ceramide to the target membrane. The efficiency of ceramide dispersal in water from the isopropanol solution and the propensity to aggregate prior to reaching the target membranes are all expected to change with alterations in chain length.

Changes in the degree of unsaturation and especially changes in the location and nature of the *trans* double bond did not have as great an effect as was expected. Simple removal of the double bond, forming dihydroceramide, resulted in failure to form channels (Siskind and Colombini 2000; Stiban et al. 2006). However, conversion of *trans* to *cis* did not influence the channel-forming ability. Moving the double bond between the hydroxyls or moving it farther down the chain, away from the C3-hydroxyl, did not have much influence on channel-forming ability. Replacing the double bond with a hydroxyl, as in phytoceramide, also was well tolerated. However, inserting a *cis* double bond half way down the acyl chain did destabilize the channel. All this raises the question of why the double bond is needed. Perhaps it is a question of being able to reach a critical concentration of ceramide to allow for channel formation. Dihydroceramide may form aggregates in the membrane that reduce the free dihydroceramide concentration in the membrane resulting in a failure to form channels. Alternatively, the packing of dihydroceramide in the form of a channel may be inherently unstable. Indeed, mixtures of ceramide and dihydroceramide are less effective in making channels as opposed to ceramide alone. As little at 1 part in 10 of dihydroceramide reduces the ability of ceramide to form channels (Stiban et al. 2006).

Methylation of the polar residues would be expected to destabilize ceramide channels by reducing the hydrogen-bonded network proposed to form the polar inner surface of the channel and the hydrogen bonds connecting the ceramides within the columns (Fig. 1). Indeed, methylation of the C1-hydroxyl resulted in failure to measurably permeabilize the MOM to cytochrome *c*. However, this analog did result in adenylate kinase release, indicating the formation of transient, unstable channels. When tested on planar phospholipid membranes, small short-lived conductances were observed as opposed to the large channels characteristic of ceramide (Perera et al. 2012a). By contrast, methylation of the C3-hydroxyl resulted in no significant change in channel-forming ability. This difference between the modification of the two hydroxyls was recapitulated in changes in the chiral centers. Conversion of carbon 3 from R to S, a modification that would change the orientation of the C3-hydroxyl, had no effect on channel formation. However, an additional change at carbon 2 (from S to R) resulted in a large drop in channel-forming ability. The latter change would have altered the orientation of the C1-hydroxyl. The most unexpected outcome was the observation of channel formation by an analog methylated at the amide nitrogen. While displaying a marked reduction in the ability to permeabilize the MOM and the formation of unstable channels in planar phospholipid membranes, nevertheless the formation of any channels at all was unexpected because of the critical importance placed on the role of the amide linkage in forming the ceramide columns, the fundamental structures of the working model of the ceramide channel. However, molecular dynamic simulations have shown that the carbonyl group of the amide linkage could hydrogen bond with one of the hydroxyls and so would not need to hydrogen bond with the amide hydrogen. This would form a channel with a different structure (supplemental figure to Perera et al. 2012a).

Although the ability to form channels tolerates a variety of structural changes in the ceramide molecule, changes that are expected to enhance channel stability actually do so. Urea ceramide is expected to increase the hydrogen bonding that is proposed to give the ceramide columns mechanical strength and it does have an increased propensity for channel formation in the MOM. The addition of a second *trans* double bond next to the natural one in a position that allows π resonance also favors channel formation in the MOM.

The relative ability of analogs to form channels in phospholipid membranes was in harmony with the results obtained with isolated mitochondria. Phytoceramide formed channels with properties similar to those of ceramide, including sensitivity to disassembly by the addition of La^{3+}. Methylation of the C1-hydroxyl destabilized the channel but small transient conductances were still formed in planar membranes.

The observation of synergy and antagonism between ceramide and the analogs indicates the existence of both compatible and incompatible interactions. Synergy was observed with analogs whose structure may be expected to complement that of ceramide. For example, combining short-chain and long-chain ceramides might result in complementary structures and, indeed, those showed synergy. Molecules with very similar structures such that one was essentially replaceable by the other showed neither synergy nor antagonism. Analogs expected to be poorly compatible with the structure of the ceramide channel showed antagonism. This was the case for phytoceramide and the methylated amide analog. Surprisingly this cooperativity was found to be in the membrane insertion process rather than just in channel formation, indicating that insertion of ceramide into the MOM may be an autocatalytic process.

6 Disruption of Ceramide Channels by Other Lipids

The dynamic nature of ceramide channels would be expected to make them susceptible to disruption by similar but fundamentally incompatible lipids. Random collisions result in the growth of ceramide channels by the insertion of one or more ceramides into an existing channel. Similarly, thermal vibrations should result in single ceramides or organized groups of ceramides to leave the channels for a location on one of the monolayers of the membrane. As discussed above, even large fragments can leave the channel resulting in large conductance drops. Throughout this dynamic motion, a variety of molecules, in addition to ceramide, will be colliding with the ceramide channel and perhaps inserting transiently or perhaps not so transiently into the ceramide structure. If one of these molecules were to interact poorly with the existing structure it would likely return rapidly to one of the monolayers, propelled by a net repulsive interaction. However, if the molecule were to interact favorably with the channel, it might become incorporated even if such an incorporation were to destabilize the structure leading to disassembly. Perhaps this is how one might interpret how and why both sphingosine and dihydroceramide destabilize ceramide channels. Indeed sphingosine by itself is capable of forming small channels but when added to a ceramide channel it disassembles that channel (Elrick et al. 2006). Each, alone, forms structures with some order but the combination may form poorly organized structures that result in instability. In sharp contrast, phospholipids serve as a medium for ceramide channel formation. Perhaps the very different structure results in exclusion, at least exclusion of phospholipids from the ceramide channel, and thus stability.

7 Channel Formation Shows Specificity for the Mitochondrial Outer Membrane

Ceramide exists in various cellular membranes, especially the E.R., Golgi, and plasma membrane, yet the formation of large channels seems to be restricted to the mitochondrial outer membrane. Certainly if large channels were formed in the plasma membrane, uncontrolled cell death would follow rapidly. Using erythrocytes as model plasma membranes, the addition of ceramide to these does not result in channel formation and therefore cell lysis (Siskind et al. 2002). Even channels small enough to allow the translocation of NaCl would result in lysis and thus even small channels were not formed. By measuring the amount of ceramide that had been delivered to the erythrocyte membrane, Siskind and coworkers (2006) demonstrated that a molar ratio of ceramide to membrane lipids that was 50-fold greater than that needed to permeabilize the MOM still did not cause detectable erythrocyte lysis. The molecular basis for the increased sensitivity of the MOM or refractoriness of the plasma membrane has not been determined.

8 Visualization of Ceramide Channels Formed in Phospholipid Membranes

The large size of the ceramide channel allows individual channels to be resolved by negative stain electron microscopy. Figure 3 shows four electron micrographs of ceramide channels formed in phospholipid membranes. Control membranes showed no such structures. Negative stain fills both the holes in the channel and the space around the dried and flattened vesicles. This amorphous material scatters electrons in proportion to the number of osmium atoms present and thus the intensity of the stain is proportional to the thickness of the stain. Note that the density of the stain in Fig. 3b within the channel is about the same as the density at the edges of the vesicle attesting to the conclusion that the stain indeed penetrates right through the membrane. An analysis of the shape of the stain is consistent with the channel being a right cylinder because the rounding is attributable to the resolution of the technique. Higher resolution methods would be necessary to gain more insight into the shape of the pore.

The use of spectroscopic methods to gain information on the structure of the channel is problematic because the large majority of the ceramide in the membrane is not part of the channel structure.

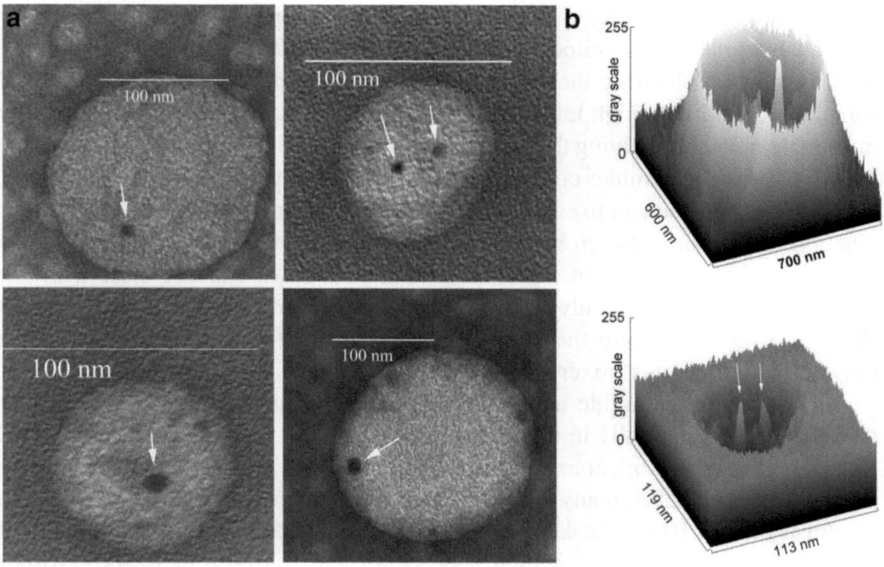

Fig. 3 (**a**) Electron micrographs of negatively stained liposomes containing ceramide channels. (**b**) Densitometry scans of two electron micrographs showing the intensity of the negative stain. The figure was revised from Samanta et al. (2011)

9 Technical Challenges and Apparent Conflicts in the Literature

Questions have been raised, both in print and in person, about various aspects of ceramide channel formation (e.g., Yuan et al. 2003; Lee et al. 2011). These have led to confusion and skepticism. Thus it is important to address these issues here.

Questions of reproducibility of the ability of ceramide to permeabilize the outer membrane of isolated mitochondria likely stem from the inherent virtual insolubility of ceramide in water. This is especially true for medium- and long-chain ceramides. It is critical that the ceramide be first solubilized in a solvent compatible with water. Ethanol, 2-propanol, and DMSO are completely miscible with water. Dodecane is not. The addition of dodecane to increase the dispersion of ceramide, although widely used (e.g., Kashkar et al. 2005; Lee et al. 2011), is problematic. Being very poorly soluble in water, dodecane either forms small droplets in the solution or dissolves in membranes, increasing the membrane thickness and changing the membrane properties. Isopropanol is a better solvent for medium- and long-chain ceramides than either DMSO or ethanol and therefore is a more useful solvent. Moreover, warming an ethanol solution to increase ceramide solubility prior to adding to the mitochondrial suspension results in concerns regarding evaporation and potential oxidation at the higher temperature.

Once a true solution is formed, the ceramide needs to be dispersed in the mitochondrial suspension in such a way as to maximize ceramide insertion into mitochondria and minimize solvent effects on mitochondria. This is achieved by vigorously vortexing the mitochondrial solution in a microfuge tube while simultaneously slowly delivering the ceramide solution. This minimizes the exposure of some mitochondria to high levels of solvent and disperses the ceramide, hopefully into single molecules, giving the latter opportunity to collide with and insert into the MOM. Clearly if ceramides collide into each other prior to colliding into the MOM, they will probably adhere to each other by hydrophobic interactions. Hence, there is kinetic competition between insertion and aggregation. Experimental results indicate that aggregates do not insert well into the MOM. Indeed, even with the procedure just described, only a few percent of the added ceramide actually inserts (Siskind et al. 2006) into the MOM and the extent of insertion varies with the operator and the vortex mixer used.

The amount of ceramide added to isolated mitochondria appears to be higher than that present naturally in the MOM. One reason for this has just been pointed out. Only the amount of ceramide that actually inserts in the MOM is relevant. The second aspect is that in many cells ceramide levels increase in the MOM early in apoptosis (Vance 1990; Matsko et al. 2001; Dai et al. 2004). The measured levels of ceramide reached (Garcia-Ruiz et al. 1997; Rodriguez-Lafrasse et al. 2002; Birbes et al. 2005) match the ones delivered to the MOM that result in 20–30 % permeabilization of the MOM (Siskind et al. 2006). This is sufficient to release proteins from mitochondria and initiate the execution phase of apoptosis. However, a final point is that the effectiveness of ceramide depends on the level and activity of Bcl-2 family proteins. In cases where apoptosis induction does not significantly elevate ceramide levels, the activation of Bax and the inhibition of Bcl-xL by BH3-only proteins could make the existing ceramide levels sufficient to form ceramide channels (Siskind et al. 2008; Ganesan et al. 2010).

It is well known that Bax/BAK double-knockout cells are highly resistant to the induction of apoptosis. Yet the MOM of mitochondria isolated from these cells is easily permeabilized by the addition of ceramide (Siskind et al. 2008). Does this represent a conflict between the cellular response and the action of ceramide on isolated mitochondria? No! The induction of apoptosis in the parental strain results in an elevation of ceramide levels in the MOM. This elevation does not happen in the double-knockout cells (Siskind et al. 2010). The reason is that BAK is necessary for the activation of the ceramide synthase responsible for the ceramide elevation (Siskind et al. 2010). Therefore, are the Bax/BAK knockout cells resistant to apoptosis because of the lack of Bax and BAK or because of the failure to elevate the mitochondrial ceramide level? The answer is unclear and likely depends on the apoptotic pathway being used by the cell. A further recent example of this entanglement between ceramide and Bcl-2 family proteins is the reported requirement for ceramide synthesis in *C. elegans* for caspase activation by Bcl-2 family proteins in response to radiation (Deng et al. 2008). Indeed, ceramide and the Bcl-2 family proteins are intertwined in their roles in apoptosis.

The extent of permeabilization of the MOM by ceramide is often assessed by measuring the release of cytochrome *c*. Alternatively, the permeability can be estimated by measuring the rate of oxidation of exogenously added reduced cytochrome *c* (Siskind et al. 2002). The two assays are complementary, since cytochrome *c* release only needs to occur once whereas the rate of oxidation of cytochrome *c* measures real-time permeabilization. An additional factor is that cytochrome *c* is one of the smallest proteins released and tends to be bound to the inner membrane: two factors that may confound the interpretation of the results (e.g., Yuan et al. 2003). Indeed, there is evidence (Ganesan et al. 2011) that Bax permeabilization of the MOM starts off in vitro as transient channels capable of causing protein release with little actual measurable permeabilization. These channels are small and grow slowly as evident by the increase in the size of the proteins that can be released. Measuring the release of small proteins, like adenylate kinase, detects the formation of the small transient channels whereas measuring the rate of cytochrome *c* oxidation detects the large channels. In vivo, protein release occurs rapidly and proteins of various sizes seem to be released at the same time (Muñoz-Pinedo et al. 2006). This indicates that large channels are formed rapidly in vivo and so the measurement of MOM permeabilization by measuring the rate of cytochrome *c* oxidation is a better assessment of permeabilization reflecting the in vivo process.

Ceramide channel formation in planar phospholipid membranes does not form the attractive "clean" conductance levels seen with the typical small channels. Rather the conductance is "noisy," i.e., fluctuates erratically, especially as the channel grows in size. There are both discrete conductance increases typical of small channels and gradual increases. The former are of varying sizes and the latter are similar to conductances observed when "defects" form in membranes. Membranes containing nonspecific conductances prior to ceramide addition are discarded by the experimenter and new membranes are formed in an effort to reconstitute well-behaved channel conductances. However, the very nature of very large channels may be responsible for the varying conductance increases observed with the formation of ceramide channels. Both the assembly and disassembly processes as well as thermally driven structural changes would result in changes in conductance as described. Given this expectation, how does one gain confidence that the channels observed are similar to the models depicted in Fig. 1? The observations described in Sect. 3, the structural specificity described in Sect. 5, and the regulation by Bcl-2 family proteins provide a strong case for the overall picture of what ceramide channels are. However, for any one individual experiment it is the sensitivity to La^{3+}-induced disassembly that provides a distinction between a ceramide channel and a membrane defect. Although La^{3+} is not a specific inhibitor, it does not affect a conductance due to the formation of a defective membrane.

10 Conclusion

The permeabilization of the MOM to proteins results in the release of intermembrane-space proteins leading to the onset of the execution phase of apoptosis. This permeabilization, generally attributed to the action of the pro-apoptotic proteins, Bax and BAK, may be far more complex. Liposome and planar membrane experiments clearly show that Bax does not need ceramide to form channels and vice versa. However, the reported synergy between these agents indicates evolutionary selection for this interaction. Also the regulation of ceramide channels by anti-apoptotic proteins reinforces the conclusion that ceramide and Bcl-2 family proteins are both involved in MOM permeabilization. The unique ability of ceramide to form large stable channels capable of translocating proteins through membranes can hardly be an accidental occurrence but rather an evolutionarily conserved process. The structure of ceramide is not merely an accident but selected for based on its functional characteristics: one of these being the ability to form channels.

Acknowledgments Ceramide channels were discovered by Leah J. Siskind and the research described was performed by her and the following in temporal order: Johnny Stiban, Matthew J. Elrick, Sharon Fluss, Wenzhi Tan, Vidyaramanan Ganesan, Meenu N. Perera, Soumya Samanta, and Kai-Ti Chang. These were aided by numerous undergraduate volunteers. I am also grateful to others who collaborated and assisted with the research: Andriy Anishkin, Sergei Sukharev, Richard N. Kolesnick, R. Blake Hill, Marie Hardwick, Alicja Bielawska, Zdzislaw M. Szulc, Robert Bittman, and Jacek Bielawski. Support from the National Science Foundation (MCB-1023008) is gratefully acknowledged.

References

Anishkin A, Sukharev S, Colombini M (2006) Searching for the molecular arrangement of transmembrane ceramide channels. Biophys J 90:2414–2426

Birbes H, El Bawab S, Hannun YA, Obeid LM (2001) Selective hydrolysis of a mitochondrial pool of sphingomyelin induces apoptosis. FASEB J 15:2669–2679

Birbes H, Luberto C, Hsu YT, El Bawab S, Hannun YA, Obeid LM (2005) A mitochondrial pool of sphingomyelin is involved in TNFalpha-induced Bax translocation to mitochondria. Biochem J 386:445–451

Colombini M (2010) Ceramide channels and their role in mitochondria-mediated apoptosis. Biochim Biophys Acta 1797:1239–1244

Dai Q, Liu J, Chen J, Durrant D, McIntyre TM, Lee RM (2004) Mitochondrial ceramide increases in UV-irradiated HeLa cells and is mainly derived from hydrolysis of sphingomyelin. Oncogene 23:3650–3658

Deng X, Yin X, Allan R, Lu DD, Maurer CW, Haimovitz-Friedman A, Fuks Z, Shaham S, Kolesnick R (2008) Ceramide biogenesis is required for radiation-induced apoptosis in the germ line of *C. elegans*. Science 322:110–115

Di Paola M, Cocco T, Lorusso M (2000) Ceramide interaction with the respiratory chain of heart mitochondria. Biochemistry 39:6660–6668

Di Paola M, Zaccagnino P, Montedoro G, Cocco T, Lorusso M (2004) Ceramide induces release of pro-apoptotic proteins from mitochondria by either a Ca^{2+}-dependent or a Ca^{2+}-independent mechanism. J Bioenerg Biomembr 36:165–170

Elrick MJ, Fluss S, Colombini M (2006) Sphingosine, a product of ceramide hydrolysis by ceramidase, disassembles ceramide channels. Biophys J 91:1749–1756

Ganesan V, Perera MN, Colombini D, Datskovskiy D, Chadha K, Colombini M (2010) Ceramide and activated Bax act synergistically to permeabilize the mitochondrial outer membrane. Apoptosis 15:553–562

Ganesan V, Samanta S, Colombini M (2011) Effect of ionic strength on Bax:tBID mediated mitochondrial outer membrane permeabilization (MOMP). Biophys J 100:43a

Garcia-Ruiz C, Colell A, Mari M, Morales A, Fernandez-Checa JC (1997) Direct effect of ceramide on the mitochondrial electron transport chain leads to generation of reactive oxygen species. Role of mitochondrial glutathione. J Biol Chem 272:11369–11377

Ghafourifar P, Klein SD, Schucht O, Schenk U, Pruschy M, Rocha S, Richter C (1999) Ceramide induces cytochrome *c* release from isolated mitochondria. Importance of mitochondrial redox state. J Biol Chem 274:6080–6084

Kashkar H, Wiegmann K, Yazdanpanah B, Haubert D, Krönke M (2005) Acid sphingomyelinase is indispensable for UV light-induced Bax conformational change at the mitochondrial membrane. J Biol Chem 280:20804–20813

Lee H, Rotolo JA, Mesicek J, Penate-Medina T, Rimner A, Liao WC, Yin X, Ragupathi G, Ehleiter D, Gulbins E, Zhai D, Reed JC, Haimovitz-Friedman A, Fuks Z, Kolesnick R (2011) Mitochondrial ceramide-rich macrodomains functionalize Bax upon irradiation. PLoS One 6 (6):e19783

Matsko CM, Hunter OC, Rabinowich H, Lotze MT, Amoscato AA (2001) Mitochondrial lipid alterations during Fas- and radiation-induced apoptosis. Biochem Biophys Res Commun 287:1112–1120

Montes LR, Ruiz-Arguello MB, Goni FM, Alonso A (2002) Membrane restructuring via ceramide results in enhanced solute efflux. J Biol Chem 277:11788–11794

Muñoz-Pinedo C, Guío-Carrión A, Goldstein JC, Fitzgerald P, Newmeyer DD, Green DR (2006) Different mitochondrial intermembrane space proteins are released during apoptosis in a manner that is coordinately initiated but can vary in duration. Proc Natl Acad Sci U S A 103:11573–11578

Pajewski R, Djedovic N, Harder E, Ferdani R, Schlesinger PH, Gokel GW (2005) Pore formation in and enlargement of phospholipid liposomes by synthetic models of ceramides and sphingomyelin. Bioorg Med Chem 13:29–37

Pastorino JG, Tafani M, Rothman RJ, Marcinkeviciute A, Hoek JB, Farber JL (1999) Functional consequences of the sustained or transient activation by Bax of the mitochondrial permeability transition pore. J Biol Chem 274:31734–31739

Perera MN, Bielawska A, Szulc ZM, Bittman R, Colombini M (2011) Bax, Bcl-xL exert their regulation on different sites on the ceramide channel. Biophys J 100:43a

Perera MN, Ganesan V, Siskind LJ, Szulc ZM, Bielawski J, Bielawska A, Bittman R, Colombini M (2012a) Ceramide channels: influence of molecular structure on channel formation in membranes. Biochim Biophys Acta 1818:1291–1301

Perera MN, Lin SH, Peterson YK, Bielawska A, Szulc ZM, Bittman R, Colombini M (2012b) BAX, Bcl-xL exert their regulation on different sites of the ceramide channel. Biochem J 445:81–91

Rodriguez-Lafrasse C, Alphonse G, Aloy MT, Ardail D, Gerard JP, Louisot P, Rousson R (2002) Increasing endogenous ceramide using inhibitors of sphingolipid metabolism maximizes ionizing radiation-induced mitochondrial injury and apoptotic cell killing. Int J Cancer 101:589–598

Samanta S, Stiban J, Maugel TK, Colombini M (2011) Visualization of ceramide channels by transmission electron microscopy. Biochim Biophys Acta 1808:1196–1201

Siskind LJ, Colombini M (2000) The lipids C_2- and C_{16}-ceramide form large stable channels: implications for apoptosis. J Biol Chem 275:38640–38644

Siskind LJ, Kolesnick RN, Colombini M (2002) Ceramide channels increase the permeability of the mitochondrial outer membrane to small proteins. J Biol Chem 277:26796–26803

Siskind LJ, Davoody A, Lewin N, Marshall S, Colombini M (2003) Enlargement and contracture of C_2-ceramide channels. Biophys J 85:1560–1575

Siskind LJ, Kolesnick RN, Colombini M (2006) Ceramide forms channels in mitochondrial outer membranes at physiologically relevant concentrations. Mitochondrion 6:118–125

Siskind LJ, Feinstein L, Yu T, Davis JS, Jones D, Choi J, Zuckerman JE, Tan W, Hill RB, Hardwick JM, Colombini M (2008) Anti-apoptotic Bcl-2 family proteins disassemble ceramide channels. J Biol Chem 283:6622–6630

Siskind LJ, Mullen TD, Rosales KR, Clarke CJ, Hernandez-Corbacho MJ, Edinger AL, Obeid LM (2010) The Bcl-2 protein BAK is required for long-chain ceramide generation during apoptosis. J Biol Chem 285:11818–11826

Stiban J, Fistere D Jr, Colombini M (2006) Dihydroceramide hinders ceramide channel formation: implications on apoptosis. Apoptosis 11:773–780

Vance JE (1990) Phospholipid synthesis in a membrane fraction associated with mitochondria. J Biol Chem 265:7248–7256

Yuan H, Williams SD, Adachi S, Oltersdorf T, Gottlieb RA (2003) Cytochrome *c* dissociation and release from mitochondria by truncated Bid and ceramide. Mitochondrion 2:237–244

Genetic Disorders of Simple Sphingolipid Metabolism

Virginie Albinet, Marie-Lise Bats, Carmen Bedia, Frédérique Sabourdy, Virginie Garcia, Bruno Ségui, Nathalie Andrieu-Abadie, Thorsten Hornemann, and Thierry Levade

Contents

V. Albinet • V. Garcia • B. Ségui • N. Andrieu-Abadie
Institut National de la Santé et de la Recherche Médicale (INSERM) UMR1037, Centre de Recherches en Cancérologie de Toulouse (CRCT), Team n°4, Université de Toulouse, CHU Rangueil, BP 84225, Toulouse cedex 4 31432, France

M.-L. Bats • F. Sabourdy • T. Levade (✉)
Institut National de la Santé et de la Recherche Médicale (INSERM) UMR1037, Centre de Recherches en Cancérologie de Toulouse (CRCT), Team n°4, Université de Toulouse, CHU Rangueil, BP 84225, Toulouse cedex 4 31432, France

Laboratoire de Biochimie Métabolique, Institut Fédératif de Biologie, CHU Purpan, Toulouse, France
e-mail: thierry.levade@inserm.fr

C. Bedia
Institut National de la Santé et de la Recherche Médicale (INSERM) UMR1037, Centre de Recherches en Cancérologie de Toulouse (CRCT), Team n°4, Université de Toulouse, CHU Rangueil, BP 84225, Toulouse cedex 4 31432, France

CSIC, Barcelona, Spain

T. Hornemann (✉)
Institute for Clinical Chemistry, University Hospital Zurich and Center for Integrative Human Physiology, University of Zurich, Zurich, Switzerland
e-mail: thorsten.hornemann@usz.ch

E. Gulbins and I. Petrache (eds.), *Sphingolipids: Basic Science and Drug Development*, Handbook of Experimental Pharmacology 215, DOI 10.1007/978-3-7091-1368-4_7,

Abstract A better understanding of the functions sphingolipids play in living organisms can be achieved by analyzing the biochemical and physiological changes that result from genetic alterations of sphingolipid metabolism. This review summarizes the current knowledge gained from studies both on human patients and mutant animals (mice, cats, dogs, and cattle) with genetic disorders of sphingolipid metabolism. Genetic alterations affecting the biosynthesis, transport, or degradation of simple sphingolipids are discussed.

Keywords Sphingolipids • Disorders • Genetics • Biosynthesis

Abbreviations

1-deoxymetSa	1-Deoxymethyl-sphinganine
1-deoxySa	1-Deoxy-sphinganine
1-deoxySL	1-Deoxy-sphingolipid
CDase	Ceramidase
FD	Farber disease
HSAN1	Hereditary sensory and autonomic neuropathy type 1
S1P	Sphingosine 1-phosphate
SK	Sphingosine kinase
SL	Sphingolipid
SM	Sphingomyelin
SPL	S1P lyase
SPT	Serine-palmitoyltransferase

1 Introduction

Since the identification in the mid 1960s of the enzymatic deficiency underlying Gaucher disease, the most prevalent lysosomal storage disease in humans (Brady et al. 1965), a number of genetic defects affecting sphingolipid (SL) metabolism have been identified. Such inherited disorders are for the vast majority autosomal recessive traits and involve the degradation of SLs. However, much more recently genetic disorders of SL biosynthesis and trafficking have also been identified. In addition, quite numerous animal models harboring a genetic defect in SL metabolism have been described. These include both spontaneous and engineered mutant animals of various species.

This review focuses on genetic defects in mammals (mostly humans and mice) and those that affect the metabolism of simple SLs (i.e., those with a simple substituent at the sphingoid base C1, such as hydrogen or phosphate), whose major pathways in mammalian cells are depicted in Fig. 1. Genetic defects of SL metabolism in yeast (Cowart and Obeid 2007; Dickson 2008), drosophila (Rao and Acharya 2008; Kraut 2011), and plants (Pata et al. 2010) have recently been reviewed. In addition, review articles on the genetic deficiencies of glycosphingolipid metabolism (mostly degradation) in humans and mice have recently been published (Sabourdy et al. 2008; Schulze et al. 2009; Wennekes et al. 2009; Xu et al. 2010).

Except for one disease, i.e., *SPTLC* mutations, all these disorders are due to a loss of function. In general, two classifications for such genetic defects can be considered. First, according to the nature of the metabolic pathway, defects in biosynthesis, catabolism, or transport can be distinguished. A second classification, based on the nature of the metabolic abnormality, will distinguish defects characterized by (1) insufficient production of one or more SL (for instance, defects in 3-ketosphinganine reductase, ceramide synthase, or dihydroceramide desaturase); (2) accumulation of one or more undegraded SL (for instance, ceramidase deficiency); (3) abnormal trafficking (such as defects in ceramide transporter (CERT) and Niemann–Pick C1 (NPC1)); and (4) production of an abnormal/toxic metabolite (for example, *SPTLC* mutations). Since for some genetic diseases or conditions presented here the pathogenesis is still unclear, the first mode of classification has been selected.

2 Defects in the Biosynthesis of Simple Sphingolipids

2.1 Serine-Palmitoyltransferases

Serine-palmitoyltransferase (SPT, [EC 2.3.1.50]) catalyzes the first and rate-limiting step in the cellular *de novo* synthesis of SLs, which is the pyridoxal phosphate-dependent condensation of L-serine and palmitoyl-CoA (Fig. 1). SPT is a heteromeric enzyme composed of the three subunits SPTLC1 (55 kDa), SPTLC2 (65 kDa), and SPTLC3 (63 kDa). SPTLC1 and SPTLC2 are ubiquitously expressed whereas SPTLC3 shows a more tissue-specific distribution with high expression levels in placenta and heart. SPTLC2 and SPTLC3 share about 70 % homology, including the presence of a pyridoxal phosphate-binding motif which is absent in SPTLC1. The active SPT has a molecular weight of 450–480 kDa and was suggested to be an octamer composed of four SPTLC1–SPTLC2 or SPTLC1–SPTLC3 dimers (Hornemann et al. 2007). However, additional putative regulatory components of the SPT complex have been reported recently (Han et al. 2009; Breslow et al. 2010). Functional studies revealed that SPTLC2 and SPTLC3 exhibit differences in respect

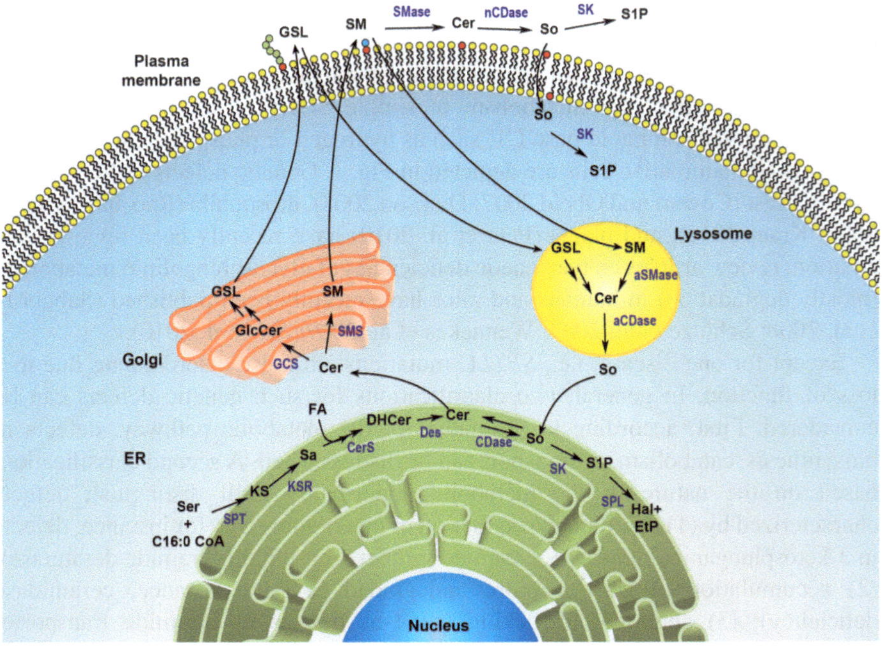

Fig. 1 Schematic view of the subcellular compartmentalization of sphingolipid metabolism. *Abbreviations*: *CDase* ceramidase, *aCDase* acid CDase, *nCDase* neutral CDase, *Cer* ceramide, *CerS* ceramide synthase, *Des* desaturase, *DHCer* dihydroceramide, *EtP* ethanolamine phosphate, *FA* fatty acid, *GlcCer* glucosylceramide, *GSL* glycosphingolipid, *Hal* hexadecenal, *KS* ketosphinganine, *KSR* KS reductase, *Sa* sphinganine, *Ser* L-serine, *SK* sphingosine kinase, *SM* sphingomyelin, *SMase* sphingomyelinase, *aSMase* acid SMase, *SMS* SM synthase, *So* sphingosine, *SPL* S1P lyase, *SPT* serine-palmitoyltransferase, *S1P* sphingosine 1-phosphate

to their substrate specificity. SPTLC2 primarily metabolizes palmitoyl (C_{16})-CoAs whereas SPTLC3 has a higher activity towards shorter acyl-CoAs, such as lauroyl (C_{12})- and myristoyl (C_{14})-CoA.

SPT is essential for development, since homozygosity for SPT deficiency in mice is embryonically lethal. Heterozygous *Sptlc1* and *Sptlc2* knockout mice are viable and show a reduced *in vitro* SPT activity but no reduction in plasma sphingomyelin (SM), total cholesterol, phospholipids, or liver SM levels (Hojjati et al. 2005). However, *Sptlc1+/−* mice show a reduced intestinal cholesterol absorption (Li et al. 2009b) whereas *Sptlc2+/−* mice have increased insulin sensitivity (Li et al. 2011). Liver-specific *Sptlc2−/−* mice show decreased plasma SM levels and increased hepatic apoE secretion (Li et al. 2009a).

Various missense mutations in the human *SPTLC1* and *SPTLC2* genes are causing the hereditary sensory and autonomic neuropathy type 1 (HSAN1, OMIM#162400) (Bejaoui et al. 2001; Dawkins et al. 2001; Verhoeven et al. 2004) (see Table 1). Thus far, no mutation in the SPTLC3 subunit has been associated with HSAN1 (Rotthier et al. 2010). HSAN1 is a rare, autosomal, and dominantly inherited axonal neuropathy with a progressive loss of pain and temperature sensation. Sense for vibration and joint position is preserved. First symptoms start usually

Table 1 Mammalian genetic defects of simple sphingolipid biosynthesis

Gene	Protein	Species	Disease	Gene defect	Changes in SL	Clinical and pathological phenotype	Age of death	References
SPTLC1	SPT subunit 1	Human	HSAN1	See text	Increased deoxySa and deoxymetSa	Hereditary sensory neuropathy type 1	Up to normal	(Rotthier et al. 2009; Penno et al. 2010)
Sptlc1	SPT subunit 1	Mouse		SPTLC1^{C133W}	Elevated deoxySLs	Peripheral neuropathy with preceding hyperpathia		(McCampbell et al. 2005; Eichler et al. 2009)
				Knockout	In heterozygotes, no changes in plasma SM levels, but reduced SM in apical membranes of enterocytes	Reduced intestinal cholesterol absorption in *Sptlc1*−/+ mice	Embryonic lethality of homozygotes	(Hojjati et al. 2005; Li et al. 2009b)
SPTLC2	SPT subunit 2	Human	HSAN1		Increased deoxySLs	Hereditary sensory neuropathy type 1		(Rotthier et al. 2010)
Sptlc2	SPT subunit 2	Mouse		Knockout	In heterozygotes, no changes in plasma SM levels		Embryonic lethality of homozygotes	(Hojjati et al. 2005)
				Liver-specific knockout	Reduced plasma and liver SM levels, and increased plasma apoE		ND	(Li et al. 2009a)
FVT1	3-Ketosphinganine reductase	Cattle		p.A175T	Not reported	Spinal muscular atrophy	A few weeks	(Krebs et al. 2007)
Cers1	Ceramide synthase 1	Mouse	Flincher Toppler	Single nucleotide deletion or substitution	Reduced total, C18:0, C18:1 but increased C14 and C16:0 Cer levels in brain; increased So, Sa, S1P, and dihydroS1P	Cerebellar ataxia at 3 weeks of age; Purkinje cell degeneration	As normal mice	(Zhao et al. 2011)

(continued)

Table 1 (continued)

Gene	Protein	Species	Disease	Gene defect	Changes in SL	Clinical and pathological phenotype	Age of death	References
Cers2	Ceramide synthase 2	Mouse		Gene trap in intron 1	Reduced C22–C24 Cer in brain, liver, and kidney; reduced C22–C24 SM in liver and kidney; reduced GalCer and sulfatide levels in sciatic nerves and myelin	Abnormal myelin, impaired initiation of motor activity, liver tumors (between 7 and 9 months)	ND	(Imgrund et al. 2009)
				Gene trap in intron 1	Reduced C22–C24 levels of Cer, SM, and HexCer in liver; increased C16 Cer and Sa levels in liver; reduced C22–C24 non-hydroxy and 2-hydroxy-GalCer in brain	Weight loss; hepatomegaly, liver nodules, and hepatocarcinoma; astrogliosis, microglial activation, and lysosomal storage of undefined material; myoclonic jerks	About 16 months	(Pewzner-Jung et al. 2010a; b; Ben-David et al. 2011)
Degs1	Dihydroceramide desaturase	Mouse		Knockout	Reduced Cer and increased DH-Cer in homozygous 1-day-old pups	Small size, decreased body weight, scaly skin, tremor, failure to thrive of homozygous mice. Enhanced insulin sensitivity of heterozygous mice	8–10 weeks	(Holland et al. 2007)

Abbreviations: *Cer* ceramide, *DH-Cer* dihydroceramide, *GalCer* galactosylceramide, *HexCer* monohexosylceramide, *ND* not defined in the lower limbs and then progress further to the upper extremities (Houlden et al. 2006). There are frequent positive sensory symptoms of severe shooting or burning pain in the limbs and an early but transient period of allodynia in some patients. Due to profound sensory impairments, many affected individuals develop neuropathic ulcers and Charcot joints requiring amputation. This distinguishes HSAN1 from other dominantly inherited sensory neuropathies such as Charcot–Marie–Tooth type 2. There is also prominent and often early motor involvement in most patients. Autopsy reports from HSAN1 patients show pronounced degeneration of dorsal root ganglion cells with depletion of myelinated axons in peripheral nerves. Loss of unmyelinated axons is present, but reported to be less severe (Denny-Brown 1951; Reimann et al. 1958; Houlden et al. 2006; Lindahl et al. 2006). To date, six mutations in the *SPTLC1* (C133W, C133Y, C133R, V144D, S331F, and A352V) and four mutations in the *SPTLC2* (V359M, G382V, T409M, and I505F) genes were convincingly associated with HSAN1. However, not all reported SPT mutations are causing the disease. The G387A mutation (Verhoeven et al. 2004) appears to be a benign variant not associated with HSAN1 (Hornemann et al. 2009).

In vitro enzyme activity of the mutant SPT is generally reduced. This was shown in various mutant overexpressing cell lines including transformed lymphocytes from HSAN1 patients (Bejaoui et al. 2002; Gable et al. 2002; Dedov et al. 2004). However, total SL levels were not altered in plasma of HSAN1 patients indicating that HSAN1 is not directly caused by haplo-insufficiency. This was further confirmed in a transgenic mouse model for HSAN1. Mice expressing the $SPTLC1^{C133W}$ mutation develop an age-dependent peripheral neuropathy with motor and sensory impairments. Although *in vitro* SPT activity is reduced in these mice, the total plasma SL levels are not altered (McCampbell et al. 2005). A reduced *in vitro* SPT activity is also seen in heterozygous *Sptlc1* and *Sptlc2* knockout mice which do not develop neuropathic symptoms (Hojjati et al. 2005). Therefore, the reduction in enzymatic activity is per se not a sufficient explanation for the pathogenesis of HSAN1.

It was shown recently that SPT does not strictly depend on serine as a substrate but can also metabolize alanine and glycine to a certain extent (Zitomer et al. 2009). This forms an atypical class of 1-deoxysphingolipids (1-deoxySLs). The use of alanine and glycine instead of serine results in the formation of the two sphingoid bases 1-deoxy-sphinganine (1-deoxySa) and 1-deoxymethyl-sphinganine (1-deoxymetSa), respectively. Both metabolites lack the C_1 hydroxyl group of regular sphingoid bases and hence cannot be converted into complex SLs (e.g., phospho- and glycosphingolipids) nor degraded by the classical pathway which requires the formation of sphingosine 1-phosphate as a catabolic intermediate (Menaldino et al. 2003). The promiscuous activity of SPT is greatly increased in case of the HSAN1 mutants resulting in highly elevated 1-deoxySL levels (Penno et al. 2010). This was demonstrated in mutant overexpressing HEK293 cells but also in lymphocytes and plasma from HSAN1 patients. Significantly elevated 1-deoxySL levels were also found in plasma and tissues of the $SPTLC1^{C133W}$ transgenic mice (Eichler et al. 2009). In contrast, double-transgenic mice which concomitantly overexpressed the $SPTLC1^{C133W}$ mutant in conjunction with the wild-type SPTLC1 showed only marginally elevated 1-deoxySL levels and did not develop neurological symptoms

(Eichler et al. 2009). Highly elevated 1-deoxySL levels were primarily detected in the sciatic nerves of HSAN1 mice, but not in the CNS or spinal cord. This fully corresponds to the pattern of pathology in HSAN1 patients who do not exhibit any mental impairment. Moderately elevated 1-deoxySL levels were also found in testes and to a lower extent in liver of the HSAN1 mice. The 1-deoxySL levels in testes correlated with a low sperm count and a significantly reduced fertility in the transgenic mice (Eichler et al. 2009). The neurotoxicity of 1-deoxySLs was confirmed *in vitro* on cultured dorsal root ganglion neurons which showed a dose-dependent reduction of neurite length, number, and branching in the presence of 1-deoxy-sphingoid bases. This neurotoxic effect was more pronounced for 1-deoxySa than for 1-deoxymetSa (Penno et al. 2010).

Interestingly, the formation of 1-deoxy-sphingoid bases is modulated in the presence of certain amino acids. The addition of alanine or glycine to the culture medium of $SPTLC1^{C133W}$ expressing HEK293 cells led to a significant increase in 1-deoxySa and 1-deoxymetSa formation whereas the addition of D-amino acids like D-alanine and D-serine markedly decreased 1-deoxySL formation. In particular, the addition of the canonical substrate L-serine significantly suppressed the formation of 1-deoxySLs, whereas in parallel the formation of sphinganine was stimulated (Garofalo et al. 2011). Thus, L-serine showed a corrective effect on both HSAN1 phenotypes—the reduced generation of sphinganine and the increased formation of neurotoxic 1-deoxySLs.

The benefit of L-serine supplementation was further explored in the HSAN1 mouse model. Mice receiving an L-serine-enriched diet (10 % w/w) showed a significant decrease in plasma 1-deoxySL levels within 4–5 days, and the levels remained low for the whole feeding period. As a consequence, L-serine-fed HSAN1 mice did not develop neurological symptoms whereas age-matched, but normally fed $SPTLC1^{C133W}$ mice developed severe neuronal deficits after 6–9 months. In contrast, supplementation with L-alanine (10 % w/w) led to significantly increased 1-deoxySL plasma levels allied with the development of severe neurological deficits in the mice already at an age of 2–3 months.

The therapeutic effect of an oral L-serine supplementation was confirmed in a 10-week pilot study with HSAN1 patients. Probands received a total dose of 200 mg or 400 mg L-serine/kg/day divided in three portions. In both groups, L-serine supplementation resulted in a significant reduction of plasma 1-deoxySL levels reaching normal levels after 5–6 weeks (Garofalo et al. 2011). After termination of the 10-week trial, 1-deoxySL plasma levels started to increase again. Improvements in the neurological symptoms were not examined and also not expected within the short time frame of this study. However, some patients reported an increase in sensation (hand tingling, increased menstrual cramps), improvements in skin robustness and wound healing, as well as faster nail and body hair growth. Based on these results, a simple and inexpensive oral L-serine supplementation might be a perspective for a future therapy in HSAN1.

2.2 Ketosphinganine Reductase

Only one condition has been described that leads to markedly reduced activity of the 3-ketosphinganine reductase, which is characterized by a point mutation discovered in cattle, believed to cause autosomal recessive neurodegenerative disease spinal muscular atrophy (Krebs et al. 2007). Whereas SL levels have not been determined in the tissues of the affected calves, this observation would emphasize the vulnerability of (bovine) neuronal cells to either the substrate or the products of 3-ketosphinganine reductase. Nevertheless, a subsequent study indicated that mutations in the 3-ketosphinganine reductase do not appear to account for spinal muscular atrophy in humans (Parkinson et al. 2008).

2.3 Ceramide Synthases

So far, six mammalian genes encoding *bona fide* ceramide synthases have been characterized (Levy and Futerman 2010). While no human disease has been reported so far to be caused by mutations in the *CERS* genes, spontaneous or engineered defects in the corresponding mouse genes have been recently described (Table 1). In line with the predominant expression of Cers1 in brain, mutations in the *Cers1* coding region result in an ataxic phenotype in mice. On the other hand, disruption of *Cers2* leads both to a hepatopathy that partially resembles that induced by fumonisin B1, a fungal inhibitor of ceramide synthase, and an encephalopathy (see Table 1).

In these mouse models, the lipid molecules acting as a culprit in the disease pathogenesis remain to be identified. Nevertheless, studies of these defects have led to the following conclusions. First, there is no metabolic redundancy between Cers1 or Cers2 and other ceramide synthases; at best, the decreased content of some ceramide species due to a defective synthase isoform is compensated by an elevation of other ceramide species. These observations also corroborate the previously reported substrate specificity of this family of enzymes. Second, Cers1 and Cers2 play key distinct tissue-specific functions that are likely ascribed to specific (or very restricted) molecular ceramide species in a given cell type or organ. Third, the substrate (chiefly, sphinganine) and/or lipid products of Cers1 and Cers2 play critical roles in liver and brain disease development. Whether similar conclusions hold true in other mammals and whether comparable diseases in humans are caused by *CERS* defects is still unknown.

2.4 Dihydroceramide Desaturase

Two genes in mammals (*DEGS1* and *DEGS2*) are involved in the introduction of a double bond in the sphingoid moiety to produce ceramides. So far, only ablation of the *Degs1* gene in mice has been described (Holland et al. 2007). Rather simple analyses of SL levels in different organs of this animal model indicated that the *Degs1* gene product plays a major role in desaturing dihydroceramide. While homozygous null mice display a complex phenotype that deserves further attention, heterozygous animals were refractory to glucocorticoid-induced insulin resistance, which led the authors to conclude about the critical function of ceramide in glucose homeostasis and the development of insulin resistance.

3 Defects in the Degradation of Simple Sphingolipids

3.1 Ceramidases

Ceramidases (CDases) are the enzymes that catalyze the hydrolysis of ceramides to form sphingosine and free fatty acids. According to their pH optima of activity, they have been classified into acid, neutral, and alkaline forms. Five human CDases encoded by five distinct genes have been identified: an acid CDase (*ASAH1*), a neutral CDase (*ASAH2*), and three alkaline CDases (*ACER1/ASAH3*, *ACER2/ ASAH3L*, and *ACER3/APHC*) (Mao and Obeid 2008).

Acid CDase is a ubiquitous lysosomal enzyme whose genetic deficiency causes Farber disease (FD, OMIM#228000), also called lipogranulomatosis (Levade et al. 2009). FD is a rare inherited lipid storage disorder characterized by accumulation of ceramide in cells and tissues of patients. Clinically, FD patients show deformed and painful joints, subcutaneous granulomas, hoarseness due to laryngeal involvement, and premature cell death (see Table 2). Also, patients may present hepatosplenomegaly and nervous system dysfunction (Levade et al. 2009). Currently, besides palliative care, there is no treatment for FD. Bone marrow or hematopoietic stem cell transplantations have been performed in a few trials in patients without neurological involvement. As a result, only partial resolution of the peripheral symptoms has occurred (Vormoor et al. 2004; Ehlert et al. 2006). Recently, a preclinical gene therapy study for FD has been carried out employing a lentiviral vector coding for acid CDase (Walia et al. 2011). This vector was used for transduction and transplantation of primary hematopoietic cells in three enzymatically normal nonhuman primates. An increase of acid CDase activity and a reduction of ceramide levels in peripheral blood and bone marrow cells, spleen, and liver throughout a 1-year study period were observed, without compromising the health status of the animals. This therapeutic approach appears as a promising strategy for the treatment of FD and other monogenic defects in future clinical gene therapy trials.

Li et al. have disrupted the mouse gene *Asah1* in embryonic stem cells by insertional mutagenesis to investigate the role of acid CDase in mammalian development (Li et al. 2002; Eliyahu et al. 2007). Homozygosity for *Asah1* disruption resulted in very early embryonic lethality suggesting that acid CDase activity is

Table 2 Mammalian genetic defects of simple sphingolipid catabolism

Gene	Protein	Species	Disease	Gene defect	Changes in SL	Clinical phenotype	Age of death	References
ASAH1	acid CDase	Human	Farber		Increased Cer	Painful and deformed joints, subcutaneous nodules, hoarseness, neurological alterations, hepatosplenomegaly	0–30 years (often childhood)	(Levade et al. 2009)
Asah1	acid CDase	Mouse		Knockout	Increased Cer	Apoptosis of embryos at the 2-cell stage	Embryonic death	(Li et al. 2002)
Asah2	acid CDase	Mouse		Knockout	Increased Cer	No apparent pathology		(Kono et al. 2006)
Sphk1	Sphingosine kinase 1	Mouse		Knockout	Reduction of serum levels of S1P to less than 50 %	Animals are viable, fertile, and without any obvious abnormalities		(Allende et al. 2004)
Sphk1	Sphingosine kinase 1	Mouse		Knockout	Reduction of serum S1P levels by 50 %	Animals are viable, fertile, and without any obvious abnormalities		(Zemann et al. 2006)
Sphk2	Sphingosine kinase 2	Mouse		Knockout	Reduction of plasma S1P levels by about 25 %	Animals are viable, fertile, and survive at least 12 months. Less homozygous mice than expected		(Kharel et al. 2005)
Sphk2	Sphingosine kinase 2	Mouse		Knockout	No difference in serum S1P level	Animals are viable and fertile. No major histological abnormalities		(Zemann et al. 2006)
Sphk2	Sphingosine kinase 2	Mouse		Knockout	N.E.	Animals are fertile and have normal longevity. No histological abnormalities		(Mizugishi et al. 2005)

(continued)

Table 2 (continued)

Gene	Protein	Species	Disease	Gene defect	Changes in SL	Clinical phenotype	Age of death	References
Sphk1+ *Sphk2*	Sphingosine kinases	Mouse		Knockout	Undetectable S1P levels in whole double-null embryos at E11.5. Accumulation of Sa and So in *Sphk1*−/− *Sphk2*+/− female mice	Abnormal cardiovascular, muscular, and neural system development. Infertility of *Sphk1*−/− *Sphk2*+/− female mice	Embryonic death (E13.5)	(Mizugishi et al. 2007)
Sphk1+ *Sphk2*	Sphingosine kinases	Mouse		Double-conditional knockout	Undetectable plasma and lymph S1P levels	Mice generated at P5 survived to adulthood and were indistinguishable from littermates. Reduction of blood and lymph B cell counts, but normal number of B cells in the spleen and lymph nodes. Accumulation of mature T cells in the thymus. Decreased number of mature B cells in the bone marrow		(Pappu et al. 2007)
Sgpl1	S1P lyase	Mouse		Knockout	N.E.	Growth failure, vascular abnormalities, hemorrhage, anemia, skeletal defects, and renal abnormalities	8 weeks	(Schmahl et al. 2007)

Sgpl1	S1P lyase	Mouse	Knockout	Elevated levels of S1P in spleen and serum. Accumulation of S1P, So, and C16-Cer in thymus	Apparently healthy at birth. Loss of weight in the postnatal period and premature death. Mature T cell sequestration in thymus leading to lymphopenia. Decreased size of spleen and thymus. Increased levels of pro-inflammatory cytokines. Pulmonary, hematologic, and cardiac defects. Urothelial lesions	Death during the first weeks (median 29 days)	(Van Veldhoven P 2005; Vogel et al. 2009; Weber et al. 2009; Allende et al. 2011)
Sgpl1	S1P lyase	Mouse	Knock-in	Elevated S1P level in spleen	Viable and fertile. Lymphopenia (B and T cells), lymphocyte retention in thymus and secondary lymphoid organs		(Vogel et al. 2009)

Abbreviations: *N.E.* not examined, *Sa* sphinganine, *So* sphingosine essential for embryo development. The heterozygous *Asah1+/−* mice survived, but quite intriguingly showed evidence of a progressive lipid storage disease in their organs.

Neutral CDase, encoded by the *ASAH2* gene, is highly expressed in the small intestine along the brush border, where it is involved in the catabolism of dietary SLs. Kono et al. generated neutral CDase-null mice that were viable, healthy, and fertile, with no apparent gross abnormalities (Kono et al. 2006). This was possibly due to the control of ceramide levels by other CDases, such as the acid or the alkaline forms. However, increased concentrations of C16-ceramide were observed in intestine and feces, suggesting that neutral CDase is important for the digestion of dietary ceramides. In contrast, Yoshimura et al. (2004) observed that the knock-down of neutral CDase in zebrafish increased the number of embryos with severe morphological and cellular abnormalities. This indicated that in zebrafish and possibly other vertebrates, neutral CDase is essential for early development (Yoshimura et al. 2004).

With respect to the alkaline CDases, no genetic deficiency or knockout studies in animals have been reported to date.

3.2 Sphingosine Kinases

Sphingosine kinases (SK) catalyze the synthesis of sphingosine 1-phosphate (S1P) via the phosphorylation of sphingoid bases. Two isoforms of mammalian SKs (termed 1 and 2), encoded by *SPHK1* and *SPHK2* genes, have been cloned and characterized (Kohama et al. 1998; Liu et al. 2000). SK1 localizes predominantly in the cytoplasm but upon stimulation can relocate to the inner leaflet of the plasma membrane. Interestingly, in endothelial cells SK1 is secreted and able to produce S1P extracellularly (Ancellin et al. 2002). Although SK1 and SK2 are expressed in numerous tissues, they show a distinct tissue distribution and developmental expression pattern (Liu et al. 2000; Fukuda et al. 2003). Indeed, while SK1 is predominantly expressed in lung, spleen, and thymus, SK2 is highly expressed in liver, heart, and kidney. In order to better understand the role of SK in mammalian development, deficient mouse models have been generated and characterized. Homozygosity for disruption of a single SK isoform leads to viable mice that are fertile and display no obvious dysfunctions (Allende et al. 2004). Interestingly, *Sphk1−/−* mice have reduced S1P levels in serum but not in tissues. Otherwise, S1P levels in *Sphk2−/−* mice differ depending on the model described. Whereas one model, produced on a 129/Sv × C57BL/6 mixed background, exhibited a 25 % reduction of plasma S1P levels (Kharel et al. 2005), another on Balb/c mice did not show a decrease but rather an increase as compared to wild-type animals (Zemann et al. 2006). In sharp contrast, *Sphk1−/− Sphk2−/−* double-knockout mice, which exhibit strongly reduced levels of S1P, die at E13.5 due to severe defects in neurogenesis and angiogenesis (Mizugishi et al. 2007). However, the conditional *Sphk1−/− Sphk2−/−* double-knockout mouse model generated by Pappu et al. (in which SK expression is abolished at postnatal day 5 in hematopoietic and endothelial cells) was viable and showed that plasma S1P is mainly hematopoietic in origin.

Up to now, no mutation in the *SPHK* genes has been described in human disease. Nevertheless, SK and S1P are viewed as critical players in several pathological states, such as anaphylaxis, inflammation, cancer, diabetes, or atherosclerosis (Fyrst and Saba 2010).

3.3 Sphingosine 1-Phosphate Lyase

S1P lyase (SPL) is a pyridoxal 5′-phosphate-dependent aldehyde-lyase that irreversibly degrades S1P to phosphoethanolamine and hexadecenal in the final step of SL catabolism (Bandhuvula and Saba 2007). By regulating intracellular S1P levels, SPL participates in the so-called sphingolipid rheostat and has the ability to shift the balance towards cell death. While no mutation of the *SGPL1* gene has been reported so far to cause human disease, its expression appears to be altered in some cancers; SPL also plays a critical role in regulating immune functions. Indeed, SPL expression and activity were found to be downregulated in human colon cancer tissues and in adenomatous lesions of the $APC^{Min/+}$ mouse model of intestinal tumorigenesis compared to normal adjacent tissues (Oskouian et al. 2006). SPL expression has also been shown to be downregulated in human melanoma cell lines compared to normal melanocytes (Colié et al. 2009).

To further investigate the role of SPL in development, knockout and humanized mouse models have been recently created. Homozygous *Sgpl1* knockout mice die within 8 weeks after birth and exhibit significant growth failure, vascular abnormalities, anemia, skeletal defects, and renal abnormalities (Schmahl et al. 2007). Beyond the reduced life span, SPL-deficient mice also display lesions in lungs, heart, urinary tract, and bone, and develop myeloid cell hyperplasia. Lymphopenia due to alterations in lymphocyte development and egress from the thymus and secondary lymphoid organs has also been reported (Vogel et al. 2009). This phenotype is reminiscent of that caused by FTY720 (Mandala et al. 2002); in addition, the apoptosis observed in thymus might be mediated by increased levels of ceramide in this tissue (Weber et al. 2009). The humanized knock-in mice generated by the insertion of the human *SGPL1* cDNA in the *Sgpl1* null background resulted in SPL expression at 10–20 % of normal mouse SPL levels, yet failed to restore normal T-cell development and trafficking (Vogel et al. 2009). In summary, these animal models indicate that the absence of SPL leads to significant developmental and functional defects. However, whether *SGPL1* mutations are compatible with life and/or are responsible for human disease remains to be determined.

4 Defects in Transport or Trafficking of Simple Sphingolipids

4.1 Ceramide Transporter

CERT is a cytosolic protein that mediates the ATP-dependent ER-to-Golgi transfer of ceramide in a non-vesicular manner. Human CERT is encoded by the *COL4A3BP* gene which was initially isolated in 1999 as the gene for Goodpasture antigen-binding protein (GPBP) (Raya et al. 1999). Goodpasture's syndrome is a strictly human disorder caused by antibodies directed against the non-collagenous domain of the α3-chain of type IV collagen. In 2003, Hanada and coworkers showed that a spliced variant of GPBP which lacks a serine-rich domain composed of 26 amino acids was responsible for the cytosolic trafficking of ceramide from the ER to the Golgi apparatus (Hanada et al. 2003). GPBPs are phylogenetically highly conserved during evolution between lower vertebrates and mammals at the amino acid level.

CERT was identified in the Chinese hamster ovary LY-A cell line that displays a severe defect in the synthesis of sphingomyelin. Its reduced level of SM makes this cell line resistant to lysenin, a cytolytic toxin derived from the earthworm *Eisenia foetida* (Hanada et al. 2003). CERT is a hydrophilic 68 kDa protein which contains three domains:

(a) The amino terminal region (≈120 aa) forms a pleckstrin homology domain which specifically binds to phosphatidylinositol 4-phosphate, mainly distributed to the Golgi apparatus. In the mutant LY-A cell line, the PH-domain of CERT has one point mutation, G67E, which destroys the phosphoinositide-binding activity, resulting in an impaired ER-to-Golgi ceramide transport (Levine and Munro 2002; Hanada et al. 2003).
(b) The middle region (≈250 aa) contains an FFAT (two phenylalanine residues in an acidic tract) motif which interacts with vesicule-associated ER proteins (VAPs A and B) and targets CERT to ER membranes (Kawano et al. 2006).
(c) The carboxy terminal region (≈230 aa) is a START (steroidogenic acute regulatory protein-related protein transfer) domain which forms a deep lipid-binding pocket that could extract ceramide from membranes and transfer the bound ceramide to other membranes (Kumagai et al. 2005).

Gene disruption of the corresponding *CERT* in mice resulted in death around embryonic day 11.5. Cells of the mutant embryos (analyzed at E10.5) showed a 60 % reduction in the total SM content and a twofold increase in the ceramide content of the ER as compared to wild-type controls, leading to an abnormal dilation of the ER and degenerating mitochondria, probably affecting organogenesis (Wang et al. 2009). Whether mutations in the *COL4A3BP* gene cause developmental (or other) defects in humans has never been reported.

4.2 Niemann–Pick C1

NPC1 is a 13 *trans*-membrane domain glycoprotein mainly residing in late endosomes (Higgins et al. 1999; Davies and Ioannou 2000). The gene (*NPC1*) that encodes this protein is located on chromosome 18q11-q12 (Greer et al. 1997). The exact role of NPC1 is still unclear but mutations in *NPC1* are responsible for a genetic disorder called Niemann–Pick disease type C (MIM #257220) (Carstea et al. 1997). About 5 % of patients affected with this disease carry mutations in the *NPC2* gene, encoding a soluble lysosomal protein. Niemann–Pick type C is an autosomal recessive lysosomal storage disorder characterized by the accumulation of a large variety of lipids in late endosomes and lysosomes (Table 3). While unesterified cholesterol and SM accumulate in peripheral tissues, brain storage is dominated by GM2 and GM3 gangliosides. Niemann–Pick type C is a neurovisceral disease of variable severity (for a review, see Vanier 2010; Patterson et al. 2009). Disease onset (as well as death) can range from the neonatal period to adult age. Progressive neurological impairment occurs in all patients, and may manifest as cerebellar ataxia, dysarthria, dysphagia, and/or dementia. Visceral involvement, mainly splenomegaly and hepatomegaly or, more rarely, neonatal cholestasis and respiratory distress, always precedes the neurological disease, but may be absent in some patients.

Two murine, one feline, and one canine spontaneous models of Niemann–Pick type C have been reported (Morris et al. 1977; Miyawaki et al. 1982; Lowenthal et al. 1990; Kuwamura et al. 1993), showing clinical and pathological features quite similar to those in humans. Experimental analyses based on both human samples and animal models led to the conclusion that a key function of NPC1 is the transport of unesterified cholesterol out of the acidic compartments (Liscum and Faust 1989; Infante et al. 2008; Kwon et al. 2009). The exact role of NPC1, however, in the metabolism or trafficking of SLs is still unsolved. Nonetheless, its involvement in the transport of sphingosine has been postulated (Lloyd-Evans et al. 2008). Sphingosine is the final product of SL catabolism and is known as a bioactive lipid with pro-apoptotic properties. Previous work showed a marked increase in the levels of sphingosine in Niemann–Pick type C tissues (Goldin et al. 1992; Rodriguez-Lafrasse et al. 1994). It has recently been proposed that sphingosine might be the first lipid accumulating in late endosomes/lysosomes of Niemann–Pick type C cells. This storage would affect the late endosome/lysosome calcium homeostasis, inducing a global defect in vesicular trafficking and the secondary accumulation, in late endosomes, of cholesterol and SLs (Lloyd-Evans et al. 2008). Niemann–Pick type C can thus be viewed as a genetic disorder of sphingosine metabolism where NPC1 could play the role of a sphingosine transporter. Nevertheless, this view has recently been challenged by a work which showed no specific sequestration of lysosomal sphingosine in NPC1-deficient cells (Blom et al. 2012). While there is currently no effective therapy for this disorder, promising results have been obtained in affected mice and cats using the cholesterol chelators cyclodextrins (Davidson et al. 2009; Liu et al. 2009; Ward et al. 2010).

Table 3 Mammalian genetic defects of simple sphingolipid transport

Gene	Protein	Species	Disease	Gene defect	Changes in SL	Clinical and pathological phenotype	Age of death	References
Col4a3bp	Cert	Mouse		Knockout	Increased Cer levels in ER and mitochondria; decreased SM levels	Embryonic lethality; cardiac abnormalities, structural and functional defects of ER and mitochondria	E11.5	(Wang et al. 2009)
NPC1	Niemann–Pick C1 protein	Human	Niemann–Pick type C		In liver and spleen, accumulation of FC, SM, LBPA, GlcCer, LacCer, So (×20), and Sa In brain, storage of GM2, GM3, and So (×3)	Neonatal cholestasis, hepatosplenomegaly, progressive neurological deterioration, vertical supranuclear gaze palsy Neuroinflammation, neuroaxonal dystrophy, Purkinje cell death. Foamy cells and sea-blue histiocytes in visceral organs	0–60 years and more (most often 10–25 years)	(Patterson et al. 2009; Vanier 2010)
Npc1	Niemann–Pick C1 protein	Mouse		*Spm* (mutation ND) and npc^{nih} (retrotransposon-like insertion and 703 bp deletion)	In liver and spleen: SM, FC, GlcCer, LacCer accumulation. In liver: LBPA, GM3, and GA2 In brain: FC, GM1, GM2, and GM3 storage	Neurological symptoms, weight loss, lymph nodes enlargement, infertile. Brain and cerebellar atrophy. Dysmyelination, neuronal storage, and axonal spheroids. Foamy macrophages in liver and spleen. Severe Purkinje cell depletion	12–14 weeks (for *Spm*) 10 weeks (for npc^{nih})	(Morris et al. 1977; Pentchev et al. 1980; Miyawaki et al. 1982; Kuwamura et al. 1993; Taniguchi et al. 2001; Zervas et al. 2001;

							Patterson et al. 2009; Vanier 2010)
Npc1	Niemann–Pick C1 protein	Mouse	Conditional knockout in cerebellar Purkinje cells		Motor deficits by 10 weeks. Progressive loss of Purkinje cells in an anterior-to-posterior gradient	No premature death	(Elrick et al. 2010)
Npc1	Niemann–Pick C1 protein	Cat	p.C955S	In the liver, FC, GlcCer, LacCer, and SM storage In the brain, GM2 and GM3 accumulation	Progressive neurological disorders: Ataxia, tremors. Growth retardation. Neuroaxonal dystrophy. Foamy macrophages in liver, spleen, lymph nodes, and lung	About 20 weeks (euthanazied at 9 weeks)	(Lowenthal et al. 1990; Somers et al. 2003; Vite et al. 2008)
Npc1	Niemann–Pick C1 protein	Dog	ND	In liver and spleen, FC (but no SM) storage. GM2, GM3, and LacCer accumulation in brain	Progressive neurological disorder: Unable to stand at 9 months. Mild cerebellar atrophy. No hepatosplenomegaly. Foamy macrophages in liver, spleen, lymph nodes, and lung. Marked loss of Purkinje and granular cells in the cerebellum	Euthanazied at 9 months	(Kuwamura et al. 1993)

Abbreviations: *Cer* ceramide, *DH-Cer* dihydroceramide, *FC* free cholesterol, *GalCer* galactosylceramide, *HexCer* monohexosylceramide, *LBPA* lysobisphosphatidic acid, *NPC* Niemann–Pick disease type C, *Sa* sphinganine, *So* sphingosine, *ND* not defined

5 Concluding Remarks

A number of alterations in the genes encoding proteins that control the metabolism of simple SLs have already been identified or created. The cell lines derived from the human patients or animals bearing these defects, as well as the corresponding mutant mice, represent unique models for the study of SL metabolism and functions. Analysis of the metabolic and functional consequences of these genetic defects has already produced invaluable insights into the physiological role of SLs. Nevertheless, not only further gene products which remain very poorly studied (such as ketosphinganine reductase, dihydroceramide desaturases, or S1P phosphatases) need to be characterized, but also more refinements in manipulation of gene expression can be introduced to generate animal models that help dissect the functions of SLs and, perhaps, replicate reliably various human pathological conditions.

Acknowledgements Studies by T.L.'s group were supported by, Université Paul Sabatier, ANR (SphingoDR program), RITC, LNCC and the Vaincre les Maladies Lysosomales Foundation, and those of T.H.'s group by the Gebert Rüf Foundation, the Center for Integrative Human Physiology (ZIHP, University of Zurich) and the European Commission (LSHM-CT-2006-037631) (for T.H.'s group).

References

Allende ML, Sasaki T, Kawai H, Olivera A, Mi Y, van Echten-Deckert G, Hajdu R, Rosenbach M, Keohane CA, Mandala S, Spiegel S, Proia RL (2004) Mice deficient in sphingosine kinase 1 are rendered lymphopenic by FTY720. J Biol Chem 279:52487–52492

Allende ML, Bektas M, Lee BG, Bonifacino E, Kang J, Tuymetova G, Chen W, Saba JD, Proia RL (2011) Sphingosine-1-phosphate lyase deficiency produces a pro-inflammatory response while impairing neutrophil trafficking. J Biol Chem 286:7348–7358

Ancellin N, Colmont C, Su J, Li Q, Mittereder N, Chae SS, Stefansson S, Liau G, Hla T (2002) Extracellular export of sphingosine kinase-1 enzyme. Sphingosine 1-phosphate generation and the induction of angiogenic vascular maturation. J Biol Chem 277:6667–6675

Bandhuvula P, Saba JD (2007) Sphingosine-1-phosphate lyase in immunity and cancer: silencing the siren. Trends Mol Med 13:210–217

Bejaoui K, Wu C, Scheffler MD, Haan G, Ashby P, Wu L, de Jong P, Brown RH Jr (2001) SPTLC1 is mutated in hereditary sensory neuropathy, type 1. Nat Genet 27:261–262

Bejaoui K, Uchida Y, Yasuda S, Ho M, Nishijima M, Brown RH Jr, Holleran WM, Hanada K (2002) Hereditary sensory neuropathy type 1 mutations confer dominant negative effects on serine palmitoyltransferase, critical for sphingolipid synthesis. J Clin Invest 110:1301–1308

Ben-David O, Pewzner-Jung Y, Brenner O, Laviad EL, Kogot-Levin A, Weissberg I, Biton IE, Pienik R, Wang E, Kelly S, Alroy J, Raas-Rothschild A, Friedman A, Brügger B, Merrill AH Jr, Futerman AH (2011) Encephalopathy caused by ablation of very long acyl chain ceramide synthesis may be largely due to reduced galactosylceramide levels. J Biol Chem 286:30022–30033

Blom T, Li Z, Bittman R, Somerharju P, Ikonen E (2012) Tracking sphingosine metabolism and transport in sphingolipidoses: NPC1 deficiency as a test case. Traffic 13(9):1234–1243

Brady RO, Kanfer JN, Shapiro D (1965) Metabolism of glucocerebrosides. II. Evidence of an enzymatic deficiency in Gaucher's disease. Biochem Biophys Res Commun 18:221–225

Breslow DK, Collins SR, Bodenmiller B, Aebersold R, Simons K, Shevchenko A, Ejsing CS, Weissman JS (2010) Orm family proteins mediate sphingolipid homeostasis. Nature 463:1048–1053

Carstea ED, Morris JA, Coleman KG, Loftus SK, Zhang D, Cummings C, Gu J, Rosenfeld MA, Pavan WJ, Krizman DB, Nagle J, Polymeropoulos MH, Sturley SL, Ioannou YA, Higgins ME, Comly M, Cooney A, Brown A, Kaneski CR, Blanchette-Mackie EJ, Dwyer NK, Neufeld EB, Chang TY, Liscum L, Strauss JF III, Ohno K, Zeigler M, Carmi R, Sokol J, Markie D, O'Neill RR, van Diggelen OP, Elleder M, Patterson MC, Brady RO, Vanier MT, Pentchev PG, Tagle DA (1997) Niemann-Pick C1 disease gene: homology to mediators of cholesterol homeostasis. Science 277:228–231

Colié S, Van Veldhoven PP, Kedjouar B, Bedia C, Albinet V, Sorli SC, Garcia V, Djavaheri-Mergny M, Bauvy C, Codogno P, Levade T, Andrieu-Abadie N (2009) Disruption of sphingosine 1-phosphate lyase confers resistance to chemotherapy and promotes oncogenesis through Bcl-2/Bcl-xL upregulation. Cancer Res 69:9346–9353

Cowart LA, Obeid LM (2007) Yeast sphingolipids: recent developments in understanding biosynthesis, regulation, and function. Biochim Biophys Acta 1771:421–431

Davidson CD, Ali NF, Micsenyi MC, Stephney G, Renault S, Dobrenis K, Ory DS, Vanier MT, Walkley SU (2009) Chronic cyclodextrin treatment of murine Niemann-Pick C disease ameliorates neuronal cholesterol and glycosphingolipid storage and disease progression. PLoS One 4:e6951

Davies JP, Ioannou YA (2000) Topological analysis of Niemann-Pick C1 protein reveals that the membrane orientation of the putative sterol-sensing domain is identical to those of 3-hydroxy-3-methylglutaryl-CoA reductase and sterol regulatory element binding protein cleavage-activating protein. J Biol Chem 275:24367–24374

Dawkins JL, Hulme DJ, Brahmbhatt SB, Auer-Grumbach M, Nicholson GA (2001) Mutations in SPTLC1, encoding serine palmitoyltransferase, long chain base subunit-1, cause hereditary sensory neuropathy type I. Nat Genet 27:309–312

Dedov VN, Dedova IV, Merrill AH Jr, Nicholson GA (2004) Activity of partially inhibited serine palmitoyltransferase is sufficient for normal sphingolipid metabolism and viability of HSN1 patient cells. Biochim Biophys Acta 1688:168–175

Denny-Brown D (1951) Hereditary sensory radicular neuropathy. J Neurol Neurosurg Psychiatry 14:237–252

Dickson RC (2008) Thematic review series: sphingolipids. New insights into sphingolipid metabolism and function in budding yeast. J Lipid Res 49:909–921

Ehlert K, Roth J, Frosch M, Fehse N, Zander N, Vormoor J (2006) Farber's disease without central nervous system involvement: bone-marrow transplantation provides a promising new approach. Ann Rheum Dis 65:1665–1666

Eichler FS, Hornemann T, McCampbell A, Kuljis D, Penno A, Vardeh D, Tamrazian E, Garofalo K, Lee HJ, Kini L, Selig M, Frosch M, Gable K, von Eckardstein A, Woolf CJ, Guan G, Harmon JM, Dunn TM, Brown RH Jr (2009) Overexpression of the wild-type SPT1 subunit lowers desoxysphingolipid levels and rescues the phenotype of HSAN1. J Neurosci 29:14646–14651

Eliyahu E, Park JH, Shtraizent N, He X, Schuchma EH (2007) Acid ceramidase is a novel factor required for early embryo survival. FASEB J 21:1403–1409

Elrick MJ, Pacheco CD, Yu T, Dadgar N, Shakkottai VG, Ware C, Paulson HL, Lieberman AP (2010) Conditional Niemann-Pick C mice demonstrate cell autonomous Purkinje cell neurodegeneration. Hum Mol Genet 19:837–847

Fukuda Y, Kihara A, Igarashi Y (2003) Distribution of sphingosine kinase activity in mouse tissues: contribution of SPHK1. Biochem Biophys Res Commun 309:155–160

Fyrst H, Saba JD (2010) An update on sphingosine-1-phosphate and other sphingolipid mediators. Nat Chem Biol 6:489–497

Gable K, Han G, Monaghan E, Bacikova D, Natarajan M, Williams R, Dunn TM (2002) Mutations in the yeast LCB1 and LCB2 genes, including those corresponding to the hereditary sensory neuropathy type I mutations, dominantly inactivate serine palmitoyltransferase. J Biol Chem 277:10194–10200

Garofalo K, Penno A, Schmidt BP, Lee HJ, Frosch MP, von Eckardstein A, Brown RH, Hornemann T, Eichler FS (2011) Oral L-serine supplementation reduces production of neurotoxic deoxysphingolipids in mice and humans with hereditary sensory autonomic neuropathy type 1. J Clin Invest 121:4735–4745

Goldin E, Roff CF, Miller SP, Rodriguez-Lafrasse C, Vanier MT, Brady RO, Pentchev PG (1992) Type C Niemann-Pick disease: a murine model of the lysosomal cholesterol lipidosis accumulates sphingosine and sphinganine in liver. Biochim Biophys Acta 1127:303–311

Greer WL, Riddell DC, Byers DM, Welch JP, Girouard GS, Sparrow SM, Gillan TL, Neumann PE (1997) Linkage of Niemann-Pick disease type D to the same region of human chromosome 18 as Niemann-Pick disease type C. Am J Hum Genet 61:139–142

Han G, Gupta SD, Gable K, Niranjanakumari S, Moitra P, Eichler F, Brown RH Jr, Harmon JM, Dunn TM (2009) Identification of small subunits of mammalian serine palmitoyltransferase that confer distinct acyl-CoA substrate specificities. Proc Natl Acad Sci U S A 106:8186–8191

Hanada K, Kumagai K, Yasuda S, Miura Y, Kawano M, Fukasawa M, Nishijima M (2003) Molecular machinery for non-vesicular trafficking of ceramide. Nature 426:803–809

Higgins ME, Davies JP, Chen FW, Ioannou YA (1999) Niemann-Pick C1 is a late endosome-resident protein that transiently associates with lysosomes and the trans-Golgi network. Mol Genet Metab 68:1–13

Hojjati MR, Li Z, Jiang XC (2005) Serine palmitoyl-CoA transferase (SPT) deficiency and sphingolipid levels in mice. Biochim Biophys Acta 1737:44–51

Holland WL, Brozinick JT, Wang LP, Hawkins ED, Sargent KM, Liu Y, Narra K, Hoehn KL, Knotts TA, Siesky A, Nelson DH, Karathanasis SK, Fontenot GK, Birnbaum MJ, Summers SA (2007) Inhibition of ceramide synthesis ameliorates glucocorticoid-, saturated-fat-, and obesity-induced insulin resistance. Cell Metab 5:167–179

Hornemann T, Wei Y, von Eckardstein A (2007) Is the mammalian serine palmitoyltransferase a high-molecular-mass complex? Biochem J 405:157–164

Hornemann T, Penno A, Richard S, Nicholson G, van Dijk FS, Rotthier A, Timmerman V, von Eckardstein A (2009) A systematic comparison of all mutations in hereditary sensory neuropathy type I (HSAN I) reveals that the G387A mutation is not disease associated. Neurogenetics 10:135–143

Houlden H, King R, Blake J, Groves M, Love S, Woodward C, Hammans S, Nicoll J, Lennox G, O'Donovan DG, Gabriel C, Thomas PK, Reilly MM (2006) Clinical, pathological and genetic characterization of hereditary sensory and autonomic neuropathy type 1 (HSAN I). Brain 129:411–425

Imgrund S, Hartmann D, Farwanah H, Eckhardt M, Sandhoff R, Degen J, Gieselmann V, Sandhoff K, Willecke K (2009) Adult ceramide synthase 2 (CERS2)-deficient mice exhibit myelin sheath defects, cerebellar degeneration, and hepatocarcinomas. J Biol Chem 284:33549–33560

Infante RE, Radhakrishnan A, Abi-Mosleh L, Kinch LN, Wang ML, Grishin NV, Goldstein JL, Brown MS (2008) Purified NPC1 protein. II. Localization of sterol binding to a 240-amino acid soluble luminal loop. J Biol Chem 283:1064–1075

Kawano M, Kumagai K, Nishijima M, Hanada K (2006) Efficient trafficking of ceramide from the endoplasmic reticulum to the Golgi apparatus requires a VAMP-associated protein-interacting FFAT motif of CERT. J Biol Chem 281:30279–30288

Kharel Y, Lee S, Snyder AH, Sheasley-O'neill SL, Morris MA, Setiady Y, Zhu R, Zigler MA, Burcin TL, Ley K, Tung KSK, Engelhard VH, Macdonald TL, Pearson-White S, Lynch KR (2005) Sphingosine kinase 2 is required for modulation of lymphocyte traffic by FTY720. J Biol Chem 280:36865–36872

Kohama T, Olivera A, Edsall L, Nagiec MM, Dickson R, Spiegel S (1998) Molecular cloning and functional characterization of murine sphingosine kinase. J Biol Chem 273:23722–23728

Kono M, Dreier JL, Ellis JM, Allende ML, Kalkofen DN, Sanders KM, Bielawski J, Bielawska A, Hannun YA, Proia RL (2006) Neutral ceramidase encoded by the Asah2 gene is essential for the intestinal degradation of sphingolipids. J Biol Chem 281:7324–7331

Kraut R (2011) Roles of sphingolipids in Drosophila development and disease. J Neurochem 116:764–778

Krebs S, Medugorac I, Röther S, Strässer K, Förster M (2007) A missense mutation in the 3-ketodihydrosphingosine reductase FVT1 as candidate causal mutation for bovine spinal muscular atrophy. Proc Natl Acad Sci U S A 104:6746–6751

Kumagai K, Yasuda S, Okemoto K, Nishijima M, Kobayashi S, Hanada K (2005) CERT mediates intermembrane transfer of various molecular species of ceramides. J Biol Chem 280:6488–6495

Kuwamura M, Awakura T, Shimada A, Umemura T, Kagota K, Kawamura N, Naiki M (1993) Type C Niemann-Pick disease in a boxer dog. Acta Neuropathol 85:345–348

Kwon HJ, Abi-Mosleh L, Wang ML, Deisenhofer J, Goldstein JL, Brown MS, Infante RE (2009) Structure of N-terminal domain of NPC1 reveals distinct subdomains for binding and transfer of cholesterol. Cell 137:1213–1224

Levade T, Sandhoff K, Schulze H, Medin JA (2009) Acid ceramidase deficiency: Farber lipogranulomatosis. In: Valle D et al (eds) Scriver's OMMBID (Online metabolic & molecular bases of inherited disease). McGraw-Hill, New York, pp 3573–3588

Levine TP, Munro S (2002) Targeting of Golgi-specific pleckstrin homology domains involves both PtdIns 4-kinase-dependent and -independent components. Curr Biol 12:695–704

Levy M, Futerman AH (2010) Mammalian ceramide synthases. IUBMB Life 62:347–356

Li CM, Park JH, Simonaro CM, He X, Gordon RE, Friedman AH, Ehleiter D, Paris F, Manova K, Hepbildikler S, Fuks Z, Sandhoff K, Kolesnick R, Schuchman EH, Hepbiloikler S (2002) Insertional mutagenesis of the mouse acid ceramidase gene leads to early embryonic lethality in homozygotes and progressive lipid storage disease in heterozygotes. Genomics 79:218–224

Li Z, Li Y, Chakraborty M, Fan Y, Bui HH, Peake DA, Kuo MS, Xiao X, Cao G, Jiang XC (2009a) Liver-specific deficiency of serine palmitoyltransferase subunit 2 decreases plasma sphingomyelin and increases apolipoprotein E levels. J Biol Chem 284:27010–27019

Li Z, Park TS, Li Y, Pan X, Iqbal J, Lu D, Tang W, Yu L, Goldberg IJ, Hussain MM, Jiang XC (2009b) Serine palmitoyltransferase (SPT) deficient mice absorb less cholesterol. Biochim Biophys Acta 1791:297–306

Li Z, Zhang H, Liu J, Liang CP, Li Y, Li Y, Teitelman G, Beyer T, Bui HH, Peake DA, Zhang Y, Sanders PE, Kuo MS, Park TS, Cao G, Jiang XC (2011) Reducing plasma membrane sphingomyelin increases insulin sensitivity. Mol Cell Biol 31:4205–4218

Lindahl AJ, Lhatoo SD, Campbell MJ, Nicholson G, Love S (2006) Late-onset hereditary sensory neuropathy type I due to SPTLC1 mutation: autopsy findings. Clin Neurol Neurosurg 108:780–783

Liscum L, Faust JR (1989) The intracellular transport of low density lipoprotein-derived cholesterol is inhibited in Chinese hamster ovary cells cultured with 3-beta-[2-(diethylamino)ethoxy] androst-5-en-17-one. J Biol Chem 264:11796–11806

Liu H, Sugiura M, Nava VE, Edsall LC, Kono K, Poulton S, Milstien S, Kohama T, Spiegel S (2000) Molecular cloning and functional characterization of a novel mammalian sphingosine kinase type 2 isoform. J Biol Chem 275:19513–19520

Liu B, Turley SD, Burns DK, Miller AM, Repa JJ, Dietschy JM (2009) Reversal of defective lysosomal transport in NPC disease ameliorates liver dysfunction and neurodegeneration in the npc1−/− mouse. Proc Natl Acad Sci U S A 106:2377–2382

Lloyd-Evans E, Morgan AJ, He X, Smith DA, Elliot-Smith E, Sillence DJ, Churchill GC, Schuchman EH, Galione A, Platt FM (2008) Niemann-Pick disease type C1 is a sphingosine storage disease that causes deregulation of lysosomal calcium. Nat Med 14:1247–1255

Lowenthal AC, Cummings JF, Wenger DA, Thrall MA, Wood PA, de Lahunta A (1990) Feline sphingolipidosis resembling Niemann-Pick disease type C. Acta Neuropathol 81:189–197

Mandala S, Hajdu R, Bergstrom J, Quackenbush E, Xie J, Milligan J, Thornton R, Shei GJ, Card D, Keohane C, Rosenbach M, Hale J, Lynch CL, Rupprecht K, Parsons W, Rosen H (2002) Alteration of lymphocyte trafficking by sphingosine-1-phosphate receptor agonists. Science 296:346–349

Mao C, Obeid LM (2008) Ceramidases: regulators of cellular responses mediated by ceramide, sphingosine, and sphingosine-1-phosphate. Biochim Biophys Acta 1781:424–434

McCampbell A, Truong D, Broom DC, Allchorne A, Gable K, Cutler RG, Mattson MP, Woolf CJ, Frosch MP, Harmon JM, Dunn TM, Brown RH Jr (2005) Mutant SPTLC1 dominantly inhibits serine palmitoyltransferase activity in vivo and confers an age-dependent neuropathy. Hum Mol Genet 14:3507–3521

Menaldino DS, Bushnev A, Sun A, Liotta DC, Symolon H, Desai K, Dillehay DL, Peng Q, Wang E, Allegood J, Trotman-Pruett S, Sullards MC, Merrill AH Jr (2003) Sphingoid bases and de novo ceramide synthesis: enzymes involved, pharmacology and mechanisms of action. Pharmacol Res 47:373–381

Miyawaki S, Mitsuoka S, Sakiyama T, Kitagawa T (1982) Sphingomyelinosis, a new mutation in the mouse: a model of Niemann-Pick disease in humans. J Hered 73:257–263

Mizugishi K, Yamashita T, Olivera A, Miller GF, Spiegel S, Proia RL (2005) Essential role for sphingosine kinases in neural and vascular development. Mol Cell Biol 25:11113–11121

Mizugishi K, Li C, Olivera A, Bielawski J, Bielawska A, Deng CX, Proia RL (2007) Maternal disturbance in activated sphingolipid metabolism causes pregnancy loss in mice. J Clin Invest 117:2993–3006

Morris M, Bhuvaneswaran C, Boothe A (1977) Tissue cholesterol storage disorder in BALB/c mice. Fed Proc 36:1158

Oskouian B, Soriyakumaran P, Borowsky AD, Crans A, Dillard-Telm L, Tam YY, Bandhuvula P, Saba JD (2006) Sphingosine-1-phosphate lyase potentiates apoptosis via p53- and p38-dependent pathways and is down-regulated in colon cancer. Proc Natl Acad Sci U S A 103:17384–17389

Pappu R, Schwab SR, Cornelissen I, Pereira JP, Regard JB, Xu Y, Camerer E, Zheng YW, Huang Y, Cyster JG, Coughlin SR (2007) Promotion of lymphocyte egress into blood and lymph by distinct sources of sphingosine-1-phosphate. Science 316:295–298

Parkinson NJ, Baumer D, Rose-Morris A, Talbot K (2008) Candidate screening of the bovine and feline spinal muscular atrophy genes reveals no evidence for involvement in human motor neuron disorders. Neuromuscul Disord 18:394–397

Pata MO, Hannun YA, Ng CK (2010) Plant sphingolipids: decoding the enigma of the Sphinx. New Phytol 185:611–630

Patterson MC, Vanier MT, Suzuki K, Morrie JA, Carstea E, Neufeld EB, Blanchette-Mackie JE, Pentchev PG (2009) Niemann-Pick disease type C: a lipid trafficking disorder. In: Valle D et al (eds) Scriver's OMMBID (Online metabolic & molecular bases of inherited disease). McGraw-Hill, New York, pp 3611–3633

Penno A, Reilly MM, Houlden H, Laurá M, Rentsch K, Niederkofler V, Stoeckli ET, Nicholson G, Eichler F, Brown RH Jr, von Eckardstein A, Hornemann T (2010) Hereditary sensory neuropathy type 1 is caused by the accumulation of two neurotoxic sphingolipids. J Biol Chem 285:11178–11187

Pentchev PG, Gal AE, Booth AD, Omodeo-Sale F, Fouks J, Neumeyer BA, Quirk JM, Dawson G, Brady RO (1980) A lysosomal storage disorder in mice characterized by a dual deficiency of sphingomyelinase and glucocerebrosidase. Biochim Biophys Acta 619:669–679

Pewzner-Jung Y, Brenner O, Braun S, Laviad EL, Ben-Dor S, Feldmesser E, Horn-Saban S, Amann-Zalcenstein D, Raanan C, Berkutzki T, Erez-Roman R, Ben-David O, Levy M, Holzman D, Park H, Nyska A, Merrill AH Jr, Futerman AH (2010a) A critical role for ceramide synthase 2 in liver homeostasis: II. insights into molecular changes leading to hepatopathy. J Biol Chem 285:10911–10923

Pewzner-Jung Y, Park H, Laviad EL, Silva LC, Lahiri S, Stiban J, Erez-Roman R, Brügger B, Sachsenheimer T, Wieland F, Prieto M, Merrill AH Jr, Futerman AH (2010b) A critical role for ceramide synthase 2 in liver homeostasis: I. alterations in lipid metabolic pathways. J Biol Chem 285:10902–10910

Rao RP, Acharya JK (2008) Sphingolipids and membrane biology as determined from genetic models. Prostaglandins Other Lipid Mediat 85:1–16

Raya A, Revert F, Navarro S, Saus J (1999) Characterization of a novel type of serine/threonine kinase that specifically phosphorylates the human goodpasture antigen. J Biol Chem 274:12642–12649

Reimann HA, McKechnie WG, Stanisavljevic S (1958) Hereditary sensory radicular neuropathy and other defects in a large family: reinvestigation after twenty years and report of a necropsy. Am J Med 25:573–579

Rodriguez-Lafrasse C, Rousson R, Pentchev PG, Louisot P, Vanier MT (1994) Free sphingoid bases in tissues from patients with type C Niemann-Pick disease and other lysosomal storage disorders. Biochim Biophys Acta 1226:138–144

Rotthier A, Baets J, De Vriendt E, Jacobs A, Auer-Grumbach M, Lévy N, Bonello-Palot N, Kilic SS, Weis J, Nascimento A, Swinkels M, Kruyt MC, Jordanova A, De Jonghe P, Timmerman V (2009) Genes for hereditary sensory and autonomic neuropathies: a genotype-phenotype correlation. Brain 132:2699–2711

Rotthier A, Auer-Grumbach M, Janssens K, Baets J, Penno A, Almeida-Souza L, Van Hoof K, Jacobs A, De Vriendt E, Schlotter-Weigel B, Löscher W, Vondráček P, Seeman P, De Jonghe P, Van Dijck P, Jordanova A, Hornemann T, Timmerman V (2010) Mutations in the SPTLC2 subunit of serine palmitoyltransferase cause hereditary sensory and autonomic neuropathy type I. Am J Hum Genet 87:513–522

Sabourdy F, Kedjouar B, Sorli SC, Colié S, Milhas D, Salma Y, Levade T (2008) Functions of sphingolipid metabolism in mammals—lessons from genetic defects. Biochim Biophys Acta 1781:145–183

Schmahl J, Raymond CS, Soriano P (2007) PDGF signaling specificity is mediated through multiple immediate early genes. Nat Genet 39:52–60

Schulze H, Kolter T, Sandhoff K (2009) Principles of lysosomal membrane degradation: cellular topology and biochemistry of lysosomal lipid degradation. Biochim Biophys Acta 1793:674–683

Somers KL, Royals MA, Carstea ED, Rafi MA, Wenger DA, Thrall MA (2003) Mutation analysis of feline Niemann-Pick C1 disease. Mol Genet Metab 79:99–103

Taniguchi M, Shinoda Y, Ninomiya H, Vanier MT, Ohno K (2001) Sites and temporal changes of gangliosides GM1/GM2 storage in the Niemann-Pick disease type C mouse brain. Brain Dev 23:414–421

Van Veldhoven P (2005) Sphingosine 1-phosphate lyase deficient mice. Presented at the 46th International Conference on the Bioscience of Lipids, Ajaccio, France. Chem Phys Lipids 136:164–165

Vanier MT (2010) Niemann-Pick disease type C. Orphanet J Rare Dis 5:16

Verhoeven K, Coen K, De Vriendt E, Jacobs A, Van Gerwen V, Smouts I, Pou-Serradell A, Martin JJ, Timmerman V, De Jonghe P (2004) SPTLC1 mutation in twin sisters with hereditary sensory neuropathy type I. Neurology 62:1001–1002

Vite CH, Ding W, Bryan C, O'Donnell P, Cullen K, Aleman D, Haskins ME, Van Winkle T (2008) Clinical, electrophysiological, and serum biochemical measures of progressive neurological and hepatic dysfunction in feline Niemann-Pick type C disease. Pediatr Res 64:544–549

Vogel P, Donoviel MS, Read R, Hansen GM, Hazlewood J, Anderson SJ, Sun W, Swaffield J, Oravecz T (2009) Incomplete inhibition of sphingosine 1-phosphate lyase modulates immune system function yet prevents early lethality and non-lymphoid lesions. PLoS One 4:e4112

Vormoor J, Ehlert K, Groll AH, Koch HG, Frosch M, Roth J (2004) Successful hematopoietic stem cell transplantation in Farber disease. J Pediatr 144:132–134

Walia JS, Neschadim A, Lopez-Perez O, Alayoubi A, Fan X, Carpentier S, Madden M, Lee CJ, Cheung F, Jaffray DA, Levade T, McCart JA, Medin JA (2011) Autologous transplantation of lentivector/acid ceramidase-transduced hematopoietic cells in nonhuman primates. Hum Gene Ther 22:679–687

Wang X, Rao RP, Kosakowska-Cholody T, Masood MA, Southon E, Zhang H, Berthet C, Nagashim K, Veenstra TK, Tessarollo L, Acharya U, Acharya JK (2009) Mitochondrial degeneration and not apoptosis is the primary cause of embryonic lethality in ceramide transfer protein mutant mice. J Cell Biol 184:143–158

Ward S, O'Donnell P, Fernandez S, Vite CH (2010) 2-Hydroxypropyl-beta-cyclodextrin raises hearing threshold in normal cats and in cats with Niemann-Pick type C disease. Pediatr Res 68:52–56

Weber C, Krueger A, Münk A, Bode C, Van Veldhoven PP, Gräler MH (2009) Discontinued postnatal thymocyte development in sphingosine 1-phosphate-lyase-deficient mice. J Immunol 183:4292–4301

Wennekes T, van den Berg RJBHN, Boot RG, van der Marel GA, Overkleeft HS, Aerts JMFG (2009) Glycosphingolipids—nature, function, and pharmacological modulation. Angew Chem Int Ed Engl 48:8848–8869

Xu YH, Barnes S, Sun Y, Grabowski GA (2010) Multi-system disorders of glycosphingolipid and ganglioside metabolism. J Lipid Res 51:1643–1675

Yoshimura Y, Tani M, Okino N, Iida H, Ito M (2004) Molecular cloning and functional analysis of zebrafish neutral ceramidase. J Biol Chem 279:44012–44022

Zemann B, Kinzel B, Müller M, Reuschel R, Mechtcheriakova D, Urtz N, Bornancin F, Baumruker T, Billich A (2006) Sphingosine kinase type 2 is essential for lymphopenia induced by the immunomodulatory drug FTY720. Blood 107:1454–1458

Zervas M, Dobrenis K, Walkley SU (2001) Neurons in Niemann-Pick disease type C accumulate gangliosides as well as unesterified cholesterol and undergo dendritic and axonal alterations. J Neuropathol Exp Neurol 60:49–64

Zhao L, Spassieva SD, Jucius TJ, Shultz LD, Shick HE, Macklin WB, Hannun YA, Obeid LM, Ackerman SL (2011) A deficiency of ceramide biosynthesis causes cerebellar purkinje cell neurodegeneration and lipofuscin accumulation. PLoS Genet 7:e1002063

Zitomer NC, Mitchell T, Voss KA, Bondy GS, Pruett ST, Garnier-Amblard EC, Liebeskind LS, Park H, Wang E, Sullards MC, Merrill AH Jr, Riley RT (2009) Ceramide synthase inhibition by fumonisin B1 causes accumulation of 1-deoxysphinganine: a novel category of bioactive 1-deoxysphingoid bases and 1-deoxydihydroceramides biosynthesized by mammalian cell lines and animals. J Biol Chem 284:4786–4795

The Role of Ceramide-1-Phosphate in Biological Functions

L. Alexis Hoeferlin, Dayanjan S. Wijesinghe, and Charles E. Chalfant

Contents

Abstract In mammalian cells, cermide-1-phosphate (C1P) is produced via the ATP-dependent mechanism of converting ceramide to C1P by the enzyme, ceramide kinase (CERK). CERK was first described as a calcium-stimulated lipid kinase that co-purified with brain synaptic vesicles, and to date, CERK is the only identified mammalian enzyme known to produce C1P in cells. C1P has steadily emerged as a bioactive sphingolipid involved in cell proliferation, macrophage migration, and inflammatory events. The recent generation of the CERK knockout mouse and the development of CERK inhibitors have furthered our current understanding of CERK-derived C1P in regulating biological processes. In this chapter, the history of C1P as well as the biological functions attributed to C1P are reviewed.

Keywords Ceramide kinase • Ceramide-1-phosphate • Eicosanoids • Proliferation • Immunity

L.A. Hoeferlin • D.S. Wijesinghe • C.E. Chalfant (✉)
Department of Biochemistry and Molecular Biology, Virginia Commonwealth University, 1101 East Marshall Street, P.O. Box 980614, Richmond, VA 23298-0614, USA
e-mail: cechalfant@vcu.edu

E. Gulbins and I. Petrache (eds.), *Sphingolipids: Basic Science and Drug Development*, Handbook of Experimental Pharmacology 215, DOI 10.1007/978-3-7091-1368-4_8,

1 Ceramide-1-Phosphate: The Early Years

Ceramide-1-phosphate (C1P) is synthesized in mammalian cells by the direct phosphorylation of ceramide by ceramide kinase (CERK). To date, CERK is the only known mammalian enzyme to produce C1P (Sugiura et al. 2002), and the enzyme was first described by Bajjalieh and coworkers (1989) as a calcium-stimulated lipid kinase co-purified with brain synaptic vesicles and possessing activity specific for the conversion of ceramide to C1P. Soon after this initial finding, the production of C1P was observed in the human pro-myelocytic leukemia cell line, HL-60 (Dressler and Kolesnick 1990). In this same study, the authors demonstrated that during stimulation, C1P was produced from ceramide derived from sphingomyelin, but not from glycosphingolipids. Follow-up studies by Kolesnick and Hemer (1990) reported a CERK activity distinguishable from diacylglycerol kinase in HL-60 cells verifying the findings of Bajjalieh and coworkers. After these initial studies, over a decade passed before successful cloning of the CERK enzyme was accomplished, and this new molecular "tool" provided researchers with the means to study the role of not only CERK in cellular functions but also C1P.

After the cloning of CERK, the mRNA for the enzyme was found to be expressed in heart, kidney, lung, brain, and hematopoietic cells (Sugiura et al. 2002). Analysis of the CERK mRNA sequence showed that human CERK protein consists of 537 amino acids, which closely resembles the amino acid homology and structure of sphingosine kinase 1 (Sphk1) and 2 (Sphk2). Specifically, CERK was found to contain the five conserved domains (C1–C5) previously identified for Sphk1 and 2. CERK also contains additional conserved regions across several species (*M. musculus*, *D. melanogaster*, *C. elegans*, and *O. sativa*) that are not homologous to SphK. These include a PH-domain at the N-terminus known to bind the β/γ subunit of heterotrimeric G-proteins, phospoinositol-4,5-bisphophate, and phosphorylated tyrosine residues (Sugiura et al. 2002). These conserved domains have been shown to play a regulatory function for the enzyme. For example, Igarashi and coworkers and Bornancin and coworkers have both demonstrated that the PH-domain is required for the activity of CERK in vitro as well as proper localization of the enzyme in cells (Kim et al. 2005; Carre et al. 2004). Interestingly, expression of the PH-domain alone also demonstrated improper localization suggesting that the catalytic domain also imparts specificity for specific internal membranes of the cell (Kim et al. 2006).

CERK also contains a calcium/calmodulin (CaM)-binding motif of the 1-8-14 type B spanning residues 422–435 [(F/I/L/V/W) XXXXXX (F/A/I/L/V/W) XXXXX (F/I/L/V/W) with a net charge of 2+ to 4+] (Sugiura et al. 2002). The functionality of this calcium/CaM-binding motif was confirmed by Igarashi and coworkers who demonstrated that CaM interacts with CERK and acts as a calcium "sensor" for the enzyme (Mitsutake and Igarashi 2005). Specifically, they showed that the CaM antagonist W-7 decreased both CERK activity and intracellular C1P formation. Additionally, exogenously added CaM enhanced CERK activity in vitro even at low concentrations of Ca^{2+}.

CERK also contains two conserved phosphorylation sites: a casein kinase II phosphorylation site [(S/T) XX (D/E)] at Ser^{340} and a cAMP-dependent phosphorylation site at Ser^{424} (Sugiura et al. 2002). There are also many putative protein kinase C (PKC) phosphorylation sites conserved in mammals: Ser^{72}, Thr^{118}, Thr^{127}, Ser^{230}, Ser^{300}, Ser^{340}, and Ser^{424}. At present, the only study investigating the role of phosphorylation in regulating the activity of CERK was carried out by Bornancin and coworkers, which demonstrated that a mutation of Ser^{340}Ala affected the stability of CERK (Chen et al. 2010). The Ser^{340} residue is located downstream of the catalytic site in a region that has been suggested to possess a regulatory role in CERK activity (Chen et al. 2010). Future identification of the kinases that are involved in CERK phosphorylation may provide additional understanding of how CERK activity is regulated.

Prior to the cloning of CERK, initial studies of the function of C1P utilized exogenous delivery of the sphingolipid followed by the examination of a biological phenotype. In this regard, the first biological activity of C1P was described by the Brindley laboratory. Specifically, Gomez-Munoz et al. (1995) demonstrated that C1P induced DNA synthesis and cell division. Since this initial study, a number of biological activities attributed to C1P have been steadily increasing, further enhancing its recognition as an important lipid-signaling molecule. Currently, C1P has been demonstrated to play a role in DNA synthesis (Gomez-Munoz et al. 1995), macrophage proliferation and migration (Gangoiti et al. 2010; Granado et al. 2009), $cPLA_2\alpha$ activation and subsequent production of inflammatory mediators (Pettus et al. 2004; Subramanian et al. 2005; Lamour et al. 2009), as well as inhibition of apoptosis via inhibition of acid sphingomyelinase (A-SMase) (Gomez-Munoz et al. 2004). More recent studies have discovered a potential role for C1P in the processing of the pro-inflammatory cytokine tumor necrosis-alpha (TNFα) (Lamour et al. 2011). These new advances in our knowledge of C1P biology have been facilitated by the development of accurate and reliable methods for detecting the relatively low cellular levels of C1P (Wijesinghe et al. 2010), as well as the availability of CERK-deficient animals, CERK inhibitors, and C1P agonists. Here, we discuss the major findings that have provided substantial evidence supporting a distinct role for C1P in cell growth and inflammatory processes.

2 C1P in Cell Growth and Survival: A Pro-survival Player

As previously stated, the first biological effect for C1P was reported by Gomez-Munoz and coworkers (1995) in regard to cellular proliferation/growth. For example, these early studies demonstrated that short-chain (not naturally found in cells) C1P induced DNA synthesis in Rat-1 fibroblasts (Gomez-Munoz et al. 1995). Additional studies demonstrated that treatment of T17 fibroblasts with natural C1P induced a potent increase in DNA synthesis and levels of proliferating cell nuclear antigen (PCNA) (Gomez-Munoz et al. 1997). Over the course of two reports Gangoiti and coworkers (2008a, b) further demonstrated that C1P stimulates macrophage proliferation through the downstream activation of the extracellularly regulated kinase 1 and 2 (ERK1/2) and c-Jun N-terminal kinase (JNK) pathways. The mechanisms behind the growth promoting role of C1P have recently become a bit more complex, as accompanying reports also implicate the activation of protein kinase C-alpha (PKCα) in C1P-stimulated macrophage proliferation (Gangoiti et al. 2010). Interestingly, these results suggested that the C1P induces the translocation of PKCα from the cytosol to the cell membrane, an event that was shown to be required for the mitogenic effect of C1P in macrophages (Gangoiti et al. 2010). Most recently, the Gomez-Munoz group has demonstrated that C1P also stimulates proliferation in C2C12 myoblasts, a skeletal muscle cell model (Gangoiti et al. 2012). Overall, one of the best described and characterized biological functions for C1P is the role of this lipid in promoting cellular proliferation and growth, which was also corroborated by Bornancin and coworkers using cells from the CERK ablation model (Graf et al. 2008).

C1P has also been implicated as an anti-apoptotic lipid; specifically, a later report from the Gomez-Munoz laboratory demonstrated that C1P also prevented cell death in bone marrow-derived macrophages (BMDMs) after withdrawal of macrophage colony-stimulating factor (M-CSF) (Gomez-Munoz et al. 2004). Treatment of BMDMs with C1P effectively blocked the activation of caspases and prevented DNA fragmentation upon serum removal. In the same study, this laboratory also demonstrated that C1P treatment inhibited ceramide generation from A-SMase. Furthermore, A-SMase was shown to be a direct target of C1P, consequently inducing inhibition of this enzyme (Gomez-Munoz et al. 2004). A follow-up study by Gomez-Munoz et al. (2005) demonstrated that C1P enhanced DNA binding to transcription factor NF-κB via stimulation of phosphatidylinositol 3-kinase (PI3-K) activity and protein kinase B (PKB)/(AKT) phosphorylation. Additionally, C1P treatment resulted in the upregulation of the anti-apoptotic regulator Bcl-X_L (Gomez-Munoz et al. 2005). Hence, C1P can activate a number of pro-survival pathways and antagonize the pro-apoptotic effects of ceramide.

Along these same lines, the Gomez-Munoz group has also presented evidence that supports a role for C1P in macrophage migration. Macrophage recruitment is a key event in mediating the inflammatory response as these cells are necessary for the release of cytokines, prostaglandins, and a variety of additional enzymes involved in the innate immune system. This recruitment process is highly dependent on the rate of macrophage proliferation, as well as the rate of migration and efflux (Pollard 2004). In this regard, the Gomez-Munoz laboratory demonstrated that addition of natural C1P stimulated the migration of macrophages (Granado et al. 2009). Interestingly, this finding of Granado and colleagues strongly suggested that C1P-induced migration was independent of intracellular C1P synthesis via CERK activation implicating the existence of cell-surface receptor for C1P. Furthermore, this study demonstrated that migration effects of exogenous C1P did not act through the currently known S1P receptors (Granado et al. 2009), a closely related sphingolipid known to induce cell survival and migration. Hence, the findings of the Gomez-Munoz laboratory suggest that C1P-stimulated macrophage migration is coupled to an, as of yet, unidentified G_i protein receptor. Indeed, a study by Zor and coworkers corroborated the possible existence of receptors for C1P and C1P analogs. Specifically, this laboratory demonstrated that incubation of RAW 264.7 macrophages with the phospho-ceramide analogue-1 (PCERA-1) reduced TNFα production at the mRNA and protein level in response to LPS-stimulation (Goldsmith et al. 2009). Structure–function studies of PCERA-1 show that the phosphate group and the lipid moiety are mutually required for its activity (Matsui et al. 2002a, b). Likewise, this study by the Zor group showed that the anti-inflammatory effect of PCERA-1 was dependent on the presence of PCERA-1 in the cell media during LPS treatment of RAW264.7 macrophages, which suggests that this C1P analogue acts extracellularly by binding a protein target present at the cell membrane (Goldsmith et al. 2009). Although additional biological and biophysical studies are required to validate the existence of a C1P receptor, the biological implications of a new class of lipid receptors are exciting and should be vehemently explored. Regardless, the culmination of these studies further emphasizes the role of C1P in the regulation of cellular homeostasis in macrophages.

3 C1P in Immunity and Inflammation: "The Missing Link"

Eicosanoids are one of the most important classes of lipids, which include prostaglandins, prostacyclins, leukotrienes, and thromboxanes. These eicosanoids give rise to the classical features of inflammation, which are necessary to defend the organism against infection and injury. In spite of these protective efforts, these processes can produce an overwhelming response, which leads to an excess of these molecules. As a result, unnecessary levels of eicosanoids can promote a wide range of disease states including chronic inflammation, allergy, cardiovascular disease, and cancer (Yedgar et al. 2007). The mechanism of prostaglandin synthesis begins with the rate-limiting step, the formation of arachidonic acid (AA), via the activity of phospholipase A_2 (Murakami et al. 1996). In many cases, inflammatory cytokines (e.g., TNFα) induce the activation and translocation of Group IVA cytosolic phospholipase A_2 (cPLA$_2$α), which requires the association of cPLA$_2$α with membranes in a Ca^{2+}-dependent manner via a C2/CALB domain (Fig. 1).

The hypothesis that C1P regulated the activation of a phospholipase A_2 (PLA$_2$) and eicosanoid synthesis came from an unlikely source. A link existed between inflammation/eicosanoid synthesis and the venom from *Loxosceles recluse* (brown recluse spider). Specifically, the main component of this venom is sphingomyelinase D (SMase D), which hydrolyzes sphingomyelin to produce C1P. The pathology of a wound generated from the bite of this spider is that of an intense inflammatory response mediated by AA and prostaglandins. The production of endogenous C1P by the action of SMase D suggested the possibility of C1P acting as a pathophysiologic link in the activation of cPLA$_2$ and the inflammatory response mediated by AA and prostaglandins. Initial evidence for the validation of this hypothesis came from an investigation focused on the regulation of prostanoid synthesis, which showed that the CERK/C1P pathway was

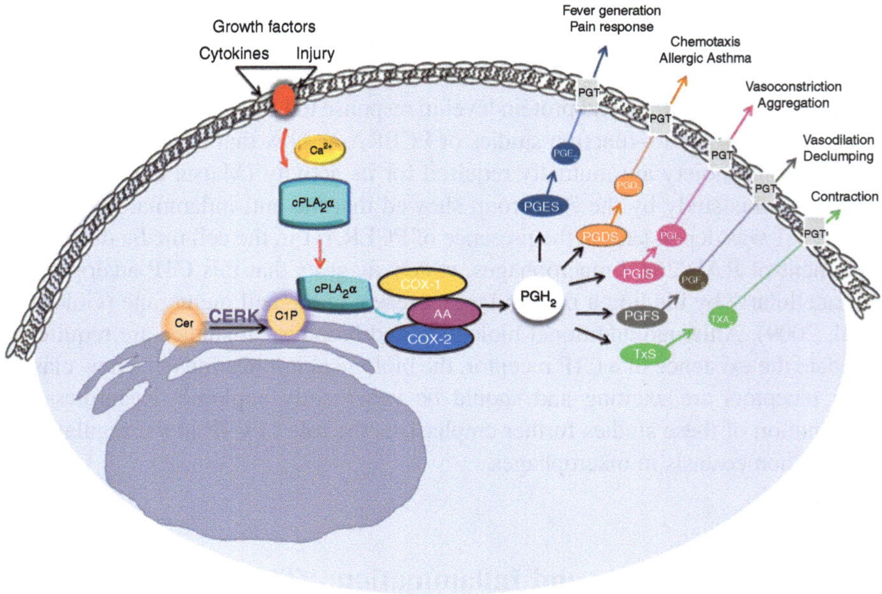

Fig. 1 The interaction between C1P and cPLA$_2\alpha$ is a crucial link in eicosanoid synthesis. Following an inflammatory stimulus, Ca^{2+} activated cPLA$_2\alpha$ translocates to the golgi membrane where it binds phosphatidylcholine (PC). CERK-derived C1P directly interacts with cPLA$_2\alpha$, thereby enhancing the association of cPLAα to the PC-rich membrane. cPLA$_2\alpha$ hydrolyzes PC to produce arachidonic acid (AA), which is further metabolized to several different eicosanoids, one of which is prostaglandin (PGH$_2$). Prostaglandins are involved in various biological processes associated with the inflammatory response

required for PLA$_2$ activation in response to calcium ionophore and cytokines (Pettus et al. 2003). Subsequent in vitro studies by our laboratory confirmed these findings, pointing to C1P as a direct activator of cPLA$_2$ through interaction with the C2/CaLB domain (Pettus et al. 2004). Collectively, these findings provided evidence for C1P as the "missing link" in the eicosanoid synthetic pathway (Fig. 1).

The Chalfant group forged forward with mechanistic studies demonstrating that the interaction of C1P and cPLA$_2\alpha$ was very specific, as closely related lipids and metabolites were unable to activate cPLA$_2\alpha$ in vitro and in cells (Pettus et al. 2004; Subramanian et al. 2005; Wijesinghe et al. 2009). The interaction site for C1P within cPLA$_2\alpha$ was characterized in depth over the course of several years, and these investigations provided evidence that C1P interacted with cPLA$_2\alpha$ at the C2 domain via a novel and previously undescribed interaction site (Pettus et al. 2004; Subramanian et al. 2005; Stahelin et al. 2007) (Fig. 1). C1P activates cPLA$_2\alpha$ by decreasing the dissociation constant of cPLA$_2\alpha$ with membranes, and by acting in a manner similar to a positive allosteric activator (Subramanian et al. 2005, 2007). The C1P/cPLA$_2\alpha$ paradigm was subsequently tested in cells, which confirmed that this interaction is required for translocation of cPLA$_2\alpha$ and subsequent production of eicosanoids in response to several inflammatory agonists (Lamour et al. 2009) (Fig. 1).

In addition to affecting the biochemical pathways through direct interaction with effector proteins like $cPLA_2\alpha$, C1P has also been surmised to induce indirect effects via changes to the structure of resident membranes. Biophysical studies by Kooijman et al. (2008, 2009) have demonstrated that C1P has the potential to alter membrane curvature, membrane fluidity, and membrane electrostatics. Specifically, negative curvature of membranes was observed upon incorporation of C1P into glycerophospholipids, which in turn induced the formation of non-lamellar structures (Kooijman et al. 2008). This property was observed even at membrane concentrations of C1P as low as <1 % with large portions of membranes in non-lamellar formations at 5 mol% (Kooijman et al. 2008). Formation of such non-lamellar structures was found to be important in events such as membrane protein insertion (Alonso et al. 2000; Martin et al. 2004) and membrane fusion (Goni and Alonso 2000; Chernomordik et al. 2006), and also has the potential to influence lipid signaling (van Blitterswijk et al. 2003; Kolesnick et al. 2000). Further studies by Koojiman et al. (2009) demonstrated that C1P affected membrane structure in a pH-dependent fashion. Specifically, C1P has a pKa_2 of 7.39, and at a slightly acidic pH of around 6 (which is found in many subcellular locations), C1P was observed to form a highly ordered crystalline structure via extensive intermolecular hydrogen bonding (Koojiman et al. 2009). However, at physiologic pH approaching its pKa_2, a high proportion of the phosphomonoester moieties of C1P are di-anionic resulting in significant repulsion among C1P molecules leading to a more diffused arrangement (Koojiman et al. 2009). In addition to pH, Ca^{2+} was also shown to affect the membrane organization of C1P by masking the negative charge, dehydrating the phosphomonoester head group, or linking different C1P molecules together (Koojiman et al. 2009). Additional biophysical studies have demonstrated that C1P formation has the potential to inhibit or reverse the formation of gel-like ceramide domains (Morrow et al. 2009). This finding has implications for a role of C1P in the destabilization of ceramide-rich lipid rafts. Thus C1P, although a relatively simple sphingolipid, has the potential to influence multiple aspects of biological membranes in a pH- and Ca^{2+}-dependent manner, which in turn has the potential to affect a significant array of cellular functions. Indeed, these membrane effects and/or changes in C1P structure may explain the stoichiometry of $cPLA_2\alpha$ activation by C1P of >4 molecules per micelle. The possibility of $cPLA_2\alpha$ or C1P interacting proteins associating with a specific C1P structure induced by localized pH or Ca^{2+} changes is an intriguing hypothesis to explore.

The connection between CERK/C1P and $cPLA_2\alpha$ activation is also of great interest with regard to disease states involving unnecessary eicosanoid production. In opposition to this link, recent studies published by the Bornancin group using an in vivo disease model for rheumatoid arthritis showed that CERK−/− mice were unaffected and responded similar to wild-type (WT) mice. These studies also suggested that $cPLA_2\alpha$-dependent pathways are unchanged in CERK-deficient mice (Graf et al. 2008). However, this study also found that while the CERK-deficient mice lacked CERK activity, $C_{18:1/16:0}$C1P was still present at significant levels with minor effects on total C1P. In an independent study utilizing a separate CERK−/− mouse, Igarashi and coworkers confirmed a minor effect on total C1P levels (Mitsutake et al. 2007). These data implicate an alternate mechanism for C1P synthesis and the possibility that the CERK-deficient mice have developed a compensatory system adapting biologically to the lack of CERK.

In contrast to the idea that $cPLA_2\alpha$-dependent pathways are completely functional in the CERK−/− mouse, Niwa et al. (2010) demonstrated that PGE_2 levels were reduced in the bronchoalveolar fluid of CERK-deficient mice compared to WT mice, thus providing initial in vivo evidence supporting a role for CERK and its product, C1P, in the regulation of eicosanoid synthesis in vivo. Furthermore, recent studies from our laboratory show that ablation of CERK results in a significant dysregulation/dysfunction in basal eicosanoid synthesis in ex vivo cells (Mietla et al. unpublished observation). In light of these latest findings, the CERK-deficient mouse model may only be partially adapted when cells are removed from the animal model. Why the CERK−/− mouse does not show the same phenotype observed in the $cPLA_2\alpha$−/− mouse is a conundrum, but the relatively unaffected levels of C1P are a likely rationale. Indeed, CERK siRNA-mediated knockout in cultured cell lines results in a significant and major reduction in total C1P levels in contrast to cell from the CERK−/− mouse (Wijesinghe et al. 2010). The phenotypic adaptation of CERK−/− mice in vivo may also be due to the higher levels of C1P found in the serum of these animals, which corresponds to the normal C1P levels found in the liver cells from the CERK−/− mouse. Regardless, the Chalfant laboratory has now created a $cPLA_2\alpha$ knock-in mouse, to overcome the controversy surrounding the CERK−/− model and the role of CERK-derived C1P in eicosanoid synthesis/inflammation disorders. This new mouse model expresses a $cPLA_2\alpha$ dysfunctional for the C1P interaction and will hopefully determine whether the C1P/$cPLA_2\alpha$ interaction plays a role in inflammatory phenotypes in vivo. By examining the C1P/$cPLA_2\alpha$ interaction directly in an in vivo model the compensatory mechanisms for C1P production activated in the CERK−/− mouse will be circumvented.

While the C1P/$cPLA_2\alpha$ interaction has been a recent major focus in the C1P research field, it is not the only function for C1P in the inflammatory response. Recent work from the Chalfant laboratory has implicated an additional role for C1P as a potential regulator of cytokine secretion, specifically TNFα, as described below (Lamour et al. 2011). TNFα is a major mediator of systemic and acute inflammation, and TNFα secretion is a pro-inflammatory event occurring in response to invading microbes. Unfortunately, dysregulation of this process results in hyper-activation of the immune response accompanied by an unnecessary amount of TNFα production and lethal tissue damage most commonly described as septic shock or sepsis (Lin and Yeh 2005). In addition to sepsis, excessive TNFα production has been linked to other diseases such as rheumatoid arthritis and cancer (Feldmann and Maini 2008; Sethi et al. 2008).

The TNFα protein is synthesized as a membrane-bound pro-peptide (Pro-TNFα) (Kriegler et al. 1988). Proteolytic cleavage of TNFα releases the active C-terminal portion from the cell surface, thus producing the secreted/soluble form of TNFα, which mediates the recruitment of subsequent activation of inflammatory cells to infected tissues or to the site of injury (Old 1988). Several enzymes have been

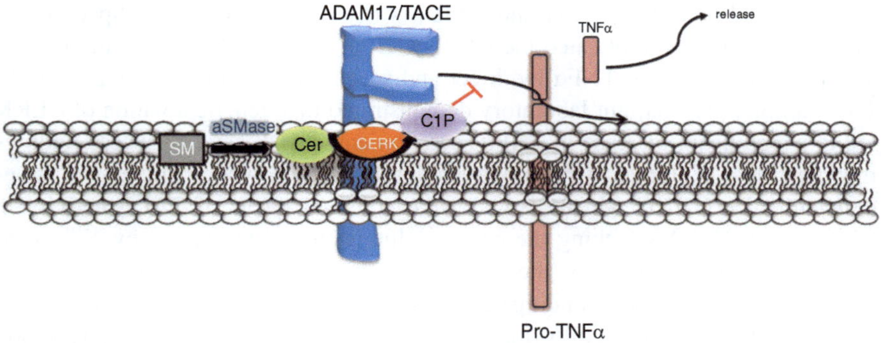

Fig. 2 Prospective role of C1P in TNFα production via direct inhibition of TACE. ADAM17/TACE is the major metalloprotease responsible for cleaving or "shedding" mature TNFα (pro-TNFα) to release the active soluble form. Recent studies have demonstrated a direct interaction between C1P and TACE, which inhibits the sheddase activity and hinders the ability of TACE to release active TNFα. Current investigations are focused on identifying the TACE residues that mediate the interaction of the enzyme with C1P

implicated in the processing of TNF, a posttranslational protease-mediated mechanism that has been described as "ectodomain shedding" (Blobel 2000). Specifically, a member of A disintegrin and metalloprotease family (ADAM), ADAM17, has extensively been shown to act as the "sheddase" for TNFα, hence the more common name, TNFα-converting enzyme (TACE) (Moss et al. 1997; reviewed in Black 2002). The sheddase activity of TACE plays a critical role in the regulation of TNFα activation via the direct cleavage of pro-TNFα, thus releasing TNFα (Fig. 2). Indeed, TACE was demonstrated to be the major TNFα sheddase in response to endotoxin stimulation (Horiuchi et al. 2007). Moreover, Blobel and coworkers demonstrated that mice bearing a temporal and conditional inactivation of TACE resulted in significantly decreased serum TNFα levels and were protected from LPS-stimulated endotoxin shock (Horiuchi et al. 2007). Due to the involvement of excessive TNFα release in sepsis and several inflammatory-associated diseases, there is a significant amount of interest in developing therapeutic strategies that can alter TNFα shedding. Thus, TACE sheddase activity has become an attractive target for novel anti-TNFα therapies. Along these lines, the mechanisms regarding the induction or termination of TACE activity following a cellular insult, such as LPS-stimulation, are presently unclear.

In regard to C1P playing a role in TNFα secretion, two recent studies have reported findings that suggest a regulatory role for the sphingolipids, ceramide and C1P, in TNFα secretion following LPS stimulation. Initially, studies by Rozenova et al. (2010) demonstrated that A-SMase may act as a regulator of posttranslational processing of TNFα via inhibition of TACE in LPS-stimulated macrophages, providing mechanistic insight for the previous finding that A-SMase-deficient mice were partially protected from the tissue-damaging effects of LPS (Haimovitz-Friedman et al. 1997). The involvement of the sphingolipid pathway in the generation of soluble TNFα was further corroborated in immortalized mouse embryonic fibroblasts (MEFs) and BMDMs (Lamour et al. 2011). Specifically, these recent studies implicated the CERK/C1P pathway in the processing of pro-TNFα to soluble/active TNFα by direct and specific inhibition of TACE (Fig. 2). For example, the Chalfant laboratory demonstrated that genetic ablation of CERK led to a significant increase in TNFα secretion and TACE activity. These findings mirrored those of studies by Nikolova-Karakashian and coworkers in the A-SMase-deficient mice (2010), which also showed increased levels of TACE activity in LPS-treated BMDMs lacking the enzyme. Interestingly, the report by Nikolova-Karakashian and coworkers showed that ceramide did not directly affect the activity of TACE. These findings led to the hypothesis that C1P may directly regulate TACE activity. Following the C1P/cPLA$_2$α model, plausible interaction sites in TACE for C1P were identified, demonstrating that C1P potently and directly inhibited TACE enzyme activity (Lamour et al. 2011). This effect of C1P was specific, as closely related lipids such as ceramide and S1P could not recapitulate this inhibitory effect. While these studies propose a different sphingolipid as an essential component of TNFα production, they clearly depict the same theory, which is the existence of a sphingolipid that regulates the posttranslational processing and secretion of TNFα via the interruption of TACE enzymatic activity. Furthermore, C1P is a direct metabolite of ceramide suggesting that CERK utilizes ceramide derived from A-SMase. Future characterization of the C1P/TACE interaction may explain the anti-inflammatory effects of some C1P analogs. For example, the C1P analogs, pCERA1 and ONO-SM-362, inhibit the production of TNFα in animal models (Avni et al. 2009; Ogata et al. 2008; Goldsmith et al. 2009). Validation of the C1P interaction site within TACE would be a substantial achievement and would allow for testing of the hypothesis that C1P analogs are blocking TNFα production in cells and in vivo via direct inhibition of TACE enzymatic activity. This interaction could perhaps be a future therapeutic target for treating sepsis in addition to chronic inflammatory diseases.

Our current understanding clearly emphasizes the role of the CERK/C1P pathway in the regulation of membrane-bound proteins with a considerable amount of involvement in fundamental inflammatory processes. The establishment and characterization of the C1P/cPLA$_2$α interaction have provided investigators with a representative mechanism that can be used to study additional potential C1P interacting proteins. Furthermore, in vivo relevance of this interaction in inflammatory disease phenotypes is still unclear, but the current subject is under intense investigation by many laboratory groups.

References

Alonso A, Goni FM, Buckley JT (2000) Lipids favoring inverted phase enhance the ability of aerolysin to permeabilize liposome bilayers. Biochemistry 39(46):14019–14024

Avni D, Goldsmith M, Ernst O, Mashiach R, Tuntland T, Meijler MM, Gray NS, Rosen H, Zor T (2009) Modulation of TNFalpha, IL-10 and IL-12p40 levels by a ceramide-1-phosphate analog, PCERA-1, in vivo and ex vivo in primary macrophages. Immunol Lett 123(1):1–8

Bajjalieh SM, Martin TF, Floor E (1989) Synaptic vesicle ceramide kinase. A calcium-stimulated lipid kinase that co-purifies with brain synaptic vesicles. J Biol Chem 264(24):14354–14360

Black RA (2002) Tumor necrosis factor-alpha converting enzyme. Int J Biochem Cell Biol 34(1): 1–5

Blobel CP (2000) Remarkable roles of proteolysis on and beyond the cell surface. Curr Opin Cell Biol 12(5):606–612

Carre A, Graf C, Stora S, Mechtcheriakova D, Csonga R, Urtz N, Billich A, Baumruker T, Bornancin F (2004) Ceramide kinase targeting and activity determined by its N-terminal pleckstrin homology domain. Biochem Biophys Res Commun 324(4):1215–1219

Chernomordik LV, Zimmerberg J, Kozlov MM (2006) Membranes of the world unite! J Cell Biol 175(2):201–207

Chen WQ, Graf C, Zimmel D, Rovina R, Krapenbauer K, Jaritz M, Parker PJ, Lubec G, Bornancin F (2010) Ceramide kinase profiling by mass spectrometry reveals a conserved phosphorylation pattern downstream of the catalytic site. J Proteome Res 9(1):420–9

Dressler KA, Kolesnick RN (1990) Ceramide 1-phosphate, a novel phospholipid in human leukemia (HL-60) cells. Synthesis via ceramide from sphingomyelin. J Biol Chem 265(25): 14917–14921

Feldmann M, Maini SR (2008) Role of cytokines in rheumatoid arthritis: an education in pathophysiology and therapeutics. Immunol Rev 223:7–19

Gangoiti P, Granado MH, Arana L, Ouro A, Gomez-Munoz A (2008a) Involvement of nitric oxide in the promotion of cell survival by ceramide 1-phosphate. FEBS Lett 582(15):2263–2269

Gangoiti P, Granado MH, Wang SW, Kong JY, Steinbrecher UP, Gomez-Munoz A (2008b) Ceramide 1-phosphate stimulates macrophage proliferation through activation of the PI3-kinase/PKB, JNK and ERK1/2 pathways. Cell Signal 20(4):726–736

Gangoiti P, Granado MH, Arana L, Ouro A, Gomez-Munoz A (2010) Activation of protein kinase C-alpha is essential for stimulation of cell proliferation by ceramide 1-phosphate. FEBS Lett 584(3):517–524

Gangoiti P, Bernacchioni C, Donati C, Cencetti F, Ouro A, Gomez-Munoz A, Bruni P (2012) Ceramide 1-phosphate stimulates proliferation of C2C12 myoblasts. Biochimie 94(3):597–607

Goldsmith M, Avni D, Levy-Rimler G, Mashiach R, Ernst O, Levi M, Webb B, Meijler MM, Gray NS, Rosen H, Zor T (2009) A ceramide-1-phosphate analogue, PCERA-1, simultaneously suppresses tumour necrosis factor-alpha and induces interleukin-10 production in activated macrophages. Immunology 127(1):103–115

Gomez-Munoz A (1998) Modulation of cell signalling by ceramides. Biochim Biophys Acta 1391 (1):92–109

Gomez-Munoz A, Duffy PA, Martin A, O'Brien L, Byun HS, Bittman R, Brindley DN (1995) Short-chain ceramide-1-phosphates are novel stimulators of DNA synthesis and cell division: antagonism by cell-permeable ceramides. Mol Pharmacol 47(5):833–839

Gomez-Munoz A, Frago LM, Alvarez L, Varela-Nieto I (1997) Stimulation of DNA synthesis by natural ceramide 1-phosphate. Biochem J 325(Pt 2):435–440

Gomez-Munoz A, Kong JY, Salh B, Steinbrecher UP (2004) Ceramide-1-phosphate blocks apoptosis through inhibition of acid sphingomyelinase in macrophages. J Lipid Res 45(1): 99–105

Gomez-Munoz A, Kong JY, Parhar K, Wang SW, Gangoiti P, Gonzalez M, Eivemark S, Salh B, Duronio V, Steinbrecher UP (2005) Ceramide-1-phosphate promotes cell survival through activation of the phosphatidylinositol 3-kinase/protein kinase B pathway. FEBS Lett 579(17): 3744–3750

Goni FM, Alonso A (2000) Membrane fusion induced by phospholipase C and sphingomyelinases. Biosci Rep 20(6):443–463

Graf C, Zemann B, Rovina P, Urtz N, Schanzer A, Reuschel R, Mechtcheriakova D, Muller M, Fischer E, Reichel C, Huber S, Dawson J, Meingassner JG, Billich A, Niwa S, Badegruber R, Van Veldhoven PP, Kinzel B, Baumruker T, Bornancin F (2008) Neutropenia with impaired immune response to streptococcus pneumoniae in ceramide kinase-deficient mice. J Immunol 180(5):3457–66

Granado MH, Gangoiti P, Ouro A, Arana L, Gonzalez M, Trueba M, Gomez-Munoz A (2009) Ceramide 1-phosphate (C1P) promotes cell migration Involvement of a specific C1P receptor. Cell Signal 21(3):405–412

Haimovitz-Friedman A, Cordon-Cardo C, Bayoumy S, Garzotto M, McLoughlin M, Gallily R, Edwards CK 3rd, Schuchman EH, Fuks Z, Kolesnick R (1997) Lipopolysaccharide induces disseminated endothelial apoptosis requiring ceramide generation. J Exp Med 186(11): 1831–1841

Horiuchi K, Kimura T, Miyamoto T, Takaishi H, Okada Y, Toyama Y, Blobel CP (2007) Cutting edge: TNF-alpha-converting enzyme (TACE/ADAM17) inactivation in mouse myeloid cells prevents lethality from endotoxin shock. J Immunol 179(5):2686–2689

Kim TJ, Mitsutake S, Kato M, Igarashi Y (2005) The leucine 10 residue in the pleckstrin homology domain of ceramide kinase is crucial for its catalytic activity. FEBS Lett 579(20):4383–4388

Kim TJ, Mitsutake S, Igarashi Y (2006) The interaction between the pleckstrin homology domain of ceramide kinase and phosphatidylinositol 4,5-bisphosphate regulates the plasma membrane targeting and ceramide 1-phosphate levels. Biochem Biophys Res Commun 342(2):611–617

Kolesnick RN, Hemer MR (1990) Characterization of a ceramide kinase activity from human leukemia (HL-60) cells. Separation from diacylglycerol kinase activity. J Biol Chem 265(31): 18803–18808

Kolesnick RN, Goni FM, Alonso A (2000) Compartmentalization of ceramide signaling: physical foundations and biological effects. J Cell Physiol 184(3):285–300

Kooijman EE, Sot J, Montes LR, Alonso A, Gericke A, de Kruijff B, Kumara S, Koni FM (2008) Membrane organization and ionization behavior of the minor but crucial lipid ceramide-1-phosphate. Biophys J 94(11):4320–4330

Koojiman EE, Vaknin D, Bu W, Joshi L, Kang SW, Gericke A, Mann EK, Kumar S (2009) Structure of ceramide-1-phosphate at the air-water solution interface in the absence and presence of Ca^{2+}. Biophys J 96(6):2204–2215

Kriegler M, Perez C, DeFay K, Albert I, Lu SD (1988) A novel form of TNF/cachectin is a cell surface cytotoxic transmembrane protein: ramifications for the complex physiology of TNF. Cell 53(1):45–53

Lamour NF, Subramanian P, Wijesinghe DS, Stahelin RV, Bonventre JV, Chalfant CE (2009) Ceramide 1-phosphate is required for the translocation of group IVA cytosolic phospholipase A2 and prostaglandin synthesis. J Biol Chem 284(39):26897–26907

Lamour NF, Wijesinghe DS, Mietla JA, Ward KE, Stahelin RV, Chalfant CE (2011) Ceramide kinase regulates the production of tumor necrosis factor alpha (TNFalpha) via inhibition of TNFalpha-converting enzyme. J Biol Chem 286(50):42808–42817

Lin WJ, Yeh WC (2005) Implication of toll-like receptor and tumor necrosis factor alpha signaling in septic shock. Shock 24(3):206–209

Martin C, Requero MA, Masin J, Konopasek I, Goni FM, Sebo P, Ostolaza H (2004) Membrane restructuring of Bordetella pertusis adenylatecyclase toxin, a member fo the RTX toxin family. J Bacteriol 186(12):3760–3765

Matsui T, Kondo T, Nishita Y, Itadani S, Nakatani S, Omawari N, Sakai M, Nakazawa S, Ogata A, Mori H, Terai K, Kamoshima W, Ohno H, Obata T, Nakai H, Toda M (2002a) Highly potent inhibitors of TNF-alpha production. Part I: discovery of new chemical leads and their structure-activity relationships. Bioorg Med Chem 10(12):3757–3786

Matsui T, Kondo T, Nishita Y, Itadani S, Tsuruta H, Fujita S, Omawari N, Sakai M, Nakazawa S, Ogata A, Mori H, Ohno H, Obata T, Nakai H, Toda M (2002b) Highly potent inhibitors of TNF-alpha production. Part 2: identification of drug candidates. Bioorg Med Chem Lett 12(6): 907–910

Mitsutake S, Igarashi Y (2005) Calmodulin is involved in the Ca2 + -dependent activation of ceramide kinase as a calcium sensor. J Biol Chem 280(49):40436–40441

Mitsutake S, Yokose U, Kato M, Matsuoka I, Yoo JM, Kim TJ, Yoo HS, Fujimoto K, Ando Y, Sugiura M, Kohama T, Igarashi Y (2007) The generation and behavioral analysis of ceramide kinase-null mice, indicating a function in cerebellar purkinje cells. Biochem Biophys Res Comm 363(3):519–524

Morrow MR, Helle A, Perry J, Vattulainen I, Weidmer SK, Holopainen JM (2009) Ceramide-1-phosphate, in contrast to ceramide, is not segregrated into lateral lipid domains in phosphatidylcholine bilayers. Biophys J 96(6):2216–2226

Moss ML, Jin SL, Milla ME, Bickett DM, Burkhart W, Carter HL, Chen WJ, Clay WC, Didsbury JR, Hassler D, Hoffman CR, Kost TA, Lambert MH, Leesnitzer MA, McCauley P, McGeehan G, Mitchell J, Moyer M, Pahel G, Rocque W, Overton LK, Schoenen F, Seaton T, Su JL, Becherer JD et al (1997) Cloning of a disintegrin metalloproteinase that processes precursor tumour-necrosis factor-alpha. Nature 385(6618):733–736

Murakami M, Nakatani Y, Kudo I (1996) Type II secretory phospholipase A2 associated with cell surfaces via C-terminal heparin-binding lysine residues augments stimulus-initiated delayed prostaglandin generation. J Biol Chem 271(47):30041–30051

Niwa S, Urtz N, Baumruker T, Billich A, Bornancin F (2010) Ovalbumin-induced plasma interleukin-4 levels are reduced in ceramide kinase-deficient DO11.10 RAG1−/− mice. Lipids Health Dis 9:1

Ogata T, Yamashita K, Horiuchi H, Okuda K, Todo S (2008) A novel tumor necrosis factor-alpha suppressant, ONO-SM362, prevents liver failure and promotes liver regeneration after extensive hepatectomy. Surgery 143(4):545–555

Old LJ (1988) Tumor necrosis factor. Sci Am 258(5):59–60, 69–75.

Pettus BJ, Bielawska A, Spiegel S, Roddy P, Hannun YA, Chalfant CE (2003) Ceramide kinase mediates cytokine- and calcium ionophore-induced arachidonic acid release. J Biol Chem 278(40):38206–38213

Pettus BJ, Bielawska A, Subramanian P, Wijesinghe DS, Maceyka M, Leslie CC, Evans JH, Freiberg J, Roddy P, Hannun YA, Chalfant CE (2004) Ceramide 1-phosphate is a direct activator of cytosolic phospholipase A2. J Biol Chem 279(12):11320–11326

Pollard JW (2004) Tumour-educated macrophages promote tumour progression and metastasis. Nat Rev Cancer 4(1):71–78

Rozenova KA, Deevska GM, Karakashian AA, Nikolova-Karakashian MN (2010) Studies on the role of acid sphingomyelinase and ceramide in the regulation of tumor necrosis factor alpha (TNFalpha)-converting enzyme activity and TNFalpha secretion in macrophages. J Biol Chem 285(27):21103–21113

Sethi G, Sung B, Aggarwal BB (2008) TNF: a master switch for inflammation to cancer. Front Biosci 13:5094–5107

Stahelin RV, Subramanian P, Vora M, Cho W, Chalfant CE (2007) Ceramide-1-phosphate binds group IVA cytosolic phospholipase a2 via a novel site in the C2 domain. J Biol Chem 282(28): 20467–20474

Subramanian P, Stahelin RV, Szulc Z, Bielawska A, Cho W, Chalfant CE (2005) Ceramide 1-phosphate acts as a positive allosteric activator of group IVA cytosolic phospholipase A2 alpha and enhances the interaction of the enzyme with phosphatidylcholine. J Biol Chem 280(18):17601–17607

Subramanian P, Vora M, Gentile LB, Stahelin RV, Chalfant CE (2007) Anionic lipids activate group IVA cytosolic phospholipase A2 via distinct and separate mechanism. J Lipid Res 48(12):2701–2708

Sugiura M, Kono K, Liu H, Shimizugawa T, Minekura H, Spiegel S, Kohama T (2002) Ceramide kinase, a novel lipid kinase. Molecular cloning and functional characterization. J Biol Chem 277(26):23294–23300

van Blitterswijk WJ, van der Luit AH, Veldman RJ, Verheij M, Borst J (2003) Ceramide: second messenger or modulator of membrane structure and dynamics? Biochem J 369(Pt 2):199–211

Wijesinghe DS, Subramanian P, Lamour NF, Gentile LB, Granado MH, Bielawska A, Szulc Z, Gomez-Munoz A, Chalfant CE (2009) Chain length specificity for activation of cPLA2alpha by C1P: use of the dodecane delivery system to determine lipid-specific effect. J Lipid Res 50(10):1986–1995

Wijesinghe DS, Allegood JC, Gentile LB, Fox TE, Kester M, Chalfant CE (2010) Use of high performance liquid chromatography-electrospray ionization-tandem mass spectrometry for the analysis of ceramide-1-phosphate levels. J Lipid Res 51(3):641–651

Yedgar S, Krimsky M, Cohen Y, Flower RJ (2007) Treatment of inflammatory diseases by selective eicosanoid inhibition: a double-edged sword? Trends Pharmacol Sci 28(9):459–464

Part II
Drug Development

Functional Inhibitors of Acid Sphingomyelinase (FIASMAs)

Johannes Kornhuber, Philipp Tripal, Erich Gulbins, and Markus Muehlbacher

Contents

Abstract Sphingolipids are not only structural components of biological membranes, but also play an important role in cellular signalling and, thus, are involved in cell proliferation and differentiation but also stress and cell death. It is therefore of great clinical relevance to define inhibitors of the enzymes involved in sphingolipid metabolism. Here, we describe the state of the art of functional inhibitors of the acid sphingomyelinase. The acid sphingomyelinase converts sphingomyelin to ceramide, a compound often involved in cell stress.

J. Kornhuber (✉) • P. Tripal • M. Muehlbacher
Department of Psychiatry and Psychotherapy, Friedrich-Alexander-University of Erlangen, Schwabachanlage 6, D 91054 Erlangen, Germany
e-mail: johannes.kornhuber@uk-erlangen.de

E. Gulbins
Department of Molecular Biology, University of Duisburg-Essen, Hufelandstrasse 55, D 45122 Essen, Germany

E. Gulbins and I. Petrache (eds.), *Sphingolipids: Basic Science and Drug Development*, Handbook of Experimental Pharmacology 215, DOI 10.1007/978-3-7091-1368-4_9,

We describe the structural and physicochemical properties, the distribution, the pharmacokinetics, the pharmocodynamics and the clinical use of direct and functional inhibitors of the acid sphingomyelinase.

Keywords Acid sphingomyelinase • Functional inhibitors • Ceramide • Structure–property-activity relation

1 Introduction

Many clinically applicable drugs function via the "lock and key" principle. In this model, the specific structure of a drug is of utmost importance in determining its interaction with target proteins such as receptors or enzymes and is also important for obtaining the desired pharmacological effects. In this article, we characterize a large group of compounds with weak basic and amphiphilic properties that functionally inhibit an important lysosomal enzyme, the acid sphingomyelinase (ASM, EC 3.1.4.12), by a mechanism different to the "lock and key" principle. These drugs function via physicochemical mechanisms based on protonation state and electrostatic interactions. Consequently, these drugs are structurally diverse but share common physicochemical properties.

Eukaryotic cells contain membrane-bound acidic compartments called lysosomes, where ATP-driven proton pumps maintain the low pH within these organelles (see Fig. 1a). At physiological pH, weak basic and amphiphilic drugs consist of molecules in the protonated and unprotonated state. Only the molecules in the unprotonated state are able to cross biological membranes (Fischer et al. 2007). However, once inside the lysosomes, these drugs become highly protonated and are thus trapped inside the acidic compartment. This phenomenon is called "acidotropism" or "lysosomotropism" (De Duve et al. 1974; Trapp et al. 2008). Figure 1a illustrates how weak organic bases are enriched within the lysosome because of their basic nature. Since the enrichment of these compounds is based on their protonation equilibrium constant and their log P value, which describes the lipophilicity of the molecule in general, a physicochemical model to predict the accumulation of compounds within the lysosome has been established (Trapp et al. 2008). These weakly basic compounds may accumulate to high levels in the lysosomes because of the constant action of the proton pumps. Depending on the physicochemical properties of the drug, the concentration of drug inside the lysosome may be more than 100-fold higher than the concentration outside the lysosome, according to empirical results (Novelli et al. 1987) and the physicochemical cell-based prediction model (Trapp et al. 2008).

What are the consequences of such high lysosomal drug concentrations? Are these consequences related to the desired or undesired pharmacological effects of a drug? Only recently has the lysosomal sequestration phenomenon been exploited in the development of novel drug targeting strategies (Black and Percival 2006; Kaufmann and Krise 2007; Kornhuber et al. 2008, 2011), and further studies need to be performed to validate lysosomal sequestration as a viable target.

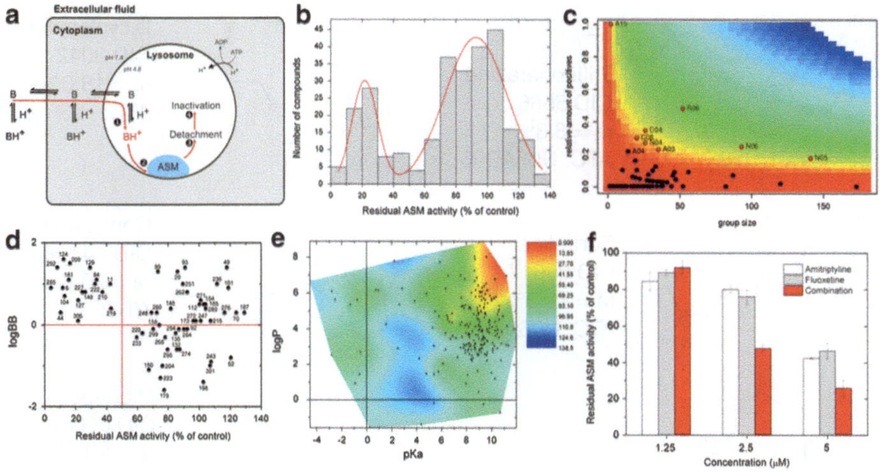

Fig. 1 (**a**) Sequential steps by which some lysosomotropic drugs result in functional inhibition of ASM. Lysosomal proton pumps maintain an acidic lysosomal pH value. Weak bases with lipophilic properties accumulate inside the lysosome; the lipophilic part of these weak bases partitions into the inner leaf of the lysosomal membrane. The basic part of the weak bases is protonated and charged due to the low pH inside the lysosome; these parts of the molecules do not partition into the membrane; instead they present their positive charge at the inner surface of the lysosomal membrane. This results in altered electrostatic properties of the inner leaf of the lysosomal membrane and detachment of the ASM. Subsequently, the ASM is inactivated in the lysosomal lumen most likely by proteolytic degradation. Reprint with permission from J Med Chem. 51; 219–237; Copyright 2008, American Chemical Society (Kornhuber et al. 2008). (**b**) The degree of functional inhibition of ASM is bimodally distributed, meaning that most drugs either act as FIASMAs or not. The figure is based on 276 compounds investigated. The red line represents a Gaussian fit to the *two peaks*. The figure is taken from (Kornhuber et al. 2011). (**c**) FIASMAs are unevenly distributed across the groups of drugs represented by the second level of the ATC drug classification system. The scatter-plot shows the relative amount of FIASMAs with a significance level coloring, where *red* indicates non-significance and *blue* extremely high significance. The significance levels are corrected for multiple testing. Only 8 of the 86 drug classes are significantly enriched with FIASMAs, namely, A03 (drugs for functional gastrointestinal disorders), A15 (appetite stimulants), C08 (calcium channel blockers), D04 (antipruritics), N04 (anti-Parkinson drugs), N05 (psycholeptics), N06 (psychoanaleptics), and R06 (antihistamines for systemic use). There are a number of other drug classes with high ratio of FIASMAs, such as A04. However, these classes do not reach significance because of the low group size. (**d**) The plot shows experimental data on functional inhibition of ASM in relation to blood–brain barrier (BBB) permeability. All compounds that do functionally inhibit ASM (i.e., residual ASM activity below 50 %) also cross the BBB (logBB values greater than 0). The figure is taken from (Kornhuber et al. 2011). (**e**) Contour-Plot for 241 compounds (*black dots*). ASM activity is color coded; *yellow*, *orange*, and *red* colors indicate high activity as functional inhibitor of ASM. It is evident that FIASMAs segregate in a narrow area of the log *P*-p*K*a-parameter space. Only those compounds combining both high log *P* and high p*K*a values do act as FIASMA. Basic p*K*a and log *P* values were calculated by ACD10 (ACD Inc., Toronto, ON). (**f**) FIASMAs work in an additive way. The combination of amitriptyline and fluoxetine, each at 2.5 μM, functionally inhibits ASM, while the drugs do not inhibit ASM when given alone. The figure is taken from (Kornhuber et al. 2011)

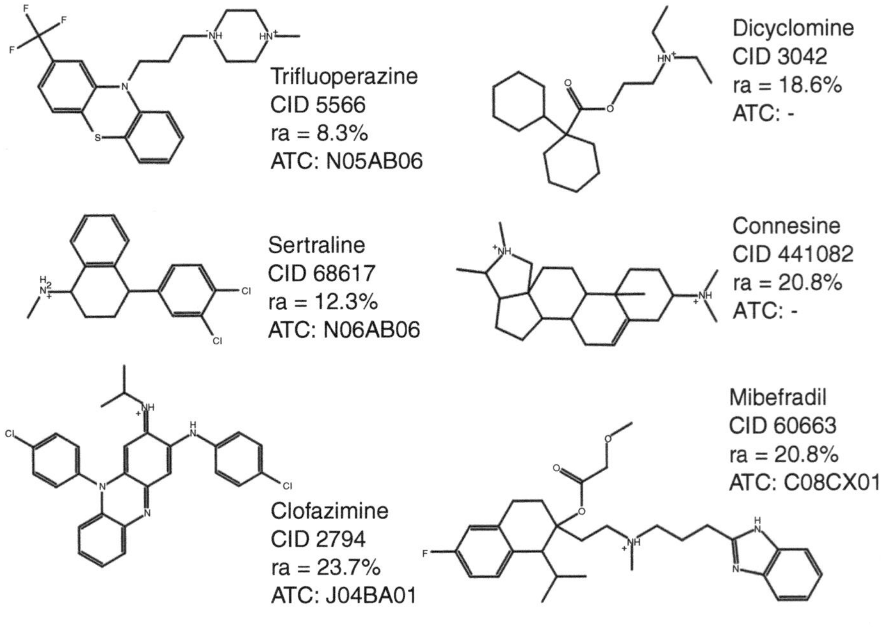

Fig. 2 FIASMAs are structurally diverse. Six randomly selected FIASMAs illustrate this diversity, as they are very different with respect to their molecular structure. It also highlights that FIASMAs occur in various ATC subgroups. *CID* PubChem's compound ID, *ra* residual ASM activity in percent of controls at 10 μM drug concentration, *ATC* anatomical therapeutic chemical drug classification system. See also Table 1 for further information

Here, we describe the accumulation of certain lysosomotropic drugs and their partition into the inner leaf of the lysosomal membrane, which results in the functional inhibition of ASM. Those compounds have the potential to reduce the activity of ASM, although they do not interact with ASM directly. Therefore, the acronym functional inhibitor of acid sphingomyelinase (FIASMA) has been proposed for this large and pharmacologically interesting group of compounds (Kornhuber et al. 2010).

The ASM is a lysosomal glycoprotein that catalyzes the hydrolysis of sphingomyelin into ceramide and phosphorylcholine. Ceramide is further metabolized to sphingosine and subsequently to sphingosine-1-phosphate. Ceramide and sphingosine-1-phosphate have antagonistic actions on cell proliferation, growth, survival, and apoptosis. High concentrations of ceramide in the cell membrane result in a reorganization of the signal transduction machinery by exclusion or the aggregation of receptors and ion channels from or into ceramide-rich membrane platforms (Bollinger et al. 2005). Both the functional inhibition of ASM and the clustering of ceramide are pure biophysical processes clearly different from the "lock and key" principle.

ASM first garnered attention for its involvement in Niemann–Pick disease, which is a lysosomal storage disorder caused by ASM deficiency (Brady et al. 1966). The pathological reduction of ASM activity in this disease is caused by a mutation in the gene coding for ASM, and the severity of Niemann–Pick disease is highly correlated with the residual activity of ASM (Schuchman 2007).

Studies using cells from patients with Niemann–Pick disease or from ASM knockout mice added novel information regarding the consequences of reduced ASM activity (Macauley et al. 2008). Remarkably, the deficiency of ASM may provide beneficial consequences, including anti-apoptotic and cytoprotective effects (Short 2010). In fact, some evidence suggests that ASM activation and ceramide accumulation play a central role in the progression of a broad range of common human diseases (Smith and Schuchman 2008). For instance, aberrant activation of ASM and/or altered levels of ceramide have been described in several psychiatric and neurological disorders, such as major depression (Kornhuber et al. 2005; Schwarz et al. 2008), morphine antinociceptive tolerance (Ndengele et al. 2009), Alzheimer's disease (Han et al. 2002; He et al. 2010; Malaplate-Armand et al. 2006), spinal cord injury (Cuzzocrea et al. 2009), and seizure disorder (Mikati et al. 2008), which suggests that these disorders may be treated by inhibiting ASM.

Activation of ASM itself is critical for several cellular responses, such as apoptosis induced by reactive oxygen and nitrogen species (Charruyer et al. 2005), chemotherapy drugs such as cisplatin (Rebillard et al. 2008), bacteria (Grassme et al. 2003), radiation (Santana et al. 1996), CD95 (Grassmé et al. 2001), or TNFα (Kim et al. 1991; Wiegmann et al. 1994). FIASMAs may therefore have broad clinical applications. For further details, please refer to the corresponding chapter within this book.

2 Direct Inhibitors of ASM

Until now, high-throughput screening assays to identify inhibitors of ASM were not successful in finding lead structures that inhibit ASM directly (Mintzer et al. 2005). It is even more difficult to find a general scaffold for drugs that are able to directly inhibit ASM because the crystal structure of the enzyme has not yet been published. Currently, the only information on the three-dimensional structure of ASM is a homology model based on the purple acid phosphatase, and this model only covers the active site of ASM (Seto et al. 2004). Consequently, only a few examples of direct ASM inhibitors have been reported (Arenz 2010). Although these compounds are able to inhibit ASM, they require localization at their target to be biologically active. Therefore, direct ASM inhibitors need to enter the lysosome to interact with ASM and be active under in vivo conditions. In contrast to functional inhibitors, direct inhibitors do not require high lysosomal drug concentrations as a prerequisite to effectively inhibit ASM. The concentration of the inhibitor needs to be of the same magnitude as the concentration of the enzyme, which is typically very low. Additional information regarding direct inhibitors of ASM is presented in the corresponding chapter and elsewhere (Arenz 2010).

3 Functional Inhibition of ASM

In addition to compounds that directly inhibit ASM, a broad group of compounds shows a high potential to inhibit ASM without evidence of direct inhibition. These compounds are called functional inhibitors and have been proposed to act via an alternative mechanism (Kornhuber et al. 2008). The inner lysosomal membrane surface is negatively charged because of the presence of the abundant phospholipid, bis(monoacylglycero)phosphate (BMP) (Gallala and Sandhoff 2011; Kolter and Sandhoff 2005; Wilkening et al. 1998). Positively charged proteins form electrostatic interactions with the negatively charged membrane and are subsequently shielded from lysosomal degradation (Hurwitz et al. 1994). In the case of native, undisturbed lysosomal metabolism, ASM is attached by electrostatic attraction to the inner lysosomal membrane. FIASMAs partition into the inner leaf of the lysosomal membrane with their lipophilic part and present their protonated and positively charged part to the lysosomal lumen, which results in a detachment of ASM from the lysosomal membrane and subsequent proteolytic degradation within the lysosomal lumen (see Fig. 1a) (Kölzer et al. 2004; Hurwitz et al. 1994). Detailed studies with one of these compounds, desipramine, showed that they facilitate the detachment of ASM from the inner lysosomal membrane. Thus, lysosomal accumulation of the drug is a prerequisite for ASM inhibition. Another prerequisite is the ability of the drug to integrate into the inner leaf of the lysosomal membrane and to detach the ASM from the lysosomal membrane.

4 Known FIASMAs

Various drugs have been tested for their ability to inhibit ASM in vitro. In a systematic approach, the conversion rate of radiolabeled ^{14}C-sphingomyelin by ASM has been investigated (Kornhuber et al. 2008, 2011). These studies showed that various drugs already approved by the Federal Drug Administration (FDA) are able to inhibit ASM, even at therapeutic concentrations. In total, 72 compounds reduced the activity of ASM by at least 50 % at 10 μM concentration and were therefore classified as FIASMAs (Kornhuber et al. 2011). Table 1 gives a comprehensive overview of FIASMAs experimentally identified to date. Remarkably, the distribution of residual ASM activity shows a bimodal distribution (see Fig. 1b). Therefore, most compounds tested clearly inhibited or did not inhibit ASM; only a few intermediate compounds were observed.

Table 1 Hitherto known FIASMAs (meaning that they reduce the ASM activity to below 50 % at a 10 μM concentration)

CID	Name	ATC short code	Log *P*	p*K*a	Residual ASM activity (%)
3678	Alverine	A03AX08	5.9	9.5	21.7
2157	Amiodarone	C01BD01	8.9	9.4	14.5
2160	Amitriptyline	N06AA09	4.9	9.2	11.7
2162	Amlodipine	C08CA01	4.2	9.0	12.0
2218	Aprindine		5.8	10.4	27.5
2247	Astemizole	R06AX11	5.8	9.0	14.3
9705	AY-9944		6.0	9.1	22.1
6832	Benztropine	N04AC01	5.0	10.5	12.7
2351	Bepridil	C08EA02	5.8	9.2	27.1
2381	Biperiden		4.0	9.8	26.2
5902	Camylofine	A03AA03	5.6	10.0	21.7
2585	Carvedilol	C07AG02	4.1	8.2	22.4
360849	Cepharanthine		5.2	7.6	9.2
2726	Chlorpromazine	N05AA01	5.2	9.4	42.4
667467	Chlorprothixene	N05AF03	6.1	9.1	22.4
1547484	Cinnarizine	N07CA02	4.6	6.5	48.9
26987	Clemastin		5.7	10.2	12.6
2794	Clofazimine	J04BA01	7.3	8.6	23.7
1548955	Clomiphene	G03GB02	8.0	9.6	13.0
2801	Clomipramine	N06AA04	5.4	9.5	21.8
2805	Cloperastine	R05DB21	5.2	8.9	26.7
441082	Connesine		5.7	10.2	20.8
2895	Cyclobenzaprine	M03BX08	5.0	9.1	26.2
2913	Cyproheptadine	A15AA01	6.4	8.9	22.2
2995	Desipramine	N06AA01	4.1	10.4	15.6
124087	Desloratadine	R06AX27	6.8	10.3	21.9
3042	Dicyclomine	A03AA07	6.1	9.2	18.6
3074	Dilazep	C01DX10	4.3	8.3	41.6
197033	Dimebon		4.0	9.1	44.1
667477	Doxepin	N06AA12	3.9	9.2	46.6
3166	Drofenine		5.6	9.2	20.8
10219	Emetine		4.9	8.9	0.4
3290	Ethopropazine	N04AA05	5.8	9.9	29.7
3336	Fendiline	C08EA01	4.9	9.5	25.2
941361	Flunarizine	N07CA03	4.7	6.4	32.7
3386	Fluoxetine	N06AB03	4.1	10.1	13.0
5281881	Flupenthixol	N05AF01	4.5	6.5	18.2
3372	Fluphenazine	N05AB02	4.8	6.8	16.5
5324346	Fluvoxamine	N06AB08	3.1	9.4	37.4
3658	Hydroxyzine		2.0	6.1	43.0
3696	Imipramine	N06AA02	4.8	9.5	32.6
3947	Lofepramine	N06AA07	7.0	6.5	19.2
3955	Loperamid	A07DA03	4.3	8.1	24.4
3957	Loratadine	R06AX13	5.9	4.8	48.5

(continued)

Table 1 (continued)

CID	Name	ATC short code	Log *P*	p*K*a	Residual ASM activity (%)
4011	Maprotiline	N06AA21	4.5	10.6	13.5
4031	Mebeverine	A03AA04	5.6	9.5	31.8
22530	Mebhydroline	R06AX15	4.1	9.1	41.9
60663	Mibefradil	C08CX01	6.3	9.3	20.8
4541	Norfluoxetine		4.4	9.1	22.5
4543	Nortriptyline	N06AA10	5.7	10.0	13.3
43815	Paroxetine	N06AB05	3.9	10.3	31.7
33630	Penfluridol	N05AG03	6.0	8.4	22.0
4746	Perhexiline	C08EX02	7.0	11.2	8.5
4748	Perphenazine	N05AB03	4.3	6.8	20.1
4822	Pimethixene	R06AX23	6.9	8.8	16.5
16362	Pimozide	N05AG02	6.4	9.4	30.6
4926	Promazin	N05AA03	4.6	9.4	33.6
4927	Promethazin	D04AA10	4.8	9.0	32.2
4976	Protriptyline	N06AA11	5.1	10.6	12.7
237	Quinacrine	P01AX05	4.8	10.5	44.3
60149	Sertindole	N05AE03	5.3	9.1	12.0
68617	Sertraline	N06AB06	4.8	9.5	12.3
442985	Solasodine		5.8	8.7	22.2
5354	Suloctidil	C04AX19	6.1	9.7	21.9
2733526	Tamoxifen	L02BA01	7.9	8.7	4.1
5405	Terfenadine	R06AX12	6.5	9.6	21.8
5452	Thioridazin	N05AC02	6.1	9.6	10.4
65576	Tomatidine		6.2	8.7	15.8
5566	Trifluoperazine	N05AB06	5.1	8.2	8.3
5568	Triflupromazine	N05AA05	5.7	9.4	29.5
5584	Trimipramine	N06AA06	5.2	9.4	13.8
91769	Zolantidine		5.0	8.9	21.6

All structures were obtained from the PubChem Web site (pubchem.ncbi.nlm.nih.gov/). Additionally, PubChem's compound ID (CID), ATC short code (taken from http://www.dimdi.de/static/de/amg/atcddd/index.htm), and the calculated log *P* and the calculated most basic p*K*a values are listed in the table. Log *P* and p*K*a values were calculated by ACD 10 (ACD Inc., Toronto, ON); ra represents the residual ASM activity. Please note that not every compound is classified by the ATC system

5 Distribution of FIASMAs Among Drugs and Natural Products

To render a detailed and objective picture of the distribution of FIASMAs with respect to their clinical application, all known FIASMAs were analyzed according to the anatomical therapeutic chemical (ATC) drug classification system, which is recommended by the World Health Organization (WHO) (World Health Organization 2011). However, not all ATC-classified drugs have been experimentally tested for their effect on ASM activity. Therefore, a prediction system has been developed to virtually screen large compound collections that is based on four computer-calculated properties and classifies compounds by a random forest model, giving a validated accuracy of 88 % (Kornhuber et al. 2011). Using this prediction system, the ATC database was virtually screened for FIASMAs. The distribution of FIASMAs with respect to their ATC code revealed a remarkable representation in specific therapeutic groups, namely, C08 (calcium channel blockers), D04 (antipruritics), N04 (anti-Parkinson drugs), A03 (drugs for functional gastrointestinal disorders), R06 (antihistamines for systemic use), N06 (psychoanaleptics), and N05 (psycholeptics) (Fig. 1c). An interesting common feature of ATC subgroups with overrepresented FIASMAs is the involvement of excitable cells, where many FIASMAs appear to primarily act. Over 40 % of the known FIASMAs belong to groups N04 (anti-Parkinson drugs), N05 (12 psycholeptics), and N06 (14 psychoanaleptics), which supports the hypothesis that FIASMAs do in fact cross the blood–brain barrier (BBB) (Fig. 1d) and that their clinical effects may be related to their functional inhibition of ASM (Kornhuber et al. 2005).

It is important to keep in mind that the size of the ATC drug groups varies widely and drug development often takes advantage of similar scaffolds that are known to be bioactive and/or have a high bioavailability. Thus, many approved drugs share common scaffolds that are known to exhibit good ADME properties. In general, groups with a high proportion of FIASMAs may be of general interest and may lead to the identification of novel FIASMAs or drugs with ASM side effects.

In contrast to drugs approved for medical use, FIASMAs are reported to be rare in natural products (Kornhuber et al. 2011). However, the chemical space of the small, drug-like molecules and natural products differs; therefore, these results must be interpreted with caution.

6 Association of FIASMAs with Other Relevant Biological and Generic Parameters

6.1 Blood–Brain Barrier

The BBB is a complex membrane system that separates the central nervous system (CNS) from the systemic blood circulation. It is formed by the endothelial cells that enclose cerebral microvessels (Bradbury 1993) and that develop tight junctions between adjacent endothelial cells. These tight junctions are able to block most molecular traffic across the BBB, which forces the traffic to take the transcellular route (Abbott and Romero 1996). Compounds entering the CNS via the transcellular route have to cross at least two membrane barriers, namely, the luminal and the abluminal plasma membranes, as well as the cytoplasm of the endothelial cells in CNS capillaries. Thus, crossing the BBB should be similar to localization within the lysosome because this process also involves the crossing of two membrane systems, the cellular and lysosomal membranes. Therefore, it can be assumed that the molecules that have to enter the CNS or the lysosomes should have similar molecular properties. In agreement with this assumption, it has been shown that FIASMAs tend to cross the BBB (Kornhuber et al. 2011). Figure 1d shows that all known FIASMAs with published quantitative data on BBB permeability, represented by their logBB values, are readily able to pass through the BBB.

It is widely known that many active transporters mediate the transport of certain substrates across the BBB (de Boer et al. 2003; de Lange 2004; Miller 2010). An interesting compound in this context is loperamide. Our experimental data classify loperamide as a FIASMA (Kornhuber et al. 2011) (Table 1). Unexpectedly, this drug does not appear in the brain (Baker 2007) because loperamide is a substrate of P-glycoprotein and is effectively transported out of the CNS (Baker 2007).

6.2 Lipinski's Rule-of-5

Lipinski's Rule-of-5 (Lo5) was established to identify properties of drugs that indicate bioavailability (Lipinski et al. 2001) by determining whether bulk molecules and drug-like molecules could be separated by a set of rules that are simple and easy to calculate. Remarkably, FIASMAs violate Lo5 more often than other drugs, mostly because of their high lipophilicity, as represented by their log *P* values (Kornhuber et al. 2011). This is noteworthy because FIASMAs are still highly effective at their target and also indicates that the requirements for general bioavailability, as estimated by Lo5 violations, differ significantly from the requirements for enrichment in an acidic compartment such as the lysosome, which is necessary for ASM inhibition.

6.3 Cationic Amphiphilic Drugs

Cationic amphiphilic drugs (*CADs*) form a special group of compounds that share at least two integral features: a cationic center and a lipophilic portion. The lipophilic part is necessary to pass through lipid bilayer membranes, but the cationic center is assumed to be a barrier against passive transport across a membrane system. In the case of weak organic bases, this hindrance is obviously linked to the degree of protonation. Altogether, CADs are assumed to be able to integrate into lipid bilayers without difficulty (Nussio et al. 2007). We suggest that a CAD should have a basic p*K*a of at least 7.4 because this value represents a functional group that is at least partially protonated at physiological pH. We also expect that a CAD has a log *P* of at least 3. Applying these two rules to the compounds listed in Table 1, the majority of compounds, i.e., 64 out of 72, are CADs. All hitherto known FIASMAs possess a basic nitrogen atom; none of the compounds without a basic nitrogen atom were able to functionally inhibit ASM. For these reasons, we examined the distribution of FIASMAs with respect to properties related to lipophilicity and the cationic character of the compound. When comparing FIASMAs to the inactive compounds (Kornhuber et al. 2011), the calculated log P and the pKa values clearly show a preferred range (FIASMAs: log $P = 5.35 \pm 1.13$, pKa $= 9.04 \pm 1.18$; inactives: log $P = 3.69 \pm 1.82$, pKa $= 7.76 \pm 3.07$). This is also evident from Fig. 1e.

6.4 Phospholipidosis

Phospholipidosis (PLD) is characterized by an excessive accumulation of phospholipids, which occurs mainly in lysosomes. Small drug-like compounds must be able to pass through two biological membranes, namely, the cellular membrane and the lysosomal membrane, to induce PLD. As this prerequisite is also found for FIASMAs, it is not unexpected that many, but not all, FIASMAs also induce PLD (Muehlbacher et al. 2012).

7 Molecular Properties Common to FIASMAs

A set of common properties can be derived from the evaluation of the known FIASMAs. There is a consensus that log P and pKa values are crucial for the functional inhibition of ASM (Kornhuber et al. 2005, 2010). A density plot of log P and pKa clearly shows that there is a preferential range for FIASMAs (see Fig. 1e) at higher log P values and more basic pKa values. Furthermore, other molecular properties, like size-intensive descriptors and the polar surface area, have been found to be important for a compound to act as a FIASMA (Kornhuber et al. 2011). For example, the molecular weight of a compound is often used to scale the effect of calculated properties with the size of the molecule (Purvis 2008). The polar surface area and properties closely correlated to the polar surface area are also frequently used in the context of biological activity (Clark 2011). In contrast, properties that avert effects on ASM can also be useful to exclude compounds within the virtual screen. An example of such a property that is incompatible with functional inhibition of ASM is an acidic pKa value; none of the acids tested experimentally so far do functionally inhibit ASM activity.

Table 2 summarizes several of the most important characteristics of FIASMAs.

Table 2 A list summarizing the general properties of FIASMAs

Structural and physicochemical properties	
Structure	FIASMAs are structurally diverse and include mono-, bi-, and tricyclic compounds (see Fig. 2). The structural diversity is also evident from hierarchical clustering of 2D-fingerprints of FIASMAs (Kornhuber et al. 2011).
Physicochemical properties	Functional inhibition of ASM requires specific physicochemical properties of a drug, namely, high log *P* and high basic p*K*a values, to accumulate within lysosomes and some further properties in order to result in detachment of ASM from the inner leaf of the lysosomal membrane (Kornhuber et al. 2011) (see Fig. 1a, e).
Cationic amphiphilic drugs	The majority of FIASMAs can be classified as cationic amphiphilic drugs (CADs). This also refers to the requirements to enrich within the lysosome, where the ASM is located.
Chemoinformatic prediction	Functional inhibition of ASM may be accurately predicted by binary random forest machine learning algorithms (Kornhuber et al. 2011).
Single-cell biodistribution	
High intralysosomal concentration	FIASMAs feature high intralysosomal concentrations, which are due to a weak basic center but still having a high log *P* value. To unfold activity in vivo, FIASMAs need to be able to accumulate within the lysosome.
Lysosomal accumulation kinetics	FIASMAs differ in terms of lysosomal accumulation kinetics. When testing a drug's effect on ASM activity, longer incubation times need to be taken into consideration (Kornhuber et al. 2011; Trapp et al. 2008).
Pharmacokinetics and pharmacodynamics	
Active in different cell lines	Usually, there is a high concordance between the effect of FIASMAs in different cell lines (in vitro) (Kornhuber et al. 2008) as well as different tissues (in vivo).
No complete inhibition	In contrast to genetic deficiency, functional inhibition of ASM usually leaves a basal level of activity. This may be the reason for good clinical tolerability of FIASMAs. However, an exception is emetine, which nearly completely inhibits ASM (see Table 1).
No habituation	FIASMAs inhibit ASM acutely, without building up a habituation effect (Kornhuber et al. 2005).
Reversible inhibition	FIASMAs show a reversible inhibition of ASM (Kornhuber et al. 2005).
No rebound effect	Upon withdrawal of FIASMAs, there is no rebound effect (Kornhuber et al. 2005).
Additive effect	FIASMAs show an additive effect with respect to ASM inhibition (Kornhuber et al. 2011) (see Fig. 1f).

(continued)

Table 2 (continued)

Pharmacokinetics and pharmacodynamics	
Bimodal distribution	FIASMAs show bimodal distribution. There are only few compounds that have an intermediate effect on the activity of ASM (see Fig. 1b).
Blood–brain barrier (BBB)	All FIASMAs with published data on BBB permeability appear to readily penetrate the BBB (Kornhuber et al. 2011).
Phospholipidosis	Most, but not all, FIASMAs induce phospholipidosis (Muehlbacher et al. 2012).
Lipinski's rule-of-5	FIASMAs tend to violate Lipinski's rule-of-5 more often than other drugs (Kornhuber et al. 2011).
Apparent volume of distribution	FIASMAs have, compared to non-FIASMAs, a large apparent volume of distribution.
Tissue plateau concentrations	Drugs like fluoxetine reach tissue plateau concentrations only after months of continued treatment (Karson et al. 1993). It is not known whether this applies to all FIASMAs.
ASM-related effects	FIASMAs are related to a large range of biological effects, mediated by ASM.
No effect on neutral and alkaline sphingomyelinase	FIASMAs do not inhibit neutral or alkaline sphingomyelinase (Kornhuber et al. 2010).
No general effect on lysosomal hydrolases	Although FIASMAs work by unspecific physicochemical mechanisms, they apparently do not functionally inhibit all lysosomal hydrolases (Kornhuber et al. 2010).
Other molecular targets	Most of the clinically approved compounds have been developed to inhibit, activate, or modify a specific biological target. Up until now, there is no compound available acting specifically as a FIASMA.
Clinical use	
Clinically available	A large number of FIASMAs are available as licensed and FDA-approved drugs.
ADME/Tox properties	Many of the hitherto identified FIASMAs have desirable ADME/Tox properties.
Active in therapeutic concentration	FIASMAs are active in lower concentration levels, even in therapeutic concentrations.
Distribution across pharmacological groups	FIASMAs are distributed widely among bioavailable drugs. Moreover, they occur among numerous classes of medicinal drugs (Kornhuber et al. 2011) (see Fig. 1c).

8 Clinical Implications of FIASMAs

Theoretically, FIASMAs should have therapeutic value in all situations where there is increased activity of the ASM/ceramide system, including clinical situations where increased activity of ASM/ceramide has already been observed, such as major depression (Gracia-Garcia et al. 2011; Kornhuber et al. 2005), alcohol abuse (Reichel et al. 2010, 2011), cystic fibrosis (Becker et al. 2010), septic shock (Claus et al. 2005), and liver fibrosis (Fernández et al. 2008). However, until now, only one clinical study has been performed that chose a FIASMA a priori as a therapeutic drug (Becker et al. 2010; Riethmüller et al. 2009). In this study, amitriptyline was used to treat cystic fibrosis (Becker et al. 2010; Riethmüller et al. 2009). Additionally, many clinical studies have used drugs that are now known to be FIASMAs. The results of these studies may be reinterpreted when the action of these drugs on ASM/ceramide is considered. One such example is multiple sclerosis. Fluoxetine has been shown to have a beneficial effect on the course of this disease (Mostert et al. 2008) and the clinical effect may be related to its effect on ASM. Another example is the action of fluoxetine in experimental models of stroke where fluoxetine has beneficial effects on motor performance and cerebral activation (Pariente et al. 2001). However, considering that fluoxetine acts on multiple neurobiological targets, other interpretations are also possible.

In addition to the clinical examples given above, several disease conditions such as cystic fibrosis (Teichgräber et al. 2008), inflammatory bowel disease (Sakata et al. 2007), Wilson's disease (Lang et al. 2007), and pulmonary edema (Göggel et al. 2004) have been examined in animal models. Thus, FIASMAs show a broad range of medicinal applications and promise to be useful in the treatment of many disorders. Further information on the potential clinical applications of FIASMAs is given in other chapters of this book.

Many of the FIASMAs that are currently available are already FDA approved, which introduces the opportunity of using these drugs in clinical studies aimed at reducing ASM/ceramide activity. However, a major disadvantage is that there is no FIASMA currently available with specific effect on ASM; all available FIASMAs also act on other biological targets. The inhibitory effect of currently licensed drugs on ASM has been identified years after the initial development of the drug. Therefore, an important goal is to develop drugs that functionally inhibit ASM without interfering with other biological targets. Such drugs may be also used as tools in identifying biological effects that are clearly related to the functional inhibition of ASM.

Pharmacological inhibition of ASM is a novel and only recently used treatment option. Many remaining open questions are related to the medical effects of FIASMAs, including the following:

- Which FIASMA should be chosen to treat human diseases?
- Are there any clinical side effects, such as obesity (Bikman and Summers 2011), that are common to all FIASMAs?
- Are there contraindications common to all FIASMAs that would proscribe their use?
- Do FIASMAs present in food constituents, such as tomatidine (Kornhuber et al. 2011), have an impact on health? In particular, is it possible that tomatidine contributes to the positive effect of mediterranean diets on health (Féart et al. 2009) via the functional inhibition of ASM?

Acknowledgments A portion of this work was supported by funding from the Deutsche Forschungsgemeinschaft (DFG grants GU 335/23-1, KO 947/11-1) and the German Ministry of Education and Research (BMBF Grant 01EX1015B).

References

Abbott NJ, Romero IA (1996) Transporting therapeutics across the blood–brain barrier. Mol Med Today 2(3):106–113

Arenz C (2010) Small molecule inhibitors of acid sphingomyelinase. Cell Physiol Biochem 26(1): 1–8

Baker DE (2007) Loperamide: a pharmacological review. Rev Gastroenterol Disord 7(suppl 3): S11–S18

Becker KA, Riethmüller J, Lüth A, Döring G, Kleuser B, Gulbins E (2010) Acid sphingomyelinase inhibitors normalize pulmonary ceramide and inflammation in cystic fibrosis. Am J Respir Cell Mol Biol 42(6):716–724

Bikman BT, Summers SA (2011) Ceramides as modulators of cellular and whole-body metabolism. J Clin Invest 121(11):4222–4230

Black WC, Percival MD (2006) The consequences of lysosomotropism on the design of selective cathepsin K inhibitors. ChemBioChem 7(10):1525–1535

Bollinger CR, Teichgräber V, Gulbins E (2005) Ceramide-enriched membrane domains. Biochim Biophys Acta 1746(3):284–294

Bradbury MWB (1993) The blood–brain barrier. Exp Physiol 78(4):453–472

Brady RO, Kanfer JN, Mock MB, Fredrickson DS (1966) The metabolism of sphingomyelin. II. Evidence of an enzymatic deficiency in Niemann–Pick diseae. Proc Natl Acad Sci 55(2): 366–369

Charruyer A, Grazide S, Bezombes C, Müller S, Laurent G, Jaffrézou J-P (2005) UV-C light induces raft-associated acid sphingomyelinase and JNK activation and translocation independently on a nuclear signal. J Biol Chem 280(19):19196–19204

Clark DE (2011) What has polar surface area ever done for drug discovery? Future Med Chem 3: 469–484

Claus RA, Bunck AC, Bockmeyer CL, Brunkhorst FM, Lösche W, Kinscherf R, Deigner H-P (2005) Role of increased sphingomyelinase activity in apoptosis and organ failure of patients with severe sepsis. FASEB J 19(12):1719–1721

Cuzzocrea S, Deigner H-P, Genovese T, Mazzon E, Esposito E, Crisafulli C, Di Paola R, Bramanti P, Matuschak G, Salvemini D (2009) Inhibition of ceramide biosynthesis ameliorates pathological consequences of spinal cord injury. Shock 31(6):635–645

de Boer AG, van der Sandt ICJ, Gaillard PJ (2003) The role of drug transporters at the blood–brain barrier. Annu Rev Pharmacol Toxicol 43:629–656

De Duve C, De Barsy T, Poole B, Trouet A, Tulkens P, Van Hoof F (1974) Lysosomotropic agents. Biochem Pharmacol 23(18):2495–2531

de Lange ECM (2004) Potential role of ABC transporters as a detoxification system at the blood-CSF barrier. Adv Drug Deliv Rev 56(12):1793–1809

Féart C, Samieri C, Rondeau V, Amieva H, Portet F, Dartigues J-F, Scarmeas N, Barberger-Gateau P (2009) Adherence to a mediterranean diet, cognitive decline, and risk of dementia. JAMA 302(6):638–648

Fernández A, Colell A, Garcia-Ruiz C, Fernandez-Checa JC (2008) Cholesterol and sphingolipids in alcohol-induced liver injury. J Gastroenterol Hepatol 23:S9–S15

Fischer H, Kansy M, Avdeef A, Senner F (2007) Permeation of permanently positive charged molecules through artificial membranes—influence of physico-chemical properties. Eur J Pharm Sci 31(1):32–42

Gallala HD, Sandhoff K (2011) Biological function of the cellular lipid BMP—BMP as a key activator for cholesterol sorting and membrane digestion. Neurochem Res 36(9):1594–1600

Göggel R, Winoto-Morbach S, Vielhaber G, Imai Y, Lindner K, Brade L, Brade H, Ehlers S, Slutsky AS, Schütze S, Gulbins E, Uhlig S (2004) PAF-mediated pulmonary edema: a new role for acid sphingomyelinase and ceramide. Nat Med 10(2):155–160

Gracia-Garcia P, Rao V, Haughey NJ, Ratnam Banduru VV, Smith G, Rosenberg PB, Lobo A, Lyketsos CG, Mielke MM (2011) Elevated plasma ceramides in depression. J Neuropsychiatry Clin Neurosci 23(2):215–218

Grassmé H, Jekle A, Riehle A, Schwarz H, Berger J, Sandhoff K, Kolesnick R, Gulbins E (2001) CD95 signaling via ceramide-rich membrane rafts. J Biol Chem 276(23):20589–20596

Grassmé H, Jendrossek V, Riehle A, von Kürthy G, Berger J, Schwarz H, Weller M, Kolesnick R, Gulbins E (2003) Host defense against pseudomonas aeruginosa requires ceramide-rich membrane rafts. Nat Med 9(3):322–330

Guidelines for ATC classifications and DDD assignment (2011) World Health Organization (WHO) Collaborating Centre for Drug Statistics Methodology. http://www.whocc.no/filearchive/publications/2010guidelines.pdf. Accessed 19 Oct 2009

Han X, Holtzman DM, McKeel DW, Kelley J, Morris JC (2002) Substantial sulfatide deficiency and ceramide elevation in very early Alzheimer's disease: potential role in disease pathogenesis. J Neurochem 82(4):809–818

He X, Huang Y, Li B, Gong C-X, Schuchman EH (2010) Deregulation of sphingolipid metabolism in Alzheimer's disease. Neurobiol Aging 31(3):398–408

Hurwitz R, Ferlinz K, Sandhoff K (1994) The tricyclic antidepressant desipramine causes proteolytic degradation of lysosomal sphingomyelinase in human fibroblasts. Biol Chem Hoppe Seyler 375(7):447–450

Inc. ACD ACD/PhysChem. version 10.0 edn., Toronto, ON, Canada. http://www.acdlabs.com

Karson CN, Newton JE, Livingston R, Jolly JB, Cooper TB, Sprigg J, Komoroski RA (1993) Human brain fluoxetine concentrations. J Neuropsychiatry Clin Neurosci 5(3):322–329

Kaufmann AM, Krise JP (2007) Lysosomal sequestration of amine-containing drugs: analysis and therapeutic implications. J Pharm Sci 96(4):729–746

Kim M-Y, Linardic C, Obeid L, Hannun Y (1991) Identification of sphingomyelin turnover as an effector mechanism for the action of tumor necrosis factor alpha and gamma-interferon. Specific role in cell differentiation. J Biol Chem 266(1):484–489

Kolter T, Sandhoff K (2005) Principles of lysosomal membrane digestion: stimulation of sphingolipid degradation by sphingolipid activator proteins and anionic lysosomal lipids. Annu Rev Cell Dev Biol 21(1):81–103

Kölzer M, Werth N, Sandhoff K (2004) Interactions of acid sphingomyelinase and lipid bilayers in the presence of the tricyclic antidepressant desipramine. FEBS Lett 559(1–3):96–98

Kornhuber J, Medlin A, Bleich S, Jendrossek V, Henkel AW, Wiltfang J, Gulbins E (2005) High activity of acid sphingomyelinase in major depression. J Neural Transm 112(11):1583–1590

Kornhuber J, Tripal P, Reichel M, Terfloth L, Bleich S, Wiftfang J, Gulbins E (2008) Identification of new functional inhibitors of acid sphingomyelinase using a structure-property-activity relation model. J Med Chem 51(2):219–237

Kornhuber J, Tripal P, Reichel M, Mühle C, Rhein C, Muehlbacher M, Groemer TW, Gulbins E (2010) Functional inhibitors of acid sphingomyelinase (FIASMAs): a novel pharmacological group of drugs with broad clinical applications. Cell Physiol Biochem 26(1):9–20

Kornhuber J, Muehlbacher M, Trapp S, Pechmann S, Friedl A, Reichel M, Mühle C, Terfloth L, Groemer TW, Spitzer GM, Liedl KR, Gulbins E, Tripal P (2011) Identification of novel functional inhibitors of acid sphingomyelinase. PLoS One 6(8):e23852

Lang PA, Schenck M, Nicolay JP, Becker JU, Kempe DS, Lupescu A, Koka S, Eisele K, Klarl BA, Rübben H, Schmid KW, Mann K, Hildenbrand S, Hefter H, Huber SM, Wieder T, Erhardt A, Häussinger D, Gulbins E, Lang F (2007) Liver cell death and anemia in Wilson disease involve acid sphingomyelinase and ceramide. Nat Med 13(2):164–170

Lipinski CA, Lombardo F, Dominy BW, Feeney PJ (2001) Experimental and computational approaches to estimate solubility and permeability in drug discovery and development settings. Adv Drug Deliv Rev 46(1–3):3–26

Macauley SL, Sidman RL, Schuchman EH, Taksir T, Stewart GR (2008) Neuropathology of the acid sphingomyelinase knockout mouse model of Niemann–Pick A disease including structure-function studies associated with cerebellar Purkinje cell degeneration. Exp Neurol 214(2):181–192

Malaplate-Armand C, Florent-Béchard S, Youssef I, Koziel V, Sponne I, Kriem B, Leininger-Muller B, Olivier J-L, Oster T, Pillot T (2006) Soluble oligomers of amyloid-β peptide induce neuronal apoptosis by activating a cPLA2-dependent sphingomyelinase-ceramide pathway. Neurobiol Dis 23(1):178–189

Mikati MA, Zeinieh M, Habib RA, El Hokayem J, Rahmeh A, El Sabban M, Usta J, Dbaibo G (2008) Changes in sphingomyelinases, ceramide, bax, bcl2, and caspase-3 during and after experimental status epilepticus. Epilepsy Res 81(2):161–166

Miller DS (2010) Regulation of P-glycoprotein and other ABC drug transporters at the blood–brain barrier. Trends Pharmacol Sci 31(6):246–254

Mintzer RJ, Appell KC, Cole A, Johns A, Pagila R, Polokoff MA, Tabas I, Snider RM, Meurer-Ogden JA (2005) A novel high-throughput screening format to identify inhibitors of secreted acid sphingomyelinase. J Biomol Screen 10(3):225–234

Mostert JP, Admiraal-Behloul F, Hoogduin JM, Luyendijk J, Heersema DJ, van Buchem MA, De Keyser J (2008) Effects of fluoxetine on disease activity in relapsing multiple sclerosis: a double-blind, placebo-controlled, exploratory study. J Neurol Neurosurg Psychiatry 79(9): 1027–1031

Muehlbacher M, Tripal P, Roas F, Kornhuber J (2012) Identification of drugs inducing phospholipidosis by novel in vitro data. ChemMedChem 7(11):1925–1934

Ndengele MM, Cuzzocrea S, Masini E, Vinci MC, Esposito E, Muscoli C, Petrusca DN, Mollace V, Mazzon E, Li D, Petrache I, Matuschak GM, Salvemini D (2009) Spinal ceramide modulates the development of morphine antinociceptive tolerance via peroxynitrite-mediated nitroxidative stress and neuroimmune activation. J Pharmacol Exp Ther 329(1):64–75

Novelli A, Lysko PG, Henneberry RC (1987) Uptake of imipramine in neurons cultured from rat cerebellum. Brain Res 411(2):291–297

Nussio MR, Sykes MJ, Miners JO, Shapter JG (2007) Characterisation of the binding of cationic amphiphilic drugs to phospholipid bilayers using surface plasmon resonance. ChemMedChem 2(3):366–373

Pariente J, Loubinoux I, Carel C, Albucher J-F, Leger A, Manelfe C, Rascol O, Chollet F (2001) Fluoxetine modulates motor performance and cerebral activation of patients recovering from stroke. Ann Neurol 50(6):718–729

Purvis GD (2008) Size-intensive descriptors. J Comput Aided Mol Des 22(6):461–468

Rebillard A, Rioux-Leclercq N, Muller C, Bellaud P, Jouan F, Meurette O, Jouan E, Vernhet L, Le Quement C, Carpinteiro A, Schenck M, Lagadic-Gossmann D, Gulbins E, Dimanche-Boitrel MT (2008) Acid sphingomyelinase deficiency protects from cisplatin-induced gastrointestinal damage. Oncogene 27(51):6590–6595

Reichel M, Greiner E, Richter-Schmidinger T, Yedibela Ö, Tripal P, Jacobi A, Bleich S, Gulbins E, Kornhuber J (2010) Increased acid sphingomyelinase activity in peripheral blood cells of acutely intoxicated patients with alcohol dependence. Alcohol Clin Exp Res 34(1):46–50

Reichel M, Beck J, Mühle C, Rotter A, Bleich S, Gulbins E, Kornhuber J (2011) Activity of secretory sphingomyelinase is increased in plasma of alcohol-dependent patients. Alcohol Clin Exp Res 35(10):1852–1859

Riethmüller J, Anthonysamy J, Serra E, Schwab M, Döring G, Gulbins E (2009) Therapeutic efficacy and safety of amitriptyline in patients with cystic fibrosis. Cell Physiol Biochem 24 (1–2):65–72

Sakata A, Ochiai T, Shimeno H, Hikishima S, Yokomatsu T, Shibuya S, Toda A, Eyanagi R, Soeda S (2007) Acid sphingomyelinase inhibition suppresses lipopolysaccharide-mediated release of inflammatory cytokines from macrophages and protects against disease pathology in dextran sulphate sodium-induced colitis in mice. Immunology 122(1):56–64

Santana P, Peña LA, Haimovitz-Friedman A, Martin S, Green D, McLoughlin M, Cordon-Cardo C, Schuchman EH, Fuks Z, Kolesnick R (1996) Acid sphingomyelinase deficient human lymphoblasts and mice are defective in radiation-induced apoptosis. Cell 86(2):189–199
Schuchman EH (2007) The pathogenesis and treatment of acid sphingomyelinase-deficient Niemann–Pick disease. J Inherit Metab Dis 30(5):654–663
Schwarz E, Prabakaran S, Whitfield P, Major H, Leweke FM, Koethe D, McKenna P, Bahn S (2008) High throughput lipidomic profiling of schizophrenia and bipolar disorder brain tissue reveals alterations of free fatty acids, phosphatidylcholines, and ceramides. J Proteome Res 7(10):4266–4277
Seto M, Whitlow M, McCarrick MA, Srinivasan S, Zhu Y, Pagila R, Mintzer R, Light D, Johns A, Meurer-Ogden JA (2004) A model of the acid sphingomyelinase phosphoesterase domain based on its remote structural homolog purple acid phosphatase. Protein Sci 13(12):3172–3186
Short B (2010) Acid sphingomyelinase deals the seal. J Cell Biol 189(6):920
Smith EL, Schuchman EH (2008) The unexpected role of acid sphingomyelinase in cell death and the pathophysiology of common diseases. FASEB J 22(10):3419–3431
Teichgräber V, Ulrich M, Endlich N, Riethmüller J, Wilker B, De Oliveira-Munding CC, van Heeckeren AM, Barr ML, von Kürthy G, Schmid KW, Weller M, Tümmler B, Lang F, Grassme H, Döring G, Gulbins E (2008) Ceramide accumulation mediates inflammation, cell death and infection susceptibility in cystic fibrosis. Nat Med 14(4):382–391
Trapp S, Rosania GR, Horobin RW, Kornhuber J (2008) Quantitative modeling of selective lysosomal targeting for drug design. Eur Biophys J 37(8):1317–1328
Wiegmann K, Schütze S, Machleidt T, Witte D, Krönke M (1994) Functional dichotomy of neutral and acidic sphingomyelinases in tumor necrosis factor signaling. Cell 78(6):1005–1015
Wilkening G, Linke T, Sandhoff K (1998) Lysosomal degradation on vesicular membrane surfaces. J Biol Chem 273(46):30271–30278

Novel Drugs Targeting Sphingolipid Metabolism

Krishna P. Bhabak and Christoph Arenz

Contents

Abstract While the evidence for an involvement of sphingolipids (SLs) in a variety of diseases is rapidly increasing, the development of sphingolipid-related drugs is still in its infancy. In fact, the recently FDA-approved fingolimod or FTY-720 (see chapter by J. Pfeilschifter for more information) is the first drug on the market to interfere with sphingolipid signaling. The reasons for this lagging are manifold and within this chapter we try to name some of them. Ceramide is in the center of sphingolipid metabolism. We describe the most important and most recent inhibitors for enzymes controlling cellular ceramide levels.

Keywords Enzyme inhibitors • Enzyme assays • Small molecules • Drugs • Natural products

1 Introduction

Historically, the first group of diseases that was related to SL metabolism is formed by the inborn lysosomal sphingolipid storage diseases also termed as sphingolipidoses. The fact that most of these diseases are very rare has certainly prevented many of the major pharmaceutical companies to make a stronger

K.P. Bhabak • C. Arenz (✉)
Institut für Chemie, Humboldt Universität zu Berlin, Brook-Taylor-Str 2, Berlin 12489, Germany
e-mail: christoph.arenz@chemie.hu-berlin.de

E. Gulbins and I. Petrache (eds.), *Sphingolipids: Basic Science and Drug Development*, Handbook of Experimental Pharmacology 215, DOI 10.1007/978-3-7091-1368-4_10,

commitment to drug development. Since these genetic diseases are characterized by a dysfunction of catabolizing enzymes, in some cases recombinant enzymes for an enzyme replacement therapy are available (imiglucerase and velaglucerase alfa, both for Gaucher's disease) and a number of gene therapy strategies are under development. Indeed, even a number of small molecules acting as site-specific chemical chaperones (Fan 2003) have been reported, some of which entered clinical studies (isofagomine, an inhibitor of glucocerebrosidase and chemical chaperone for Gaucher's disease). Miglustat is a market-approved small-molecule drug, inhibiting glycosylceramide synthase (GCS) and thus the synthesis of the main storage compound in Gaucher's disease. A more detailed review of recent therapeutic approaches to sphingolipid storage diseases can be found in the chapter by Futerman and has been published elsewhere (Hemsley and Hopwood 2011).

The second reason for a lack in sphingolipid-related drugs is the fact that detailed molecular mechanisms for the biological activity of many sphingolipids are missing. The structurally most diverse subgroup of sphingolipids is formed by the glycosphingolipids (GSL). However, many enzymes involved in GSL biosynthesis have a limited substrate specificity leading to the combinatorial biosynthesis of gangliosides (Kolter et al. 2002) arguing that there is only a minimal chance to control the function of individual GSL by inhibition of their biosynthesis. During the last years, especially those sphingolipids involved in lipid signaling processes gained increased attention. This group of signaling lipids includes ceramide, sphingosine, sphingosine-1-phosphate, and also ceramide-1-phosphate. However, despite the small number of signaling lipids, many different enzymes are involved in their regulation making a pharmacological control difficult. Ceramide is the central lipid molecule in the sphingolipid family. Ceramide levels are controlled by the de novo biosynthetic pathway, secondly by the interconversion to membrane constituents like GSL and sphingomyelin, and also by the degradation to sphingosine and sphingosine-1-phosphate. More importantly, the exact mechanisms of ceramide-mediated signaling are only partially understood and are at least of complex nature. In contrast, sphingosine-1-phosphate acts in a more classical fashion by binding to a set of clearly defined sphingosine-1-phosphate receptors. Accordingly, the enzymes and receptors involved in sphingosine-1-phosphate bioactivity have been in the focus of industrial drug development relatively early (see chapter by Pfeilschifter).

A third reason that has slowed progress in the development of sphingolipid-related drugs lies in the chemical nature of the sphingolipids. As a general rule, biological recognition is mainly mediated by hydrophobic interactions that provide high affinity but less specificity and secondly by hydrogen bonds and other electrostatic interactions mainly contributing to the specificity of interaction. Thus, it seems reasonable that in the case of GSL-metabolizing enzymes, the rational development of substrate-analog inhibitors is relatively easy, since specificity is provided by a number of hydrogen bonds. In contrast to ceramide, which forms the center part of sphingolipid metabolism, mainly consists of hydrophobic parts while it accomplishes only a total of four heteroatoms that might be able to form hydrogen

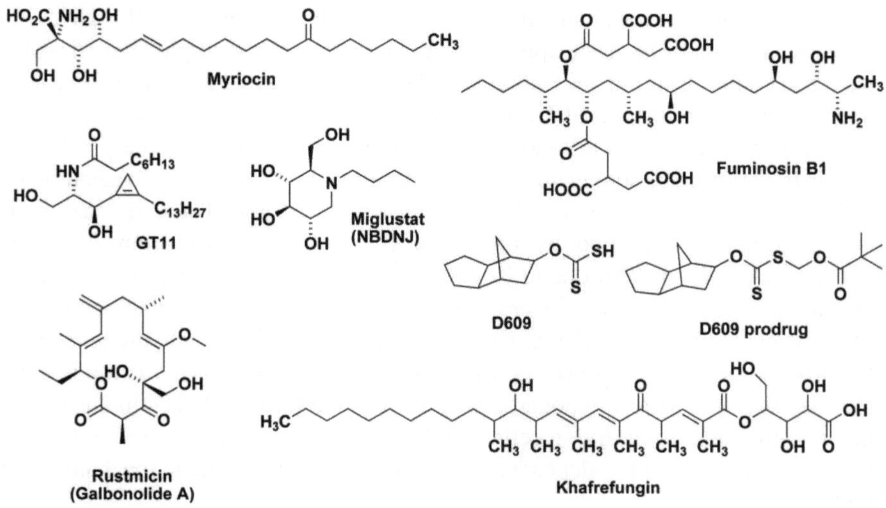

Fig. 1 Inhibitors of sphingolipid biosynthetic enzymes

bridges. It seems obvious that a potent inhibition of enzymes like the ceramidases can only be achieved by either irreversible covalently acting inhibitors or alternatively by allosteric inhibitors that bind independently of the substrate-recognizing active site of the enzyme. Inhibitors of the latter type cannot be rationally designed, but are usually identified in high-throughput-screening (HTS) formats. Indeed, we are convinced that for many sphingolipid-mediating enzymes, the unavailability of fast, specific, and cost-effective assays is one major bottleneck for the identification of novel lead structures for drug design. Thus, the development assays suitable for HTS approaches is of utmost importance. In the present chapter we concentrate on the most important or most recent developments. Further drugs and inhibitors can be found in a number of reviews (Arenz 2010; Delgado et al. 2006; Fan 2003; Kolter and Sandhoff 1999).

2 Inhibitors of Sphingolipid Biosynthetic Enzymes

Several inhibitors of ceramide de novo synthesis have been described (Fig. 1) and although some of them have very high biological activity, a market-approved drug is missing. Since sphingolipids are crucial for cellular function, the biosynthesis of ceramide is not a primary target for pharmacological strategies. However, the fact that sphingolipids play essential roles for many different organisms is obviously the main reason for the existing rich arsenal of natural secondary metabolites with potent activity against ceramide biosynthesis. Serine palmitoyl transferase (SPT), the first enzyme in ceramide de novo synthesis, is potently inhibited by the antibiotic myriocin. Due to its very potent immune-suppressive activity, myriocin was a lead structure for the development of FTY-720 (Brinkmann et al. 2010). FTY-720 however, which now has been approved for the treatment of multiple sclerosis (MS) in several countries, is not an inhibitor of SPT (see chapter by Pfeilschifter). Several other potent inhibitors of SPT like sphingofungin B have been described. For an excellent review see Delgado et al. 2006. 3-Ketosphinganine, the product of SPT, is transformed to sphinganine by 3-ketosphinganine reductase. Consecutively, sphinganine or sphingosine is acylated by ceramide synthase. The latter enzyme is most prominently inhibited by fumonisin B1, a fungal toxin from *Fusarium verticillioides*. Not at least due to its complex toxic nature, this compound is only used as a biochemical tool. Several other inhibitors of natural origin are known (Delgado et al. 2006). N-acylsphinganine or dihydroceramide is dehydrogenated by dihydroceramide desaturase (DES) to yield ceramide. This enzyme is moderately inhibited by a synthetic compound named GT11 (Triola et al. 2001). Interestingly, the inhibitory activity in living cells is in the lower nanomolar range (Triola et al. 2004), suggesting a time-dependent and thus probably covalent mechanism of inhibition.

Ceramide is substrate for the biosynthesis of sphingomyelin and glycosphingolipids. While the above enzymes all lead to a reduction of cellular ceramide concentration, inhibition of the downstream biosynthetic enzymes sphingomyelin synthase (SMS) or GCS conversely leads to an increased level of ceramide. In the light of recent research underscoring the importance of ceramide as an important regulator of cell fate, such enzymes are particularly important for the development of future drugs (see chapter by Kolesnick). The most important inhibitors are shown in Fig. 1. SMS is inhibited by the xanthogenate D609 (Luberto and Hannun 1998). Due to its ionic nature, this compound is poorly cell permeable. Its prodrug analogue X is cell permeable, twice as active, and more stable in buffer and cell media (Bai et al. 2004). D609 has some remarkable effects in a number of experimental therapies of cancer and atherosclerosis (Adibhatla et al. 2011). In most cases it remains unclear whether the observed effects can be attributed to SMS or to phosphatidyl choline-specific phospholipase C, which was originally found to be inhibited by D609 (Amtmann 1996). The other main biosynthetic route from ceramide leads to the diverse group of GSL via the synthesis of glucosyl ceramide by glucosyl ceramide synthase. As mentioned above, inhibition of this enzyme has gained importance in the context of GSL storage diseases, especially for the treatment of Gaucher's disease. Besides *N*-butyl deoxynojirimycin (NBDNJ) or Miglustat (Platt et al. 1994), which has orphan drug status in many countries, several even more potent inhibitors of this enzyme exist (Delgado et al. 2006).

In contrast to vertebrates which mainly synthesize GSL, many fungi synthesize large amounts of inositol phosphoceramide (IPC), which has essential functions in these organisms. The natural product khafrefungin is a potent inhibitor of IPC synthase in fungi and exhibits high and selective fungicidal activity (Mandala et al. 1997). The same IPC synthase inhibitory activity has been reported for rustmicin or galbonolide A. This macrolide antibiotic is active against clinically relevant microorganisms like *Candida albicans*, but it is not in clinical use, probably due to its low chemical stability (Mandala et al. 1998).

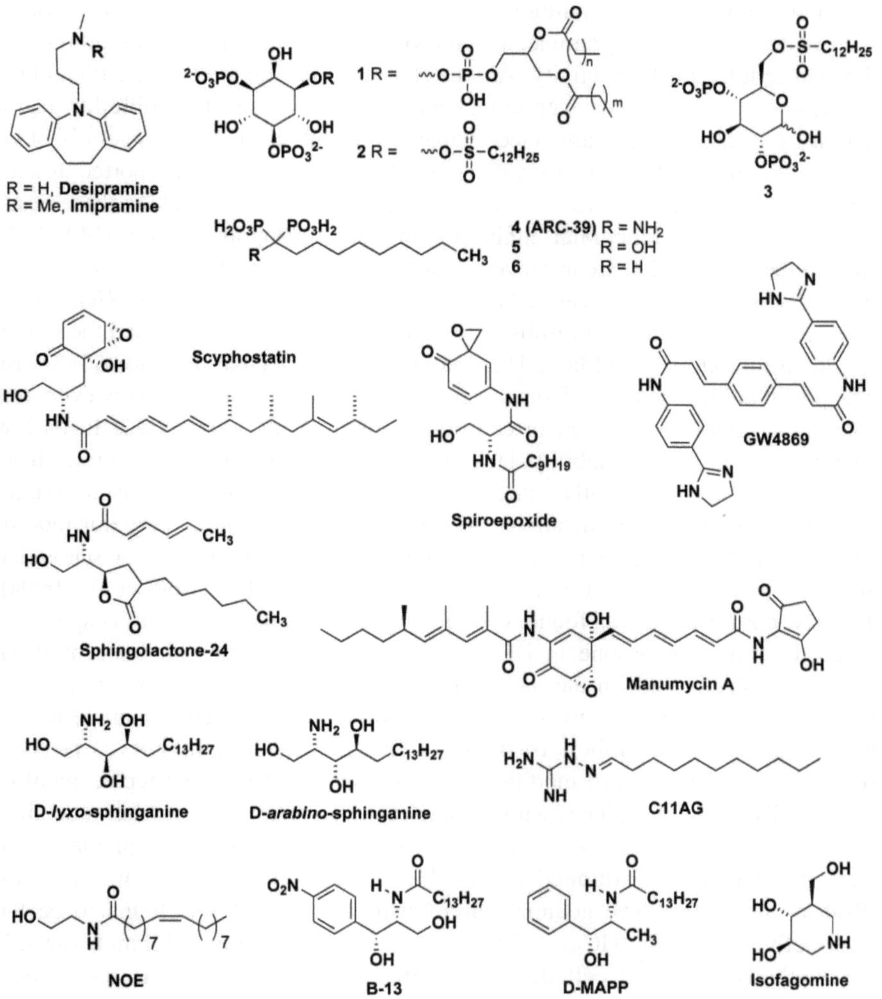

Fig. 2 Inhibitors of sphingolipid catabolic enzymes

3 Inhibitors of Catabolic Enzymes

During the last decade the interest in inhibitors targeting the catabolic enzymes of sphingolipid metabolism (Fig. 2) has steadily increased. Especially the highly abundant membrane constituent sphingomyelin is a source for the fast and inducible production of ceramide catalyzed by several sphingomyelinases that differ in pH optima and subcellular localization. The acid sphingomyelinase (aSMase) has been shown to play important roles in many diseases like cystic fibrosis (Teichgraber et al. 2008), Wilson's disease (Lang et al. 2007), and atherosclerosis (Devlin et al. 2008). Therefore, aSMase has become a potential drug target for a variety of diseases.

Though not being true inhibitors of this enzyme, the tricyclic antidepressants such as desipramine, imipramine, and amitriptyline are the most important tools to dampen cellular aSMase activity (Albouz et al. 1981). Such molecules enrich in the lysosomes to millimolar concentrations and render aSMase susceptible for degradation by lysosomal proteases (see chapter by Kornhuber). This process is obviously rather unselective and a similar type of inhibition has been reported at least for the lysosomal aCDase as well (Elojeimy et al. 2006). Nonetheless, these drugs and several other functional inhibitors of acid sphingomyelinase (FIASMA) (Kornhuber et al. 2010) are approved as antidepressants and have already successfully been used in first clinical trials for the treatment of cystic fibrosis (Riethmüller et al. 2009). The promising results with FIASMAs have also stimulated the search for direct inhibitors of aSMase. The first potent and selective inhibitor of aSMase was phosphatidyl-inositol-3,5-bisphosphate 1 (Kölzer et al. 2003). However, this compound could not be used in cell culture or in living animals due to its low chemical and biological stability, lack of cell permeability, and poor off rates from biomembranes. Very recently, our group has developed a number of bisphosphates with some important modifications over the existing compound. A major modification was the replacement of two-tailed diacylglycerol part by a one-tailed analogue with a dodecyl sulfonate ester chain (compound 2) (Roth et al. 2009a). The new compound 2 was found to be active in cell culture and inhibits apoptosis induced by dexamethasone in HEK293 cells at low micromolar concentration. A further enhancement in the inhibition towards aSMase was observed upon the replacement of inositol moiety with a carbohydrate molecule (compound 3) indicating that some further structural modification in the inhibitors is possible (Roth et al. 2010). A third modification was carried out with the replacement of biologically labile phosphates with the uncleavable phosphonates. After a detailed structure–activity correlation studies, a new series of geminal bisphosphonates 4–6 were developed that exhibited remarkable in vitro inhibition towards aSMase (Roth et al. 2009b). The geminal amino-bisphosphonate 4 was found to be the most active compound ($IC_{50} = 20$ nM) that exhibited strong inhibition towards dexamethasone-induced cell death in HEK293 cells and also exhibited dose-dependent inhibition of platelet-activating factor (PAF)-induced pulmonary edema formation in perfused rat lungs at a concentration of 100 nM (Roth et al. 2009b). Although these new sets of compounds were found to be very active in cell culture and ex vivo experiments, more data on their mechanism of action is needed. It is unclear whether these compounds are able to enter cells or whether they exclusively act on the secreted form of aSMase. Further, mostly moderate inhibitors of aSMase have been reviewed recently (Arenz 2010).

There are at least two types of neutral sphingomyelinases, which are designated as nSMase1 and nSMase2 (see chapter by Hannun). Similar to aSMase, nSMases have been implicated in many biological disorders such as inflammatory diseases, heart failure, diabetes, cell cycle arrest, etc. Therefore, the selective inhibition of nSMases has become important and challenging from a therapeutic point of view. The first potent selective inhibitor of nSMases was found to be a fungal compound scyphostatin (Nara et al. 1999). This compound exhibited much pronounced inhibition towards nSMases rather than aSMase. In contrast to scyphostatin the spiroepoxide inhibitor, a synthetic structural analogue, was found to act by an irreversible time-dependent mechanism (Arenz and Giannis 2000) and also the farnesyltransferase inhibitor manumycin A (Arenz et al. 2001) and the sphingolactone inhibitor (Wascholowski and Giannis 2006) were identified to be irreversible nSMase inhibitors. A selective noncompetitive inhibitor of purified and cellular nSMase (GW4869, $IC_{50} = 1\ \mu M$) was discovered in systematic screening approach (Luberto et al. 2002). Except scyphostatin, all these alternative nSMase inhibitors are available from commercial sources. A group of guanidinium-based lipophilic compounds was characterized as moderate nSMase inhibitors (Amtmann et al. 2000, 2003). Among those compounds, the inhibitor C11AG was found to be selective towards nSMase2 subtype (Amtmann and Zoller 2005). A recent study also reports the moderate but selective inhibition of nSMase2 by D-*lyxo*- and D-*arabino*-sphinganine (IC_{50} between 50 and 100 μM for both compounds), whereas GW4869 showed no subtype selectivity (Lee et al. 2011). Inhibitors of SMases and ceramidases (CDases) have been reviewed recently (Canals et al. 2011).

CDases catalyze the hydrolysis of the amide bond in ceramide to produce sphingosine and fatty acid. Depending on the subcellular localization and pH-optima, CDases can be divided as acid ceramidase (aCDase), neutral ceramidase (nCDase), and alkaline ceramidase (alkCDase). It should be noted that ceramide induces apoptosis, whereas its metabolite sphingosine-1-phosphate induces cell proliferation. Thus the CDases are regarded as "rheostats" of cell fate and the inhibition of CDases may be an effective therapy in cancer therapy.

Several inhibitors of CDases have been described. While acid ceramidase is selectively inhibited by B-13 with an IC_{50} of about 10 μM (Raisova et al. 2002), D-MAPP selectively inhibits alkaline ceramidase derived from HL-60 cell extracts with an IC_{50} of about 1–5 μM (Bielawska et al. 1996). However, in human melanoma and HaCaT cells, D-MAPP exhibits no activity towards neutral ceramidase, but is a moderate inhibitor of acid ceramidase (Raisova et al. 2002). At somewhat higher concentrations, even aSMase is inhibited (Schuchman and Desnick 2005). Although B-13 fairly inhibits aCDase in cell lysates, it showed nearly no cellular activity towards the lysosomal aCDase (Bai et al. 2009), obviously due to poor lysosomal penetration by this compound. A number of basic structural analogues of B-13 and D-MAPP were developed for subcellular targeting (Szulc et al. 2008; Bai et al. 2009). Inhibitors like LCL-464 can enrich in the lysosomes. *N*-oleoylethanolamine (NOE) is a very weak (IC_{50} ~0.2–0.5 M) inhibitor of aCDase (Ueda et al. 2001). However, this substance is easily available and has been shown to increase the ceramide level and thereby enhance the apoptosis in different cell lines (Strelow et al. 2000). Recently a series of structural analogues of NOE have been reported and a detailed SAR study suggested those compounds as somewhat more potent inhibitors of aCDase (Grijalvo et al. 2006).

Compared to the potential relevance of CDase inhibition for therapeutic approaches, the potency and selectivity of existing CDase inhibitors are not satisfying. As pointed out before, an HTS approach for allosteric inhibitors is needed. Recently, a homogenous assay of CDase activity has been developed that might be compatible with HTS endeavors (Bedia et al. 2007).

4 Conclusions and Outlook

A detailed review on the inhibition of sphingolipid-metabolizing enzymes reveals a number of small molecules that effectively inhibit one or more enzymes. In addition, many molecules exhibit very selective and potent inhibition towards a certain enzyme in the in vitro studies with purified enzymes but lack the activity in living cells. Therefore, the design of lipophilic molecules with some proper functionality would have some possibilities to localize in a particular cell compartment to have a selective inhibition towards a certain enzyme. Furthermore, the development of potent inhibitors could be better achieved with the discovery of some homogeneous assay methods for the determination of key enzymes involved in the sphingolipid metabolism. These advancements would definitely solve many issues related to sphingolipid-mediated pathologies.

Acknowledgments The authors gratefully acknowledge funding by the DFG priority program "Sphingolipids—Signal and Disease." K.P.B. thanks the Alexander v. Humboldt Foundation for a fellowship.

References

Adibhatla RM, Hatcher JF, Gusain A (2011) Tricyclodecan-9-yl-xanthogenate (D609) mechanism of actions: a mini-review of literature. Neurochem Res. doi:10.1007/s11064-011-0659-z

Albouz S, Hauw JJ, Berwald-Netter Y, Boutry JM, Bourdon R, Baumann N (1981) Tricyclic antidepressants induce sphingomyelinase deficiency in fibroblast and neuroblastoma cell cultures. Biomedicine 35:218–220

Amtmann E (1996) The antiviral, antitumoural xanthate D609 is a competitive inhibitor of phosphatidylcholine-specific phospholipase C. Drugs Exp Clin Res 22:287–294

Amtmann E, Zoller M (2005) Stimulation of CD95-induced apoptosis in T-cells by a subtype specific neutral sphingomyelinase inhibitor. Biochem Pharmacol 69:1141–1148

Amtmann E, Zoller M, Schilling G (2000) Neutral sphingomyelinase-inhibiting guanidines prevent herpes simplex virus-1 replication. Drugs Exp Clin Res 26:57–65

Amtmann E, Baader W, Zoller M (2003) Neutral sphingomyelinase inhibitor C11AG prevents lipopolysaccharide-induced macrophage activation. Drugs Exp Clin Res 29:5–13

Arenz C (2010) Small molecule inhibitors of acid sphingomyelinase. Cell Physiol Biochem 26:01–08

Arenz C, Giannis A (2000) Synthesis of the first selective irreversible inhibitor of neutral sphingomyelinase. Angew Chem Int Ed 39:1440–1442

Arenz C, Thutewohl M, Block O, Waldmann H, Altenbach HJ, Giannis A (2001) Manumycin A and its analogues are irreversible inhibitors of neutral sphingomyelinase. Chembiochem 2:141–143

Bai A, Meier GP, Wang Y, Luberto C, Hannun YA, Zhou D (2004) Prodrug modification increases potassium tricyclo[5.2.1.0(2,6)]-decan-8-yl dithiocarbonate (D609) chemical stability and cytotoxicity against U937 leukemia cells. J Pharmacol Exp Ther 309:1051–1059

Bai A, Szulc ZM, Bielawski J, Mayroo N, Liu X, Norris J, Hannun YA, Bielawska A (2009) Synthesis and bioevaluation of ω-N-amino analogs of B13. Bioorg Med Chem 17:1840–1848

Bedia C, Casas J, Garcia V, Levade T, Fabriàs G (2007) Synthesis of a novel ceramide analogue and its use in a high-throughput fluorogenic assay for ceramidases. Chembiochem 8:642–648

Bielawska A, Greenberg MS, Perry D, Jayadev S, Shayman JA, McKay C, Hannun YA (1996) (1S,2R)-D-erythro-2-(N-myristoylamino)-1-phenyl-1-propanol as an inhibitor of ceramidase. J Biol Chem 271:12646–12654

Brinkmann V, Billich A, Baumruker T, Heining P, Schmouder R, Francis G, Aradhye S, Burtin P (2010) Fingolimod (FTY720): discovery and development of an oral drug to treat multiple sclerosis. Nat Rev Drug Discov 9:883–897

Canals D, Perry DM, Jenkins RW, Hannun YA (2011) Drug targeting of sphingolipid metabolism: sphingomyelinases and ceramidases. Br J Pharmacol 163:694–712

Delgado A, Casas J, Llebaria A, Abad JS, Fabrias G (2006) Inhibitors of sphingolipid metabolism enzymes. Biochim Biophys Acta 1758:1957–1977

Devlin CM, Leventhal AR, Kuriakose G, Schuchman EH, Williams KJ, Tabas I (2008) Acid sphingomyelinase promotes lipoprotein retention within early atheromata and accelerates lesion progression. Arterioscler Thromb Vasc Biol 28:1723–1730

Elojeimy S, Holman DH, Liu X, El-Zawahry A, Villani M, Cheng JC, Mahdy A, Zeidan Y, Bielawska A, Hannun YA, Norris JS (2006) New insights on the use of desipramine as an inhibitor for acid ceramidase. FEBS Lett 580:4751–4756

Fan JQ (2003) A contradictory treatment for lysosomal storage disorders: inhibitors enhance mutant enzyme activity. Trends Pharmacol Sci 24:355–360

Grijalvo S, Bedia C, Triola G, Casas J, Llebaria A, Teixido J, Rabal O, Levade T, Delgado A, Fabriàs G (2006) Design, synthesis and activity as acid ceramidase inhibitors of 2-oxooctanoyl and N-oleoylethanolamine analogues. Chem Phys Lipids 144:69–84

Hemsley KM, Hopwood JJ (2011) Emerging therapies for neurodegenerative lysosomal storage disorders—from concept to reality. J Inherit Metab Dis 34:1003–1012

Kolter T, Sandhoff K (1999) Sphingolipids-their metabolic pathways and the pathobiochemistry of neurodegenerative diseases. Angew Chem Int Ed 11:1532–1568

Kolter T, Proia RL, Sandhoff K (2002) Combinatorial ganglioside biosynthesis. J Biol Chem 277:25859–25862

Kölzer M, Arenz C, Ferlinz K, Werth N, Schulze H, Klingenstein R, Sandhoff K (2003) Phosphatidylinositol-3,5-bisphosphate is a potent and selective inhibitor of acid sphingomyelinase. Biol Chem 384:1293–1298

Kornhuber J, Tripal P, Reichel M, Mühle C, Rhein C, Muehlbacher M, Groemer TW, Gulbins E (2010) Functional inhibitors of acid sphingomyelinase (FIASMAs): a novel pharmacological group of drugs with broad clinical applications. Cell Physiol Biochem 26:09–20

Lang PA, Schenck M, Nicolay JP, Becker JU, Kempe DS, Lupescu A, Koka S, Eisele K, Klarl BA, Rubben H, Schmid KW, Mann K, Hildenbrand S, Hefter H, Huber SM, Wieder T, Erhardt A, Haussinger D, Gulbins E, Lang F (2007) Liver cell death and anemia in wilson disease involve acid sphingomyelinase and ceramide. Nat Med 13:164–170

Lee DH, Kim SH, Ahn KH, Kim SK, Choi JM, Ji JE, Won JH, Park YH, Lim C, Kim S, Kim DK (2011) Identification and evaluation of neutral sphingomyelinase 2 inhibitors. Arch Pharm Res 34:229–236

Luberto C, Hannun YA (1998) Sphingomyelin synthase, a potential regulator of intracellular levels of ceramide and diacylglycerol during SV40 transformation. Does sphingomyelin synthase account for the putative phosphatidylcholine-specific phospholipase C? J Biol Chem 273:14550–14559

Luberto C, Hassler DF, Signorelli P, Okamoto Y, Sawai H, Boros E, Hazen-Martin DJ, Obeid LM, Hannun YA, Smith GK (2002) Inhibition of tumor necrosis factor-induced cell death in MCF7 by a novel inhibitor of neutral sphingomyelinase. J Biol Chem 277:41128–41139

Mandala SM, Thornton RA, Rosenbach M, Milligan J, Garcia-Calvo M, Bull HG, Kurtz MB (1997) Khafrefungin, a novel inhibitor of sphingolipid synthesis. J Biol Chem 272: 32709–32714

Mandala SM, Thornton RA, Milligan J, Rosenbach M, Garcia-Calvo M, Bull HG, Harris G, Abruzzo GK, Flattery AM, Gill CJ, Bartizal K, Dreikorn S, Kurtz MB (1998) Rustmicin, a potent antifungal agent, inhibits sphingolipid synthesis at inositol phosphoceramide synthase. J Biol Chem 273:14942–14949

Nara F, Tanaka M, Masuda-Inoue S, Yamasato Y, Doi-Yoshioka H, Suzuki-Konagai K, Kumakura S, Ojita T (1999) Biological activities of scyphostatin, a neutral sphingomyelinase inhibitor from a discomycete, Trichopeziza mollissima. J Antibiot 52:531–535

Platt FM, Neises GR, Dwek RA, Butter TD (1994) N-butyldeoxynojirimycin is a novel inhibitor of glycolipid biosynthesis. J Biol Chem 269:8362–8365

Raisova M, Goltz G, Bektas M, Bielawska A, Riebeling C, Hossini AM, Eberlea J, Hannun YA, Orfanos CE, Geilen CC (2002) Bcl-2 overexpression prevents apoptosis induced by ceramidase inhibitors in malignant melanoma and HaCaT keratinocytes. FEBS Lett 516:47–52

Riethmüller J, Anthonysamy J, Emilio Serra E, Schwab M, Döring G, Gulbins E (2009) Therapeutic efficacy and safety of amitriptyline in patients with cystic fibrosis. Cell Physiol Biochem 22:405–412

Roth AG, Redmer S, Arenz C (2009a) Potent inhibition of acid sphingomyelinase by phosphoinositide analogues. Chembiochem 10:2367–2374

Roth AG, Drescher D, Yang Y, Redmer S, Uhlig S, Arenz C (2009b) Potent and selective inhibition of acid sphingomyelinase by bisphosphonates. Angew Chem Int Ed 48:7560–7563

Roth AG, Redmer S, Arenz C (2010) Development of carbohydrate-derived inhibitors of acid sphingomyelinase. Bioorg Med Chem 18:939–944

Schuchman E, Desnick RJ (2005) Chaperone-based therapy for Niemann–Pick-disease. US Patent Application Publication, Pub. No.: US 2005/0153934 A1

Strelow A, Bernardo K, Adam-Klages S, Linke T, Sandhoff K, Kronke M, Adam D (2000) Overexpression of acid ceramidase protects from tumor necrosis factor-induced cell death. J Exp Med 192:601–612

Szulc ZM, Mayroo N, Bai A, Bielawski J, Liu X, Norris JS, Hannun YA, Bielawska A (2008) Novel analogs of d-*e*-MAPP and B13. Part 1: synthesis and evaluation as potential anticancer agents. Bioorg Med Chem 16:1015–1031

Teichgraber V, Ulrich M, Endlich N, Riethmuller J, Wilker B, De Oliveira-Munding CC, van Heeckeren AM, Barr ML, von Kurthy G, Schmid KW, Weller M, Tummler B, Lang F, Grassme H, Doring G, Gulbins E (2008) Ceramide accumulation mediates inflammation, cell death and infection susceptibility in cystic fibrosis. Nat Med 14:382–391

Triola G, Fabrias G, Llebaria A (2001) Synthesis of a cyclopropene analogue of ceramide, a potent inhibitor of dihydroceramide desaturase. Angew Chem Int Ed Engl 40:1960–1962

Triola G, Fabrias G, Dragusin M, Niederhausen L, Broere R, Llebaria A, van Echten-Deckert G (2004) Specificity of the dihydroceramide desaturase inhibitor N-[(1R,2S)-2-hydroxy-1-hydroxymethyl-2-(2-tridecyl-1-cyclopropenyl)ethyl]o ctanamide (GT11) in primary cultured cerebellar neurons. Mol Pharmacol 66:1671–1678

Ueda N, Yamanaka K, Yamamoto S (2001) Purification and characterization of an acid amidase selective for N-palmitoylethanolamine, a putative endogenous anti-inflammatory substance. J Biol Chem 276:35552–35557

Wascholowski V, Giannis A (2006) Sphingolactones: selective and irreversible inhibitors of neutral sphingomyelinase. Angew Chem Int Ed 45:827–830

The Therapeutic Potential of Nanoscale Sphingolipid Technologies

Jody L. Hankins, Ushma A. Doshi, Jeremy K. Haakenson, Megan M. Young, Brian M. Barth, and Mark Kester

Contents

Abstract Nanotechnologies, while small in size, widen the scope of drug delivery options for compounds with problematic pharmacokinetics, such as bioactive sphingolipids. We describe the development of historical sphingolipid nanotechnologies, such as nanoliposomes, and project future uses for a broad repertoire of nanoscale sphingolipid therapy formulations. In particular, we describe sphingo-nanotherapies for treatment of cancer, inflammatory disease, and cardiovascular disease. We conclude with a discussion of the challenges associated with regulatory approval, scale-up, and development of these nanotechnology therapies for clinical applications.

Keywords Sphingolipid • Nanotechnology • Ceramide • Liposome • Therapy

---

J.L. Hankins • U.A. Doshi • J.K. Haakenson • M.M. Young • B.M. Barth • M. Kester (✉)
Department of Pharmacology, R130, Penn State University College of Medicine, Hershey, PA 17033, USA
e-mail: mkester@psu.edu

E. Gulbins and I. Petrache (eds.), *Sphingolipids: Basic Science and Drug Development*, Handbook of Experimental Pharmacology 215, DOI 10.1007/978-3-7091-1368-4_11,

1 Suitability of Nanotechnologies for Sphingolipid Therapies

Spatiotemporal modulation of sphingolipid mass and metabolism is a promising avenue to treat inflammatory disease, cancer, and cardiovascular disease. Tactics for manipulation of physiologic sphingolipid mass include pharmacologic targeting of sphingolipid enzymes or transporters, or direct(ed) delivery of the desired sphingolipid to the site of action. To deliver sphingolipids as therapeutics in both a *direct* and *directed* manner, one has a number of challenges to surpass. Key considerations include improved solubility of these highly hydrophobic molecules, host response to the sphingolipid:carrier complex, and capacity to selectively target delivery to the desired physiological compartment.

Nanoscale technologies can address many of these challenges, while also capitalizing on novel properties, due to their size of 100 nm or less. Engineering of nanoparticles can facilitate alteration of size, shape, and electrostatic charge, as well as the attachment of targeting moieties. Notably, nanotherapeutics can also be designed to passively accumulate in tumor beds due to the enhanced permeability and retention effect (Matsumura and Maeda 1986). Beyond the usual pharmacokinetic profiling of any potential drug, nanotechnologies must meet additional criteria for development into drug delivery vehicles. These criteria include drug-encapsulation, -retention, -efficacy, and -stability under physiologic conditions, and the absence of toxic materials, both in the initial formulation and after metabolism (Adair et al. 2010).

2 Liposome-Based Sphingolipid Nanotherapy Development

The timeline of nanoscale sphingolipid development highlights the advancement toward a more sophisticated nanotherapeutic with predictable immunological properties (Fig. 1). While immune stimulation is desirable for nanoparticle vaccine delivery, immune escape is a key feature of nanotherapies that require long circulation times for biodistribution to the appropriate compartment. The first lipid-based nanotechnologies were phospholipid liposomes, simple lipid vesicles that enclosed an aqueous compartment with limited diffusion into or out of the vesicle (Bangham et al. 1965). Systemic administration of active pharmaceutical ingredient (API)-loaded liposomes exhibited protective encapsulation of compounds *in vivo*, albeit with a short half-life (Gregoriadis and Ryman 1972). However, the observation that the immune system could readily destroy such liposomes by complement-mediated lysis (both with and without antibody opsonization) demonstrated that further refinement was necessary for liposomes to be viable drug delivery vehicles (Haxby et al. 1969; Lachmann et al. 1970).

To optimize drug delivery, liposomes and other nanotechnologies were engineered to evade immune detection through composition and surface modification. Early liposomes were perceived as an immune challenge due to their lipid

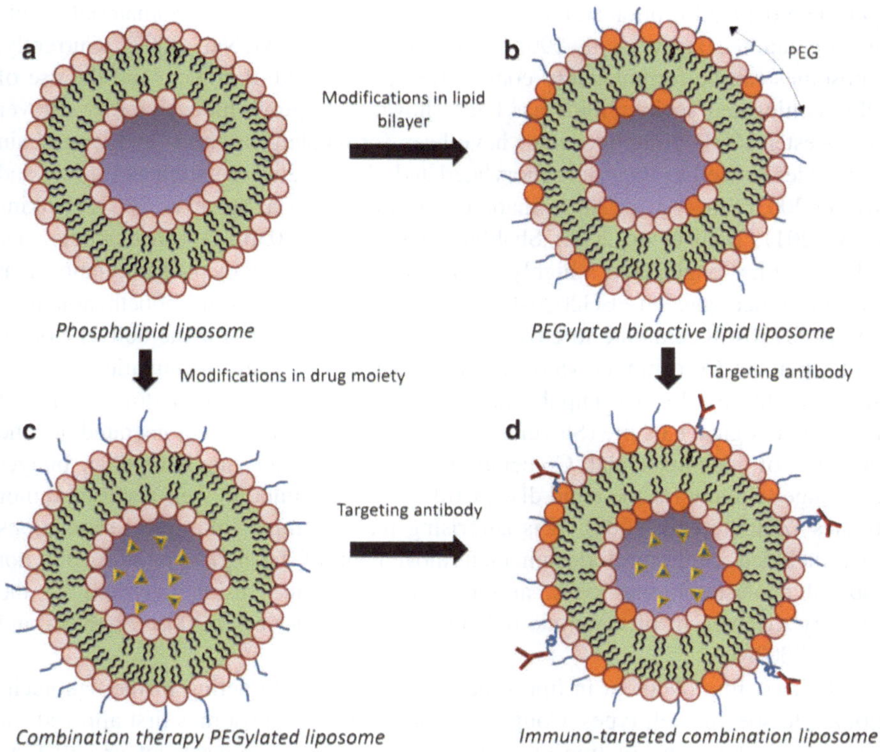

Fig. 1 The evolution of the nanoliposome. (**a**) Liposomes of various lipid mixtures (phospholipids and cholesterol) can be created at the nanoscale, less than 100 nm. (**b**) Second-generation nanoliposomes can be composed of various concentrations of bioactive lipids, including sphingolipids. (**c**) In addition, nanoliposomes can be created to contain therapeutic concentrations of active pharmaceutical ingredients (API). Nanoliposomal vehicles for bioactive lipids or APIs can then undergo surface passivation with PEG to increase biological circulation time. (**d**) In addition, discrete phospholipids can be bio-conjugated with targeting molecules, including antibodies, to create immuno-liposomes for enhanced tissue targeting

composition and tendency to aggregate into larger structures. Thus, they readily accumulated in the reticuloendothelial system, and thereby had unfavorable pharmacokinetics and distribution properties for many applications. Alterations in the lipid composition of liposomes to mimic erythrocyte membrane composition, such as incorporation of sphingomyelin and the complex sphingolipid ganglioside GM1, proved to increase liposome retention in circulation (Allen and Chonn 1987; Allen et al. 1989; Maruyama et al. 1990). Since immune system destruction is initiated by the binding of proteins such as immunoglobulins and complement to nanoparticles in the blood, surface modifications that could prevent this binding was an important development. Phospholipid-coupled polyethylene glycol (PEG) was demonstrated to effectively increase the circulation time of liposomes and protect them from degradation by preventing aggregation (Klibanov et al. 1990). Further, the addition of PEG or other surface molecules was able to effectively mask the surface, decreasing plasma protein adsorption and minimizing host responses to the nanomaterials while in circulation (Bazile et al. 1995; Dobrovolskaia and McNeil 2007). Currently, liposomes can be formulated to contain the sphingolipid of choice, and the use of PEG-conjugated sphingolipids has been shown to improve bioavailability (Stover and Kester 2003). Nanoliposomes have been formulated for short- and long-chain ceramides as well as their phosphorylated and glycosylated metabolites (Stover and Kester 2003; Stover et al. 2005; Shabbits and Mayer 2003a; Sun et al. 2008; Hankins et al. 2011; Salli et al. 2009; Shabbits and Mayer 2003b). As an example, our laboratory has engineered a highly PEGylated (14 molar percent) formulation that includes bioconjugated PEG2000-1,2-distearoyl-*sn*-glycero-3-phosphoethanolamine and PEG750-C8-ceramide to create an 80 nm-sized vehicle for the delivery of 30 molar percent C6-ceramide (Stover and Kester 2003). These formulations exhibit shelf stability for 2 years (unpublished, Nanocharacterization Laboratory, NCI) and an 11-h biological half-life (Stover and Kester 2003). We have demonstrated that the delivery of nanoliposomal C6-ceramide occurs through interlammellar bilayer exchange to allow for the rapid distribution of C6-ceramide to well-perfused tumor tissues (Zolnik et al. 2008). It is surprising that ceramide liposome formulations that included cholesterol documented movement of labeled ceramide, but not labeled cholesterol, into cells, arguing that membrane fusion may not be the primary mechanism of ceramide delivery (Stover and Kester 2003; Shabbits and Mayer 2003b).

The next improvement in liposome technology was targeting of the nanotechnology to specific cell types. Coupling of antibodies to liposomes first allowed for site-specific targeting of liposomes to the appropriate compartment or cell type (Barbet et al. 1981). Now, directed delivery can be accomplished through antibody fragments, aptamers, small molecules, peptides, or proteins targeting cell surface receptors (Yu et al. 2010).

The new paradigm in sphingolipid nanoliposome therapy is combination therapy. This includes a multipronged approach whereby one "primes" a system with exogenous sphingolipid while also delivering an enzymatic inhibitor. For example, we have shown that delivery of PDMP (D-*threo*-1-phenyl-2-decanoylamino-3-morpholino-1-propanol), a glucosylceramide inhibitor, within a C6-ceramide liposome formulation potentiated the cytotoxic effect of C6-ceramide in pancreatic cancer (Jiang et al. 2011) and glio- and neuroblastomas (Barth et al. 2010a). In addition, combination therapies may also include ceramide-containing nanoliposomes that are also delivery vehicles for ceramide-generating or sphingolipid-modulating chemotherapeutics (Barth et al. 2011). Thus, liposome formulations can be modified to modulate multiple sphingolipid pathways and provide a unique reinforcement mechanism. However, despite the ability to formulate targeted and combinational sphingolipid-containing nanoliposomes, delivery of the bioactive sphingolipids at the desired therapeutic concentration, to a specific cell type, in a complex tissue bed or tumor microenvironment remains a challenge and an area for future improvement.

3 Other Sphingolipid Nanotherapy Platforms

Nanotechnologies, other than liposomes, that incorporate sphingolipids have also been developed. These include nanoemulsions, polyethyleneoxide-modified polyepsilon-caprolactone nanoparticles (PEO-PCL), linear-dendritic nanoparticles, and calcium phosphosilicate nanoparticles (Barth et al. 2011). Nanoemulsions are suitable vehicles for topical or intravenous administration of sphingolipids. For example, delivery of 2*S*,3*S*,4*R*-2-stearoylamide-1,3,4-octadecanetriol (ceramide 3, an *N*-acylated phytosphinganine) as a topical nanoemulsion improved skin elasticity and moisture content (Yilmaz and Borchert 2006). Topical sphingolipid chemotherapies at one time appeared promising, but a clinical trial to test the efficacy of a short-chain ceramide nanoemulsion in cutaneous metastatic breast cancer patients did not demonstrate benefit (Jatoi et al. 2003). However, an intravenous ceramide and paclitaxel nanoemulsion, designed to cross the blood–brain barrier, increased apoptosis in glioblastoma cells, but has not yet been tested in vivo (Desai et al. 2008).

Other intravenous nanoformulations include PEO-PCL and thermally activated linear-dendritic nanoparticles. Each of these nanotechnologies offer unique, controlled drug delivery capabilities. PEO-PCLs are polymers that undergo hydrolysis under physiologic conditions (Kumari et al. 2010). The molecular weight of the polymer controls the release of the active ingredient, with larger structures generally exhibiting slower release characteristics. Coadministration of ceramide and either paclitaxel or tamoxifen in PEO-PCLs induced apoptosis and resensitized multidrug-resistant ovarian cancer cells (Devalapally et al. 2007, 2008; van Vlerken et al. 2007). Likewise, thermally activated linear-dendritic nanoparticles offer the opportunity for improved pharmacokinetic properties. For example, these thermally activated, linear-dendritic polymer nanoparticles can slowly release the API after reaching the target. Using linear-dendritic nanoparticles, sustained ceramide delivery induced apoptosis in breast adenocarcinoma cells (Stover et al. 2008).

Nanocolloidal suspensions are another technology platform that is conducive to intravenous administration. Calcium phosphosilicate nanoparticles (CPSNPs) are nanocolloids with unique properties. CPSNPs are approximately 20 nm in diameter and break down into nontoxic products normally found in the body, namely, calcium and phosphate ions (Morgan et al. 2008). CPSNPs escape enterohepatic circulation, which limits hepatic accumulation and toxicity because they are excreted in the feces (Barth et al. 2010b). One distinguishing feature of CPSNPs is the targeted dissolution of the nanoparticle in the late endosome (Morgan et al. 2008). CPSNPs have limited solubility at pH 7.4, but when the pH within the late endosome drops to pH 4–5, the particle dissolves (Morgan et al. 2008). Since the release of the active agent within CPSNPs is pH responsive, toxicity to normal, non-targeted cells due to accumulation of free drug in extracellular fluids is limited (Barth et al. 2010b). We speculate that release of phosphate from the particle acts to buffer the action of proton pumps to acidify the endosome, ultimately causing endosomal rupture. CPSNPs may therefore function as “bio-smart bombs.” Nanocolloidal formulations allow for controlled release of cargoes in a stable platform with limited toxicity. *In vitro* data also suggests that bioconjugated CPSNPs loaded with both fluorescent imaging agents and ceramide can potentially be developed for "theranostic" use, a strategy which combines therapeutic and diagnostic tools (Kester et al. 2008; Morgan et al. 2008; Barth et al. 2010b). Incorporation of C6-ceramide into CPSNPs has been shown to inhibit cell growth in human vascular smooth muscle cells (VSMC) much more potently than other ceramide-delivery systems (Morgan et al. 2008), arguing that CPSNPs are promising nanocolloid delivery vehicles for sphingolipids. Other nanocolloids, including silica, gold, or silver, may also allow for sphingolipid delivery, albeit as surface-decorated particles.

4 Sphingolipid Nanoscale Therapies for Cancers

Sphingolipid-centered nanotherapies have garnered attention as a strategy to selectively seek and destroy cancerous cells. The clinical success of Doxil®, a PEGylated nanoliposomal formulation of doxorubicin, is a model for the use of nanotechnology delivery systems in cancer therapy. Approved for clinical use in the treatment of ovarian cancer, AIDS-related Kaposi's sarcoma, and multiple myeloma, Doxil® significantly prolongs the half-life of doxorubicin to enhance its efficacy (Product Information Booklet). Furthermore, Doxil® limited the accumulation of doxorubicin in the myocardium, thereby reducing clinical cardiotoxicity (Rahman et al. 2007). Interestingly, incorporation of short-chain glucosylceramide into doxorubicin-containing nanoliposomes has been shown to enhance the cellular uptake of the agent and its antitumor activity in vitro and in vivo (van Lummel et al. 2011).

The ability of ceramide to selectively induce apoptosis in cancer cells versus non-transformed cells has spurred the development of several promising ceramide-based therapeutics (Barth et al. 2011). Nanoliposomes are an effective delivery vehicle for natural (Shabbits and Mayer 2003a, b) and short-chain ceramides in vivo, and have been shown to inhibit tumor growth in models of breast cancer (Stover and Kester 2003; Stover et al. 2005), J774 sarcoma (Shabbits and Mayer 2003a), melanoma (Tran et al. 2008), hepatocellular carcinoma (Tagaram et al. 2011), large granular lymphocytic leukemia (Liu et al. 2010), and pancreatic cancer (Jiang et al. 2011). As outlined, ceramide-based nanotherapies have proven efficacious in multiple solid and non-solid cancer models *in vitro* and *in vivo*. Importantly, C6-ceramide nanoliposomes selectively target cancerous cells with minimal adverse toxicity (Liu et al. 2010; Stover and Kester 2003; Stover et al. 2005; Tran et al. 2008), a result that is supported by extensive in vivo toxicology studies (see http://ncl.cancer.gov/working_technical_reports.asp).

Importantly, ceramide nanoliposomes lend themselves to combinatorial therapy by serving as delivery vehicles for cytotoxic agents. Ceramide accumulation has been observed to occur secondary to the primary effects of a wide range of chemotherapeutics, including doxorubicin, fluorouracil, sorafenib, paclitaxel, tamoxifen, and gemcitabine (Barth et al. 2011). As a result, encapsulation of these agents within ceramide-containing nanoliposomes can enhance cytotoxicity by improving pharmacokinetics and providing a multipronged mechanism for the induction of cell death. In fact, C6-ceramide nanoliposomes have been shown to synergistically enhance the preclinical therapeutic efficacy of sorafenib in melanoma and breast cancer (Tran et al. 2008) and gemcitabine in studies of pancreatic cancer (Jiang et al. 2011). Furthermore, several sphingolipid-modulating agents have been developed to target dysregulated sphingolipid metabolism in cancer and to promote cell death. These agents include inhibitors of sphingosine kinase, acid ceramidase, and glucosylceramide synthase (discussed in Barth et al. 2011). Similar to that of the chemotherapeutics discussed above, encapsulation of sphingolipid-modulating agents within ceramide nanoliposomes can greatly improve the pharmacokinetics, toxicology, and cytotoxic efficacy of these compounds. Delivery of safingol, a synthetic L-*threo* isomer of dihydrosphingosine, as an emulsion in combination with cisplatin has recently shown promising results in a phase I clinical trial for refractory adrenocortical cancer (Dickson et al. 2011). However, this formulation has been shown to cause hemolysis and dose-limiting toxicity in preclinical models (Kedderis et al. 1995). Recently, liposomal delivery of safingol has been demonstrated to improve the pharmacokinetics and cytotoxicity of the sphingolipid-modulating agent in an acute myeloid leukemia model while significantly reducing hemolysis (Tan et al. 2012). The future incorporation of safingol in ceramide nanoliposomes can be envisioned as one approach to further improve cytotoxic efficacy through combinatorial therapy.

In addition to combinatorial therapy, sphingolipid nanoliposome formulations can be specifically targeted to cancer cells. For example, the transferrin receptor is frequently overexpressed in cancer cells, making it an excellent target for drug delivery (Larrick and Cresswell 1979). Incorporation of the glycoprotein transferrin into C6-ceramide nanoliposomes has been shown to significantly increase endocytosis of the liposomes, lysosomal accumulation of ceramide, and induction of apoptosis (Koshkaryev et al. 2012). Furthermore, the glycosphingolipid, GD2, has been utilized as a target for nanotherapies. GD2 is highly expressed in cancers of neuronal origin, and the conjugation of an anti-GD2 antibody to the liposomal surface has been shown to effectively and selectively deliver siRNA targeting VEGF (Adrian et al. 2011) and doxorubicin (Pastorino et al. 2006) to neuroblastoma, as well as siRNA targeting c-myc (Pastorino et al. 2003) and fenretinide (Pagnan et al. 1999) to GD2-positive melanoma. Finally, nanotherapies engineered to target specific subcellular organelles have been developed. Ceramide is a well-known inducer of mitochondrial damage and has been shown to assemble pores in the outer mitochondrial membrane to mediate the release of cytochrome c (Siskind and Colombini 2000). C6-ceramide nanoliposomes have been targeted to mitochondria by conjugating the triphenylphosphonium (TPP) cation to a stearyl residue that is incorporated into the lipid bilayer (Boddapati et al. 2008). These targeted liposomes preferentially accumulate in the mitochondria to result in enhanced cytotoxicity in models of mouse mammary carcinoma (Boddapati et al. 2008). Collectively, these developments provide a basis for further development of sphingolipid-based nanoliposomes that target specific cancer cells and/or organelles.

Other than liposome-based formulations, we have shown that C10-ceramide-loaded CPSNPs reduced cell survival and viability of melanoma cells and breast cancer cells in vitro (Kester et al. 2008). Ceramide formulated in thermoresponsive and biodegradable linear-dendritic nanoparticles in breast adenocarcinoma cells led to prolonged release of C6-ceramide in vitro, thus displaying a great potential for the treatment of solid tumors with bioactive lipids and hyperthermia (Stover et al. 2008). Additionally, self-assembled nanoparticles containing synthetically conjugated ceramide and hyaluronic acid (HA) and the block copolymer Pluronic 85 have been reported. Docetaxel- and doxorubicin-loaded HA-ceramide nanoparticles have been shown to enhance the efficacy of each agent by effectively targeting cancer cells through the association of HA and CD44, a cell surface receptor frequently overexpressed in cancer (Cho et al. 2011, 2012). Collectively, a variety of sphingolipid-centered nanoscale therapies for cancers have shown promising results. While these therapies have focused on ceramide due to its well-documented apoptotic properties, other apoptotic sphingolipids, such as sphingosine, may be incorporated into cancer nanotherapies in the future.

5 Sphingolipid Nanotherapy in the Control of the Immune Response

Sphingolipids are important regulators of immune response, and sphingo-nanotherapies targeting inflammation are ripe for development. In a model of corneal keratitis, topical administration of C6-ceramide nanoliposomes prevented release of chemotactic cytokines from the cornea, limiting neutrophil infiltration, without inhibiting the wound healing response (Sun et al. 2008). The concept that new immunosuppressive therapies could arise from sphingolipid modulation is exemplified by the sphingolipid mimetic, FTY720, which when phosphorylated prevents egress of lymphocytes from secondary lymphoid organs through sphingosine-1-phosphate (S1P) receptor ligation and internalization (Brinkmann et al. 2002; Mandala et al. 2002; Matloubian et al. 2004). FTY720 (fingolimod) has been approved for the treatment of relapsing multiple sclerosis (Brinkmann et al. 2010). Other inhibitors of S1P signaling show promise to treat inflammatory diseases such as cancer, asthma, rheumatoid arthritis, and inflammatory bowel disease (Edmonds et al. 2011).

In addition, the acid sphingomyelinase/ceramide/ceramide-1-phosphate (C-1-P) pathway could be an exploitable target to control inflammation through nanotechnology. Acid sphingomyelinase-deficient mice displayed potentiated TNF-α secretion following LPS challenge (Rozenova et al. 2010), and both ceramide and C-1-P have been shown to dampen LPS-mediated TNF-α secretion (Jozefowski et al. 2010). Further, C-1-P directly interacts with TNF-α-converting enzyme to negatively regulate its activity (Lamour et al. 2011). Extending upon these anti-inflammatory actions of C-1-P, we have demonstrated that nanoliposomal C-1-P limited secretion of not only TNF-α but also IL-6, IL-8, and IL-1β in LPS-stimulated human peripheral blood mononuclear cells (Hankins et al. 2011).

While sphingolipid modulation is one approach to control inflammation by nanotechnology, it is also conceivable that sphingolipid moieties could be employed as engineering tools. Since pathogen-like molecules may be affixed to nanotechnologies from material impurities (or by design), and these molecules can activate Toll-like receptors, it may be desirable to modulate inflammatory responses to these nanotechnologies. Sphingolipids such as ceramide and C-1-P can limit cytokine secretion following Toll-like receptor 4 activation in some cell types (Chiba et al. 2007; Hankins et al. 2011; Jozefowski et al. 2010; Sun et al. 2008). Therefore, modification of nanotechnologies with these sphingolipids could selectively alter host detection or the response profile to these therapies. In sum, sphingolipid-targeted or sphingolipid-protected immunotherapy appears promising, and nanotherapies are uniquely qualified to deliver upon these promises.

6 Sphingolipid Nanotherapy in Cardiovascular Disease

Altered sphingolipid metabolism is critical in cardiovascular physiology and disease. We have shown that ceramide nanofilm-coated balloon catheters inhibited neointimal hyperplasia and decreased restenosis, indicating that ceramide was anti-proliferative, anti-inflammatory, and anti-atherogenic (Charles et al. 2000; O'Neill et al. 2008). In addition, ceramide delivery from balloon catheters decreased macrophage activation and fibrin matrix formation (O'Neill et al. 2008). To limit smooth muscle cell proliferation while maintaining endothelial cell wound healing response, C6-ceramide nanofilms capitalized on differential metabolism in vascular endothelial cells and smooth muscle cells (O'Neill et al. 2008). In VSMC, C6-ceramide limited proliferation and induced cell cycle arrest (Charles et al. 2000). However, in coronary artery endothelial cells, C6-ceramide was readily converted to C-1-P and glycosphingolipids, which may have facilitated wound healing (O'Neill et al. 2008). Thus, ceramide nanofilms modulated different actions in multiple cell types to orchestrate a desired response at the tissue level. Drug-eluting stents with biopolymerized PDMP, a glucosylceramide and lactosylceramide synthase inhibitor, have also been shown to prevent smooth muscle cell proliferation and prevent stent-induced hyperplasia (Tang et al. 2009), indicating that sphingolipid therapies could modulate cardiovascular disease. In the future, nanotherapeutics that target sphingolipid metabolism could be used as long-lasting, nontoxic treatments and preventative agents for restenosis, myocardial infarction, atherosclerosis, and venous thrombosis.

S1P has also been shown to have a dramatic effect on the cardiovascular system. S1P has been shown to decrease heart rate, blood pressure, and blood flow in various parts of the body, including the renal and mesenteric arteries (Bischoff et al. 2000; Peters and Alewijnse 2007). In addition, S1P mediated endothelial cell migration, proliferation, and barrier integrity (Peters and Alewijnse 2007), and it protected cardiac myocytes from hypoxic death (Alewijnse and Peters 2008). The effects of S1P on VSMC migration, proliferation, and contraction were dependent on the specific S1P receptors ($S1P_{1-5}$) expressed at a particular time (Peters and Alewijnse 2007). There may be times when angiogenesis is desirable, such as in the treatment of peripheral vascular disease, or to protect and differentiate cardiac myocytes in congestive heart failure. Thus, it may be plausible to utilize nanoformulations for the targeted delivery of S1P in these circumstances.

7 Outlook on Development and Commercialization of Sphingolipid Nanotherapies

Clinical studies with nanoscale formulations of sphingolipids have not yet commenced. Non-nanoscale topical ceramide-based gels have been tested in the clinic for cutaneous breast cancer, but did not demonstrate efficacy (Jatoi et al. 2003). Systemic delivery of non-aggregating nanoscale formulations may have more efficacy in clinical applications. Many laboratories and companies are beginning to edge toward the clinic as they continue to characterize, optimize, and validate the various pharmacological, molecular, and nanotechnological approaches to reset sphingolipid metabolism in diseased patients. Completing costly and time-consuming pharmacokinetics/ADME/toxicity studies for sphingolipid-based nanotechnologies often requires one to leverage resources from many sources. These sources include, but are not limited to, the Nanotechnology Characterization Laboratory, National Cancer Institute (Ft. Detrick, MD), or the National Institutes of Health Bridging Interventional Development Gaps Program (BRIDGs). Even with sufficient resources, critical pharmacokinetic and toxicity data for scaled-up good manufacturing practices (GMP)-grade preparation of sphingolipid-based nanoformulations remains a hurdle. As a case in point, extrusion methodologies to formulate nanoliposomes for animal experiments are not physically equivalent to a "scaled-up" formulation. To support human trials, one must employ microfluidic/emulsification methodologies to produce the required quantity (grams) of GMP-grade materials. Data from GMP-grade lipid materials are required in order to file an Investigational New Drug application to support first-in-man studies. Investigators or development companies often must negotiate and partner with organizations vertically along the supply chain to begin efficient scale-up, manufacturing, and testing of GMP-grade nanoformulations. One final consideration is that very few nanoscale drugs have been approved for human use, with Doxil® (doxorubicin-encapsulated liposomes) and Abraxane® (docetaxel-conjugated albumin) as notable exceptions. The path to market is therefore not sufficiently demarcated for one to proceed without an arsenal of resources to clear the way. The Food and Drug Administration is rightly concerned about chronic toxicities that may result from nanoscale formulations that are designed, by definition, to exhibit long biological half-lives. Yet, as more nanotechnologies, such as lipid–polymer complexes and nanocolloids (silver, gold), continue to progress through Phase 1/2 studies, it can be envisioned that variations of these nanoscale formulations engineered for systemic delivery of sphingolipids will also transition successfully to the clinic.

Conflict of Interest Statement. Penn State Research Foundation has licensed several nanoscale formulations for sphingolipids to Keystone Nano., Inc. (Boalsburg, PA). Dr. Kester is cofounder and Chief Medical Officer of Keystone Nano.

References

Adair JH, Parette MP, Altinoglu EI, Kester M (2010) Nanoparticulate alternatives for drug delivery. ACS Nano 4(9):4967–4970

Adrian JE, Wolf A, Steinbach A, Rossler J, Suss R (2011) Targeted delivery to neuroblastoma of novel siRNA-anti-GD2-liposomes prepared by dual asymmetric centrifugation and sterol-based post-insertion method. Pharm Res 28(9):2261–2272

Alewijnse AE, Peters SL (2008) Sphingolipid signalling in the cardiovascular system: good, bad or both? Eur J Pharmacol 585(2–3):292–302

Allen TM, Chonn A (1987) Large unilamellar liposomes with low uptake into the reticuloendothelial system. FEBS Lett 223(1):42–46

Allen TM, Hansen C, Rutledge J (1989) Liposomes with prolonged circulation times: factors affecting uptake by reticuloendothelial and other tissues. Biochim Biophys Acta 981(1):27–35

Bangham AD, Standish MM, Watkins JC (1965) Diffusion of univalent ions across the lamellae of swollen phospholipids. J Mol Biol 13(1):238–252

Barbet J, Machy P, Leserman LD (1981) Monoclonal antibody covalently coupled to liposomes: specific targeting to cells. J Supramol Struct Cell Biochem 16(3):243–258

Barth BM, Gustafson SJ, Young MM, Fox TE, Shanmugavelandy SS, Kaiser JM, Cabot MC, Kester M, Kuhn TB (2010a) Inhibition of NADPH oxidase by glucosylceramide confers chemoresistance. Cancer Biol Ther 10(11):1126–1136

Barth BM, Sharma R, Altinoglu EI, Morgan TT, Shanmugavelandy SS, Kaiser JM, McGovern C, Matters GL, Smith JP, Kester M, Adair JH (2010b) Bioconjugation of calcium phosphosilicate composite nanoparticles for selective targeting of human breast and pancreatic cancers in vivo. ACS Nano 4(3):1279–1287

Barth BM, Cabot MC, Kester M (2011) Ceramide-based therapeutics for the treatment of cancer. Anticancer Agents Med Chem 11(9):911–919

Bazile D, Prud'homme C, Bassoullet MT, Marlard M, Spenlehauer G, Veillard M (1995) Stealth Me.PEG-PLA nanoparticles avoid uptake by the mononuclear phagocytes system. J Pharm Sci 84(4):493–498

Bischoff A, Czyborra P, Meyer Zu Heringdorf D, Jakobs KH, Michel MC (2000) Sphingosine-1-phosphate reduces rat renal and mesenteric blood flow in vivo in a pertussis toxin-sensitive manner. Br J Pharmacol 130(8):1878–1883

Boddapati SV, D'Souza GG, Erdogan S, Torchilin VP, Weissig V (2008) Organelle-targeted nanocarriers: specific delivery of liposomal ceramide to mitochondria enhances its cytotoxicity in vitro and in vivo. Nano Lett 8(8):2559–2563

Brinkmann V, Davis MD, Heise CE, Albert R, Cottens S, Hof R, Bruns C, Prieschl E, Baumruker T, Hiestand P, Foster CA, Zollinger M, Lynch KR (2002) The immune modulator FTY720 targets sphingosine 1-phosphate receptors. J Biol Chem 277(24):21453–21457

Brinkmann V, Billich A, Baumruker T, Heining P, Schmouder R, Francis G, Aradhye S, Burtin P (2010) Fingolimod (FTY720): discovery and development of an oral drug to treat multiple sclerosis. Nat Rev Drug Discov 9(11):883–897

Charles R, Sandirasegarane L, Yun J, Bourbon N, Wilson R, Rothstein RP, Levison SW, Kester M (2000) Ceramide-coated balloon catheters limit neointimal hyperplasia after stretch injury in carotid arteries. Circ Res 87(4):282–288

Chiba N, Masuda A, Yoshikai Y, Matsuguchi T (2007) Ceramide inhibits LPS-induced production of IL-5, IL-10, and IL-13 from mast cells. J Cell Physiol 213(1):126–136

Cho HJ, Yoon HY, Koo H, Ko SH, Shim JS, Lee JH, Kim K, Kwon IC, Kim DD (2011) Self-assembled nanoparticles based on hyaluronic acid-ceramide (HA-CE) and Pluronic(R) for tumor-targeted delivery of docetaxel. Biomaterials 32(29):7181–7190

Cho HJ, Yoon IS, Yoon HY, Koo H, Jin YJ, Ko SH, Shim JS, Kim K, Kwon IC, Kim DD (2012) Polyethylene glycol-conjugated hyaluronic acid-ceramide self-assembled nanoparticles for targeted delivery of doxorubicin. Biomaterials 33(4):1190–1200

Desai A, Vyas T, Amiji M (2008) Cytotoxicity and apoptosis enhancement in brain tumor cells upon coadministration of paclitaxel and ceramide in nanoemulsion formulations. J Pharm Sci 97(7):2745–2756

Devalapally H, Duan Z, Seiden MV, Amiji MM (2007) Paclitaxel and ceramide co-administration in biodegradable polymeric nanoparticulate delivery system to overcome drug resistance in ovarian cancer. Int J Cancer 121(8):1830–1838

Devalapally H, Duan Z, Seiden MV, Amiji MM (2008) Modulation of drug resistance in ovarian adenocarcinoma by enhancing intracellular ceramide using tamoxifen-loaded biodegradable polymeric nanoparticles. Clin Cancer Res 14(10):3193–3203

Dickson MA, Carvajal RD, Merrill AH Jr, Gonen M, Cane LM, Schwartz GK (2011) A phase I clinical trial of safingol in combination with cisplatin in advanced solid tumors. Clin Cancer Res 17(8):2484–2492

Dobrovolskaia MA, McNeil SE (2007) Immunological properties of engineered nanomaterials. Nat Nanotechnol 2(8):469–478

Edmonds Y, Milstien S, Spiegel S (2011) Development of small-molecule inhibitors of sphingosine-1-phosphate signaling. Pharmacol Ther 132(3):352–360

Gregoriadis G, Ryman BE (1972) Fate of protein-containing liposomes injected into rats. An approach to the treatment of storage diseases. Eur J Biochem 24(3):485–491

Hankins JL, Fox TE, Barth BM, Unrath KA, Kester M (2011) Exogenous ceramide-1-phosphate reduces lipopolysaccharide (LPS)-mediated cytokine expression. J Biol Chem 286(52): 44357–44366

Haxby JA, Gotze O, Muller-Eberhard HJ, Kinsky SC (1969) Release of trapped marker from liposomes by the action of purified complement components. Proc Natl Acad Sci U S A 64(1): 290–295

Jatoi A, Suman VJ, Schaefer P, Block M, Loprinzi C, Roche P, Garneau S, Morton R, Stella PJ, Alberts SR, Pittelkow M, Sloan J, Pagano R (2003) A phase II study of topical ceramides for cutaneous breast cancer. Breast Cancer Res Treat 80(1):99–104

Jiang Y, DiVittore NA, Kaiser JM, Shanmugavelandy SS, Fritz JL, Heakal Y, Tagaram HRS, Cheng H, Cabot MC, Staveley-O'Carroll KF, Tran MA, Fox TE, Barth BM, Kester M (2011) Combinatorial therapies improve the therapeutic efficacy of nanoliposomal ceramide for pancreatic cancer. Cancer Biol Ther 12(7):574–585

Jozefowski S, Czerkies M, Lukasik A, Bielawska A, Bielawski J, Kwiatkowska K, Sobota A (2010) Ceramide and ceramide 1-phosphate are negative regulators of TNF-α production induced by lipopolysaccharide. J Immunol 185(11):6960–6973

Kedderis LB, Bozigian HP, Kleeman JM, Hall RL, Palmer TE, Harrison SD Jr, Susick RL Jr (1995) Toxicity of the protein kinase C inhibitor safingol administered alone and in combination with chemotherapeutic agents. Fundam Appl Toxicol 25(2):201–217

Kester M, Heakal Y, Fox T, Sharma A, Robertson GP, Morgan TT, Altinoglu EI, Tabakovic A, Parette MR, Rouse SM, Ruiz-Velasco V, Adair JH (2008) Calcium phosphate nanocomposite particles for in vitro imaging and encapsulated chemotherapeutic drug delivery to cancer cells. Nano Lett 8(12):4116–4121

Klibanov AL, Maruyama K, Torchilin VP, Huang L (1990) Amphipathic polyethyleneglycols effectively prolong the circulation time of liposomes. FEBS Lett 268(1):235–237

Koshkaryev A, Piroyan A, Torchilin VP (2012) Increased apoptosis in cancer cells in vitro and in vivo by ceramides in transferrin-modified liposomes. Cancer Biol Ther 13(1):50–60

Kumari A, Yadav SK, Yadav SC (2010) Biodegradable polymeric nanoparticles based drug delivery systems. Colloids Surf B Biointerfaces 75(1):1–18

Lachmann PJ, Munn EA, Weissmanng (1970) Complement-mediated lysis of liposomes produced by the reactive lysis procedure. Immunology 19(6):983–986

Lamour NF, Wijesinghe DS, Mietla JA, Ward KE, Stahelin RV, Chalfant CE (2011) Ceramide kinase regulates the production of tumor necrosis factor alpha (TNFalpha) via inhibition of TNFalpha-converting enzyme. J Biol Chem 286(50):42808–42817

Larrick JW, Cresswell P (1979) Modulation of cell surface iron transferrin receptors by cellular density and state of activation. J Supramol Struct 11(4):579–586

Liu X, Ryland L, Yang J, Liao A, Aliaga C, Watts R, Tan SF, Kaiser J, Shanmugavelandy SS, Rogers A, Loughran K, Petersen B, Yuen J, Meng F, Baab KT, Jarbadan NR, Broeg K, Zhang R, Liao J, Sayers TJ, Kester M, Loughran TP Jr (2010) Targeting of survivin by nanoliposomal ceramide induces complete remission in a rat model of NK-LGL leukemia. Blood 116 (20):4192–4201

Mandala S, Hajdu R, Bergstrom J, Quackenbush E, Xie J, Milligan J, Thornton R, Shei GJ, Card D, Keohane C, Rosenbach M, Hale J, Lynch CL, Rupprecht K, Parsons W, Rosen H (2002) Alteration of lymphocyte trafficking by sphingosine-1-phosphate receptor agonists. Science 296(5566):346–349

Maruyama K, Kennel SJ, Huang L (1990) Lipid composition is important for highly efficient target binding and retention of immunoliposomes. Proc Natl Acad Sci U S A 87(15): 5744–5748

Matloubian M, Lo CG, Cinamon G, Lesneski MJ, Xu Y, Brinkmann V, Allende ML, Proia RL, Cyster JG (2004) Lymphocyte egress from thymus and peripheral lymphoid organs is dependent on S1P receptor 1. Nature 427(6972):355–360

Matsumura Y, Maeda H (1986) A New concept for macromolecular therapeutics in cancer chemotherapy: mechanism of tumoritropic accumulation of proteins and the antitumor agent smancs. Cancer Res 46(12 Part 1):6387–6392

Morgan TT, Muddana HS, Altinoğlu EI, Rouse SM, Tabaković A, Tabouillot T, Russin TJ, Shanmugavelandy SS, Butler PJ, Eklund PC, Yun JK, Kester M, Adair JH (2008) Encapsulation of organic molecules in calcium phosphate nanocomposite particles for intracellular imaging and drug delivery. Nano Lett 8(12):4108–4115

O'Neill SM, Olympia DK, Fox TE, Brown JT, Stover TC, Houck KL, Wilson R, Waybill P, Kozak M, Levison SW, Weber N, Karavodin LM, Kester M (2008) C(6)-ceramide-coated catheters promote re-endothelialization of stretch-injured arteries. Vasc Dis Prev 5(3):200–210

Pagnan G, Montaldo PG, Pastorino F, Raffaghello L, Kirchmeier M, Allen TM, Ponzoni M (1999) GD2-mediated melanoma cell targeting and cytotoxicity of liposome-entrapped fenretinide. Int J Cancer 81(2):268–274

Pastorino F, Brignole C, Marimpietri D, Pagnan G, Morando A, Ribatti D, Semple SC, Gambini C, Allen TM, Ponzoni M (2003) Targeted liposomal c-myc antisense oligodeoxynucleotides induce apoptosis and inhibit tumor growth and metastases in human melanoma models. Clin Cancer Res 9(12):4595–4605

Pastorino F, Brignole C, Di Paolo D, Nico B, Pezzolo A, Marimpietri D, Pagnan G, Piccardi F, Cilli M, Longhi R, Ribatti D, Corti A, Allen TM, Ponzoni M (2006) Targeting liposomal chemotherapy via both tumor cell-specific and tumor vasculature-specific ligands potentiates therapeutic efficacy. Cancer Res 66(20):10073–10082

Peters SL, Alewijnse AE (2007) Sphingosine-1-phosphate signaling in the cardiovascular system. Curr Opin Pharmacol 7(2):186–192

Product Information Booklet. http://doxil.com/assets/DOXIL_PI_Booklet.pdf Accessed 3 Mar 2012
Rahman AM, Yusuf SW, Ewer MS (2007) Anthracycline-induced cardiotoxicity and the cardiac-sparing effect of liposomal formulation. Int J Nanomedicine 2(4):567–583
Rozenova KA, Deevska GM, Karakashian AA, Nikolova-Karakashian MN (2010) Studies on the role of acid sphingomyelinase and ceramide in the regulation of tumor necrosis factor alpha (TNFalpha)-converting enzyme activity and TNFalpha secretion in macrophages. J Biol Chem 285(27):21103–21113
Salli U, Fox TE, Carkaci-Salli N, Sharma A, Robertson GP, Kester M, Vrana KE (2009) Propagation of undifferentiated human embryonic stem cells with nano-liposomal ceramide. Stem Cells Dev 18(1):55–65
Shabbits JA, Mayer LD (2003a) High ceramide content liposomes with in vivo antitumor activity. Anticancer Res 23(5A):3663–3669
Shabbits JA, Mayer LD (2003b) Intracellular delivery of ceramide lipids via liposomes enhances apoptosis in vitro. Biochim Biophys Acta 1612(1):98–106
Siskind LJ, Colombini M (2000) The lipids C2- and C16-ceramide form large stable channels. Implications for apoptosis. J Biol Chem 275(49):38640–38644
Stover T, Kester M (2003) Liposomal delivery enhances short-chain ceramide-induced apoptosis of breast cancer cells. J Pharmacol Exp Ther 307(2):468–475
Stover TC, Sharma A, Robertson GP, Kester M (2005) Systemic delivery of liposomal short-chain ceramide limits solid tumor growth in murine models of breast adenocarcinoma. Clin Cancer Res 11(9):3465–3474
Stover TC, Kim YS, Lowe TL, Kester M (2008) Thermoresponsive and biodegradable linear-dendritic nanoparticles for targeted and sustained release of a pro-apoptotic drug. Biomaterials 29(3):359–369
Sun Y, Fox T, Adhikary G, Kester M, Pearlman E (2008) Inhibition of corneal inflammation by liposomal delivery of short-chain, C-6 ceramide. J Leukoc Biol 83(6):1512–1521
Tagaram HR, Divittore NA, Barth BM, Kaiser JM, Avella D, Kimchi ET, Jiang Y, Isom HC, Kester M, Staveley-O'Carroll KF (2011) Nanoliposomal ceramide prevents in vivo growth of hepatocellular carcinoma. Gut 60(5):695–701
Tan KB, Ling LU, Bunte RM, Chng WJ, Chiu GN (2012) In vivo efficacy of a novel liposomal formulation of safingol in the treatment of acute myeloid leukemia. J Control Release 160(2): 290–298
Tang YD, Pandey A, Kolmakova A, Wang XT, Venkatraman SS, Chatterjee S, Boey FY (2009) Use of a novel anti-proliferative compound coated on a biopolymer to mitigate platelet-derived growth factor-induced proliferation in human aortic smooth muscle cells: comparison with sirolimus. Glycoconj J 26(6):721–732
Tran MA, Smith CD, Kester M, Robertson GP (2008) Combining nanoliposomal ceramide with sorafenib synergistically inhibits melanoma and breast cancer cell survival to decrease tumor development. Clin Cancer Res 14(11):3571–3581
van Lummel M, van Blitterswijk WJ, Vink SR, Veldman RJ, van der Valk MA, Schipper D, Dicheva BM, Eggermont AM, ten Hagen TL, Verheij M, Koning GA (2011) Enriching lipid nanovesicles with short-chain glucosylceramide improves doxorubicin delivery and efficacy in solid tumors. FASEB J 25(1):280–289
van Vlerken LE, Duan Z, Seiden MV, Amiji MM (2007) Modulation of intracellular ceramide using polymeric nanoparticles to overcome multidrug resistance in cancer. Cancer Res 67(10): 4843–4850
Yilmaz E, Borchert HH (2006) Effect of lipid-containing, positively charged nanoemulsions on skin hydration, elasticity and erythema—an in vivo study. Int J Pharm 307(2):232–238
Yu B, Tai HC, Xue W, Lee LJ, Lee RJ (2010) Receptor-targeted nanocarriers for therapeutic delivery to cancer. Mol Membr Biol 27(7):286–298
Zolnik BS, Stern ST, Kaiser JM, Heakal Y, Clogston JD, Kester M, McNeil SE (2008) Rapid distribution of liposomal short-chain ceramide in vitro and in vivo. Drug Metab Dispos 36(8): 1709–1715

Novel Chemotherapeutic Drugs in Sphingolipid Cancer Research

Daniel Canals and Yusuf A. Hannun

Contents

Abstract Sphingolipid-metabolizing enzymes are becoming targets for chemotherapeutic development with an increasing interest in the recent years. In this chapter we introduce the sphingolipid family of lipids, and the role of individual species in cell homeostasis. We also discuss their roles in several rare diseases and overall, in cancer transformation. We follow the biosynthesis pathway of the sphingolipid tree, focusing on the enzymes in order to understand how using

D. Canals
Department of Medicine, University of Stony Brook, Stony Brook, New York 11794

Y.A. Hannun (✉)
Health Science Center, Stony Brook University, 100 Nicolls Road, L-4, 178, Stony Brook, NY 11794, USA
e-mail: yusuf.hannun@stonybrookmedicine.edu

E. Gulbins and I. Petrache (eds.), *Sphingolipids: Basic Science and Drug Development*, Handbook of Experimental Pharmacology 215, DOI 10.1007/978-3-7091-1368-4_12,

small molecule inhibitors makes it possible to modulate cancer progression. Finally, we describe the most used and historically significant inhibitors employed in cancer research, their relationships to sphingolipid metabolism, and some promising results found in this field.

Keywords Ceramide • Sphingosine • Glycosphingolipid • Cancer • Myriocin • Cycloserine • Fumonisin B1 • 4-HRP • Scyphostatin • GW4869 • Desipramine • NOE • B13 • E-tb • MAPP • PDMP • Safingol • PF-543 • FTY720 • ABC294640

Abbreviations

D-*e*-MAPP	(1*S*, 2*R*)-D-*erythro*-2-(*N*-Myristoylamino)-1-phenyl-1-propanol
CDase (aCDase nCDase, alkCDase)	Ceramidase (acid, neutral, and alkaline ceramidase)
CerS	Ceramide synthase
DES	Dihydroceramide desaturase
GSC	Glucosylceramide synthase
DMS	*N*, *N*-dimethylsphingosine
TMS	*N*, *N*, *N*-Trimethylsphingosine
NOE	*N*-oleoylethanolamine
PKC	Protein kinase C
SPT	Serine-palmitoyl transferase
SMase (aSMase nSMase)	Sphingomyelinase (acid, neutral sphingomyelinase)
SMS	Sphingomyelin synthase
SPP	Sphingosine 1-phosphate phosphatase
SK	Sphingosine kinase

1 Sphingolipids

1.1 A Brief Introduction of Sphingolipids

Sphingolipids are a vast family of naturally occurring lipids containing a D-*erythro*-sphingoid (Shapiro and Flowers 1962) backbone (in sphingosine, (E,2S,3R)-2-aminooctadec-4-ene-1,3-diol). The sphingolipid group encompasses a growing variety of structures such as lysosphingolipids, ceramides, cerebrosides, sulfatides, and gangliosides. Sphingolipids were discovered studying the chemical composition of the brain. The father of neurochemistry, JLW Thudichum, described for the first time sphingomyelin and sphingosine in a study published in 1884. At that time, and for a long time thereafter, sphingolipids were considered as structural compounds in biological membranes (Stoffel 1970) of eukaryotes and in plasma lipoproteins. As an exception, sphingolipids have been also found in the *Sphingomonas* bacterial genus and other bacteria (Olsen and Jantzen 2001). Working as structural components, sphingomyelin and glycosphingolipids are the most abundant sphingolipids occurring in cells, normally accounting for 10–30 % of lipids in the cellular membranes whereas other sphingolipids such as ceramide or sphingosine are much less abundant. However, as it has been found for other lipids, the role of sphingolipids was found not to be only structural. In the middle 1980s sphingosine was found to inhibit protein kinase C, suggesting a bioactive role for sphingolipids as a second messenger (Hannun et al. 1986). After that, ceramide was shown to have regulatory roles in the cell. Following these findings, their phosphorylated forms, ceramide-1-phosphate (C1P) and sphingosine 1-phospate, were also described to have roles in apoptosis, proliferation, senescence, angiogenesis, and vesicular trafficking (Hannun and Obeid 2008). Interestingly, ceramide and sphingosine 1-phosphate, separated only by two bidirectional metabolic steps, have been described to exert opposites effects in the cell. Thus, ceramide has been reported to trigger apoptosis and cell arrest whereas S1P enhances cell survival and cell proliferation.

1.2 Biosynthesis of Sphingolipids

The de novo biosynthesis of sphingolipids in mammals begins by the condensation of serine with palmitoyl-coenzyme A to form 3-ketosphinganine, catalyzed by serine-palmitoyl transferase (SPT), a pyridoxal 5′-phosphate-dependent enzyme, and then reduced to sphinganine (Fig. 1). Less abundant, and biologically less studied, glycine or alanine can be incorporated instead of serine to form 1-desoxymethyl- or 1-deoxy- derivatives, respectively. N-acylation of the amino group of sphinganine with several possible coenzyme A-activated fatty acid (normally between C16/C16:1 and C24/C24:1, although shorter and longer backbone chains have been described, as well as a variety of backbone modifications Abe et al. 1996) leads to dihydroceramides, and this step is catalyzed by at least 6 known (dihydro)ceramide synthases (CerS1-6), each one with different fatty acid length preference. These enzymes are also responsible for N-acylation of sphingosine. Ceramides are formed by desaturation of dihydroceramides by dihydroceramide desaturase (DES) (Fabrias et al. 2012).

Classically, ceramide was treated as a single biological entity; however, ceramide is a family of structurally related molecules, and recently researchers have realized the biological diversity that may accompany this structural diversity. In mammals, it is possible to identify a growing number of ceramides, which are over 300 structures. Increasing the complexity, the same species of ceramide may be found in different subcellular compartments, in biological fluids, or in different

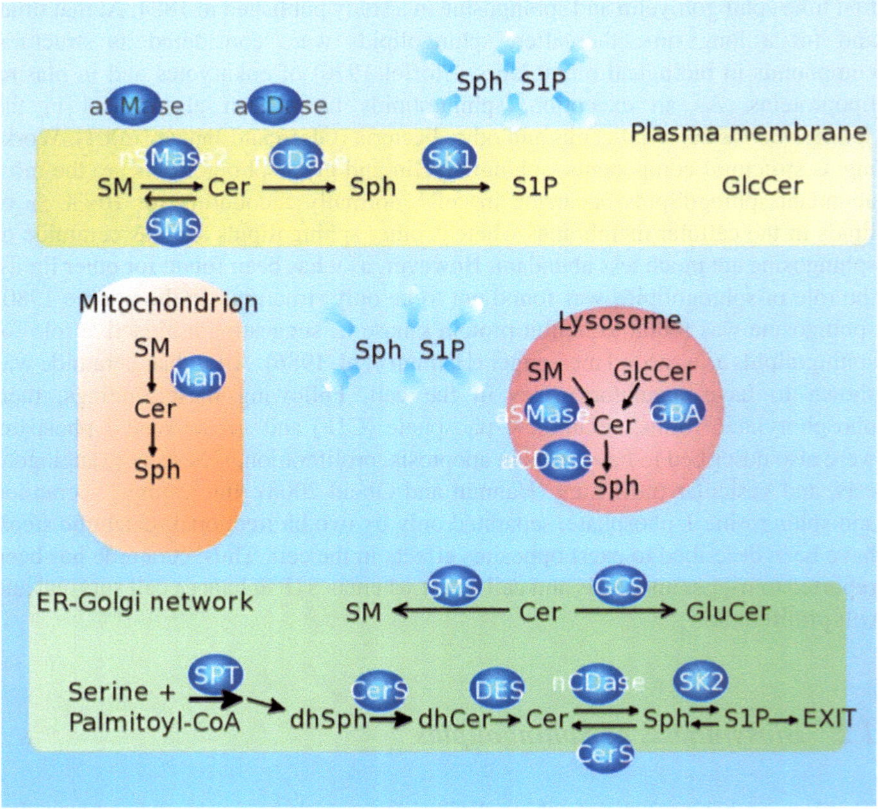

Fig. 1 Sphingolipid pathway and subcellular localization of sphingolipid enzymes. The biosynthesis of sphingolipids begins in the endoplasmatic reticulum (ER)–Golgi network. Vesicular transport distributes sphingolipids to different compartments such as plasma membrane, and lysosome. Sphingolipids and a sphingomyelinase (Man) have been found in the mitochondrion. *a/nSMase* acid/neutral sphingonyelinase, *a/nCDase* acid/neutral ceramidase, *nSMase* neutral sphingomyelinase, *SK1/2* sphingosine kinase 1/2, *Man* mitochondrial sphingomyelinase, *GBA* acid β-glucosidase, *SMS* sphingomyelinase synthase, *DES* dihydroceramide desaturase, *GCS* glucosylceramide synthase, *SPT* serine palmitoyl-CoA transferase

metabolic contexts in the cell. This variability in structure and localization raised the concept of *many ceramides* (Hannun and Obeid 2011), with different possible functions.

Ceramides are also the central hub from which diverse chemical modifications such as N-acyl hydrolysis or esterification in C1 result in hundreds of diverse structures with different potential functions. Moreover, the catabolism of more complex sphingolipids leads to ceramide formation, which can be converted to other complex sphingolipids, or continue its catabolism. That positions ceramide as a central key regulatory step in the whole sphingolipid diversity and function.

One of the fates of ceramides is its hydrolysis by ceramidases (*N*-acylsphingosine amidohydrolase, EC 3.5.1.23, here abbreviated as CDase) to form sphingosine, which is considered a bioactive molecule by itself (as mentioned before it was discovered to inhibit PKC), and it is the main sphingoid base in mammals. There are different ceramidases in human, codified by five different genes. Depending on their pH range of activity, they have been named as acid, neutral, and alkaline. Acid CDase (aCDase) is codified by the gene ASAH1. The optimum activity of aCDase is pH 4.5, and it is found in the lysosome, secreted in the media, and it could also have a role in trafficking vesicles. Another CDase has its optimum activity at pH 7–9, and it has been dubbed neutral CDase (nCDase, codified by the gene ASAH2) which has been localized bound to the plasma membrane, and also secreted to the intestinal lumen, but also may be present in mitochondria (Novgorodov et al. 2011). Finally, three different CDases with activity at pH 8.5–9.5, named as alkaline CDases (alkCDase, codified by the genes ACER1 or ASAH3, ACER2, and ACER3), are localized in the endoplasmic reticulum and the Golgi complex (Mao et al. 2001).

The potential modifications of sphingosine are N-acylation (ceramides), N-methylation (*N*, *N*-dimethylsphingosine, DMS; *N*, *N*, *N*-trimethylsphingosine, TMS), and phosphorylation. Additionally, sphingosine can be reverted/recycled to ceramide by reverse activity of CDases or by CerS activity (the same enzyme seen working on the N-acylation of dihydrosphingosine in the biosynthesis of de novo dihydroceramide/ceramide). The N-acylation by CDases does not require fatty acid activation by coenzyme A, as it is necessary for N-acylations catalyzed by CerS (El Bawab et al. 2001; Mao et al. 2000). Phosphorylation of sphingosine to sphingosine 1-phosphate is carried out by one of the two sphingosine kinases (SK1, 2). DMS also has been reported to be phosphorylated to DMS-1-phosphate; however this compound has not been further studied (Yatomi et al. 1997).

Sphingosine 1-phosphate is a bioactive lipid, triggering biological responses such as cell migration, cell proliferation, and angiogenesis through five known GPCR receptors (S1P1-5) (Hannun and Obeid 2008; Maceyka et al. 2012; Pyne and Pyne 2011; Sanchez and Hla 2004). The biological effects of sphingosine 1-phosphate might be also mediated in a receptor-independent way, and intracellular targets have been described to mediate calcium homeostasis and cell growth (Spiegel and Milstien 2011). In blood, sphingosine 1-phosphate concentration is elevated (around 400 nM), mainly associated with lipoproteins and albumin. Stimulated platelets are a main source of blood sphingosine 1-phosphate. As seen before for other sphingolipids, the conversion to sphingosine 1-phosphate is reversible, and it can return to sphingosine by the action of phosphatases (SPP1 and SPP2). Importantly, sphingosine 1-phosphate can be irreversibly hydrolyzed by sphingosine 1-phosphate lyase (SPL) to hexadecenal and ethanolamine-phosphate, which are non-sphingolipid molecules, and it is the only exit point of the sphingolipid network. Interestingly, as with SPT (the entrance point enzyme to the sphingolipid structure), also SPL (the exit point) is a pyridoxal 5′-phosphate-dependent enzyme (Bourquin et al. 2011).

Returning to ceramide, its phosphorylation by ceramide kinase (CK) results in C1P, another bioactive sphingolipid involved in cell homeostasis, inflammation, and cell migration through interaction with phospholipase A1 and possibly also by acting on a G-protein-coupled receptor (Granado et al. 2009). Growing in complexity, two sphingomyelin synthases (SMS1, 2) add a phosphocholine group, from phosphatidylcholine, to ceramide resulting in formation of sphingomyelin and diacylglycerol. Sphingomyelin is one of the main structural lipid of biological membranes, and it has been described to form defined domains in the plasma membrane together with cholesterol, offering localization domains for certain proteins.

The hydrolysis of sphingomyelin by several sphingomyelinases (SMase) generates ceramide in different compartments, which can then trigger several and distinct cellular responses. As we saw for CDase, different genes also encode for SMases: an acid SMase (aSMAse, sphingomyelin phosphodiesterase 1, codified by the gene SMPD1) which can be lysosomal or secreted to the extracellular matrix (Jenkins et al. 2009). Three neutral SMase genes (nSMAse1, nSMase2, and nSMase3) have been identified and cloned. However, mammalian nSMase1, although with bacterial neutral SMase sequence homology, and in vitro SMase activity, acts primarily as a lyso-PAF phospholipase C in vivo, with no SMase detected in cells (Marchesini et al. 2003). On the other hand, nSMase2 shows SMase activity in vitro and in vivo, whereas nSMase3 has been shown to exhibit SMase activity in some studies but not others. As such, nSMase2 has been much more studied and found to be involved in a myriad of biological effects in cells (Wu et al. 2010). All of the cloned nSMase proteins require addition of magnesium in vitro assays, although only nSMase2 has been shown to really increase sphingomyelin hydrolysis in vivo when overexpressed in breast cancer MCF-7 cell line. Moreover, an additional SMase, a magnesium-independent nSMase activity has been described in cytosolic fractions. Nevertheless, this activity has not been associated to a gene yet (Okazaki et al. 1994).

Other modifications in the ceramide structure include condensation of hexosyl units. Commonly, in the endoplasmic reticulum, units of glucose or galactose are attached in β-linkage to ceramide forming the cerebrosides glucosyl-ceramide and galactosyl-ceramide by glucosylceramide transferase and galactosylceramide transferase, respectively. The addition of more units of sugars to the glucosylceramide structure including different numbers and combinations of galactose, glucose, *N*-acetylgalactosamine, neuraminic acid (sialic acid), fucose, etc. in a precise sequence, bond configuration, and with or without ramifications defines different subcategories of glycosphingolipids, such as globo-sphingolipids, gangliosides (rich in sialic acid, forming several series: GMs, GD, GTs, GQs, GPs), and lacto-series. Sulfation of galactoceramide by galactosylceramide sulfotransferase results in another subgroup of glycosphingolipids, named sulfatides, rich in axonal myelin sheath and associated with apolipoprotein E in cerebrospinal fluid. Glycosphingolipids in general have been identified as antigens defining for example some of the blood groups, and used to define tumor-associated antigens. Glycosphingolipids are also involved in *cis* and *trans* cell interactions, contact cell growth inhibition, cell adhesion, and signal transduction (Hakomori 2008).

1.3 Sphingolipids and Disease

The enormous variety of sphingolipid structures imparts on the family a tremendous range of biological functions, which are commonly discovered in disease conditions. The misregulation of one sphingolipid metabolic enzyme, a sphingolipid receptor, or any other sphingolipid-modulated protein can develop several severe and fatal diseases. Thus, there is a collection of lysosome storage diseases due to a mutation in different lysosomal sphingolipid catabolic enzymes resulting in accumulation of one or another sphingolipid in the lysosome and provoking cellular, tissue, and organ failure. The lack of aCDase activity in the lysosome results in Farber disease. There is an accumulation of ceramide in the lysosomes, linked to a deficient neurological and general organ development, normally with short life span. A deficiency of alpha-galactosidase A (GLA) causes Fabry's disease, a multisystemic accumulation of globotriaosylceramide which results in severe complications in kidney, heart, and brain (Schaefer et al. 2009; Tarabuso 2011) commonly leading to early death. In Niemann–Pick disease, there is a lack of aSMase activity, resulting in storage of sphingomyelin in the endo-lysosomal compartment. The disease can present with different severities. Niemann–Pick type A (Ledesma et al. 2011) develops a severe neurological pathology with shorter life span. Type B is not as severe as type A but it still has a life expectancy during adulthood. Another lysosomal storage disorder is Gaucher's disease, the most common lysosomal storage disorder. Gaucher's disease presents accumulation of glucosylceramide due to lack of the lysosomal enzyme that hydrolyzes glucosylceramide to ceramide and glucose, glucocerebrosidase (acid β-glucosidase, GBA1 gene). Gaucher's disease patients often show visceral disorders, and in some severe cases neurological abnormalities (Farfel-Becker et al. 2011). Moreover, accumulation of GM2 gangliosides comprises three different disorders (Tay–Sachs disease, Sandhoff disease, and the very rare GM2A deficiency) including GM2 gangliosidosis due to deficiency in beta-hexosaminidase activity. Accumulation of GM1 by beta-galactosidase activity deficiency results in GM1 gangliosidosis. Both types of gangliosidosis are fatal. Finally, accumulations of sulfatides are described in metachromatic leukodystrophy. These diseases illustrate how important is the regulation of the sphingolipid pathway, and how a malfunction of one of their enzymes seems not to have another exit than the accumulation of the lipid. Mutations in SPT underlie hereditary autonomic neuropathy which has been mechanistically linked to formation of deoxysphingolipids.

Other diseases have been also related to malfunction of the sphingolipid pathway, including Alzheimer's disease, diabetes, atherosclerosis, cystic fibrosis, Wilson disease, and cancer. In this review, it is in cancer that we focus the drug development status involving targets in the sphingolipid pathway.

2 Sphingolipids and Cancer

Aberrations in bioactive sphingolipids, mainly ceramide, sphingosine 1-phosphate, and gangliosides, have been linked to several steps in cancer progression and response to cancer treatment, including resistance to chemotherapeutics.

In mammalian cells, accumulation of certain ceramide species in certain subcellular localizations might play a role in apoptosis (Birbes et al. 2002), cell cycle arrest, and inhibition of cell motility. Apoptosis is a natural response in tissue homeostasis, immune cell maturation, elimination of damaged or infected cells (for example exposed to UV-C light, or viral infection), or in the ontogenic process. Apoptosis can be triggered by a variety of stimuli such as cytokines (for example TNF) or DNA damage (activating p53 protein). In some cellular models the apoptotic mechanism has been shown to involve ceramide accumulation. Thus, several cancer cells and cancer models have been reported to have significant alterations in the enzymes involved in ceramide generation (aSMase, nSMAse, CerS) or degradation (aCDase Flowers et al. 2011, nCDase), often resulting in loss of ceramide, and could be one of the reasons of a cancer cell behavior, avoiding cell death.

In an opposite direction, changes in metabolism of sphingosine 1-phosphate have also been seen in cancer. S1P has been shown to induce cell proliferation, angiogenesis, cell invasion, and cell migration through five known G-protein-coupled receptors (S1P1-5). The enzymes responsible to catalyze sphingosine 1-phosphate have been reported to be overexpressed in several cancers, and thus, elevated levels of sphingosine 1-phosphate have been also reported in cancer cells and tissues. Especially relevant to cancer biology is the protein p53 which recently has been found to act upstream of SK1/sphingosine 1-phosphate pathway (Heffernan-Stroud et al. 2012).

Aberrant glycosylation has become a marker of the majority of cancers (Durrant et al. 2012). One of the recognized functions for glycosphingolipids is cell adhesion and cell motility. Changing the pattern of glycosphingolipid by expressing aberrant glycosphingolipids, and/or truncating the glycosphingolipids biosynthetic pathway, or overexpression of the glycosphingolipids, can dramatically promote or inhibit the spreading of cancer cells.

Because of these multiple contributions of sphingolipid in regulating cancer cells, their key enzymes have become targets in drug development.

3 Chemotherapeutical Drugs

Despite the observation that many sphingolipid enzymes have been implicated directly in cancer pathobiology, and even though drug inhibitors of virtually all of them have been developed, it is mainly glucosyl ceramide synthase, acid ceramidase, and sphingosine kinase that have garnered most attention as targets in recent research (Table 1).

In this chapter we review the main inhibitors of sphingolipid enzymes, and we mainly focus on those with more chemotherapeutic possibilities.

Table 1 Commonly used sphingolipid enzyme inhibitors

Enzyme targeted	Inhibitor	Structure
Serine-palmitoyl transferase	Myriocin	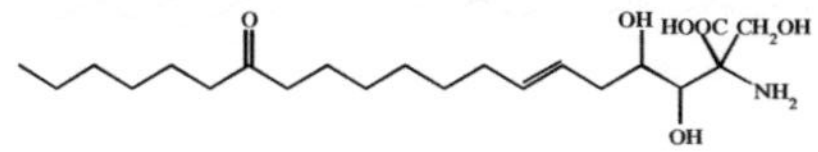
	L-cycloserine	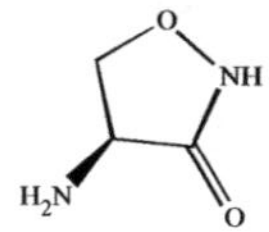
(Dihydro)ceramide desaturase	4-HPR	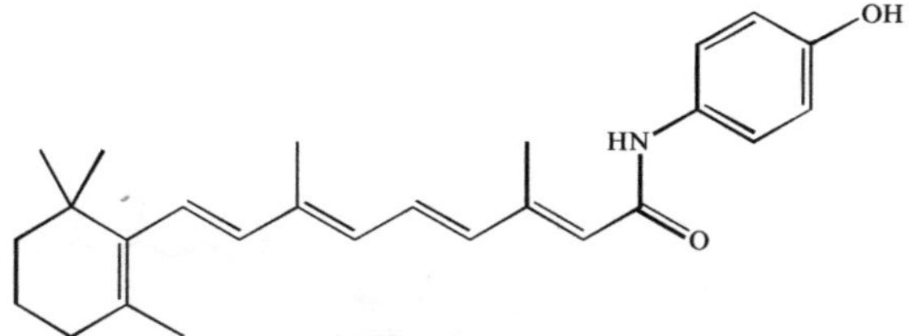 4-HPR
Ceramide synthase	Fumonisin B1	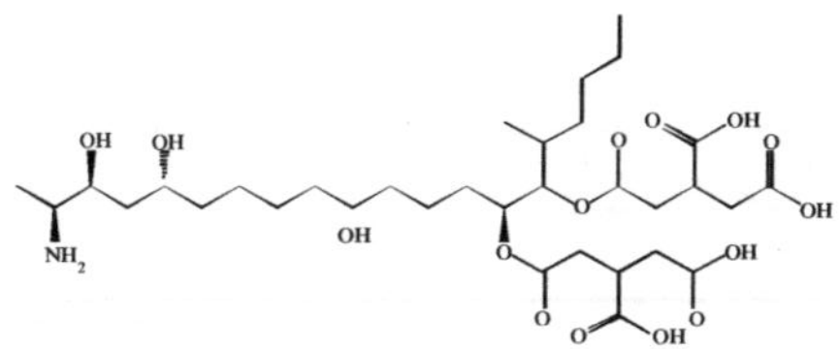

(continued)

Table 1 (continued)

Enzyme targeted	Inhibitor	Structure
Neutral sphingomyelinase2	Scyphostatin	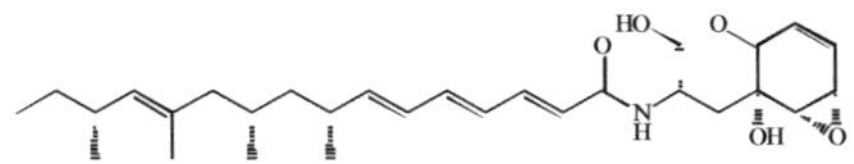
	GW4869	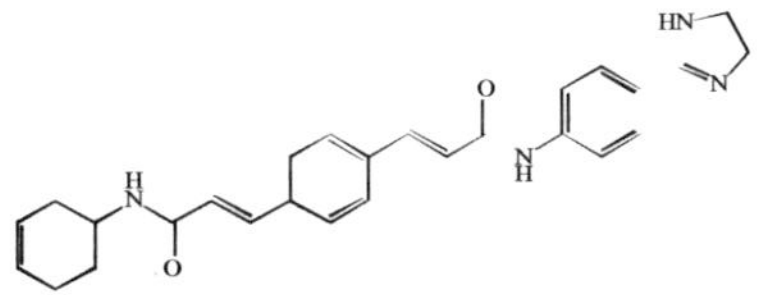
Acid sphingomyelinase/acid ceramidase	Desipramine	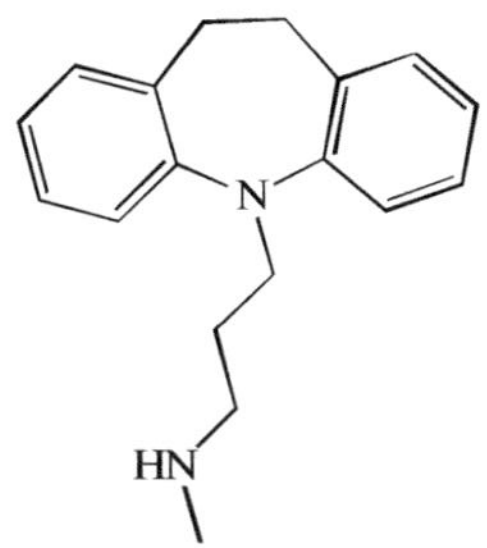
Acid ceramidase	NOE	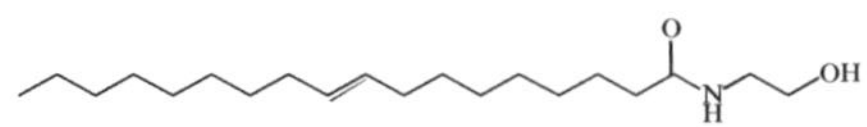

	B13	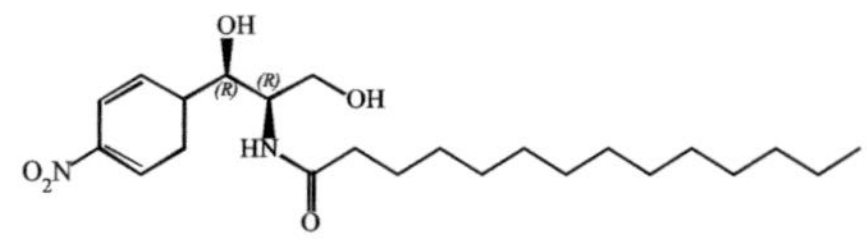
	E-tb	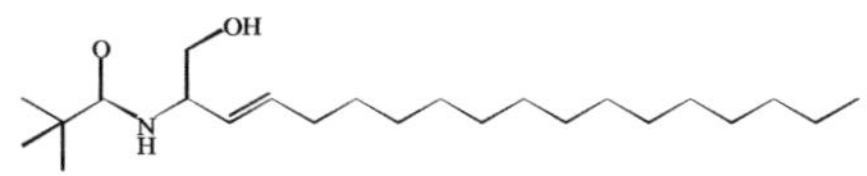
Neutral ceramidase/alkaline ceramidase	D-*e*-MAPP	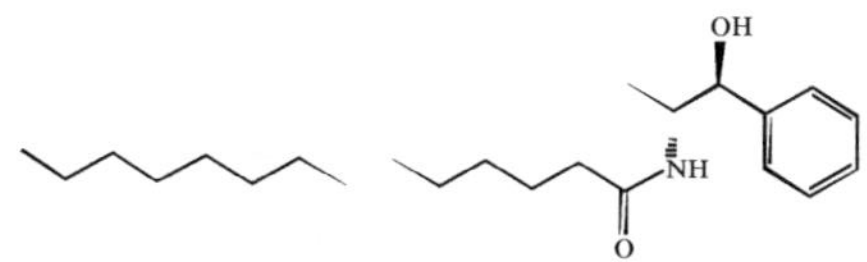
Glucosylceramide synthase	D-*threo*-PDMP	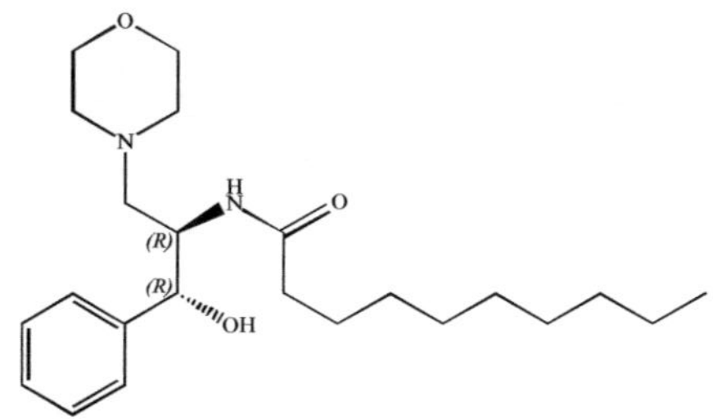

(continued)

Table 1 (continued)

Enzyme targeted	Inhibitor	Structure
Sphingosine kinase	Safingol	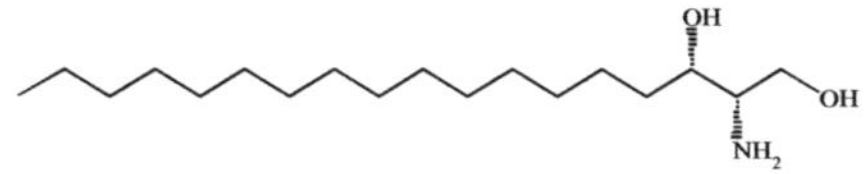
	DMS	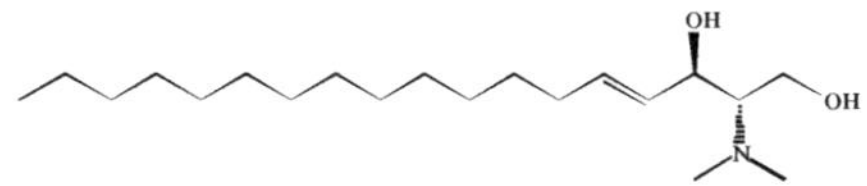
	PF-543	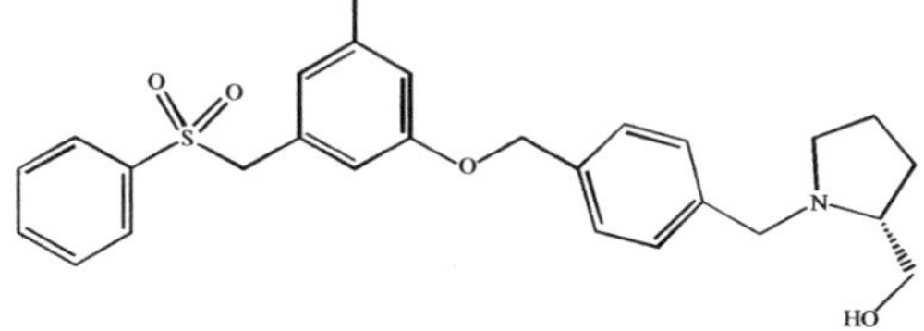
	SK1-I	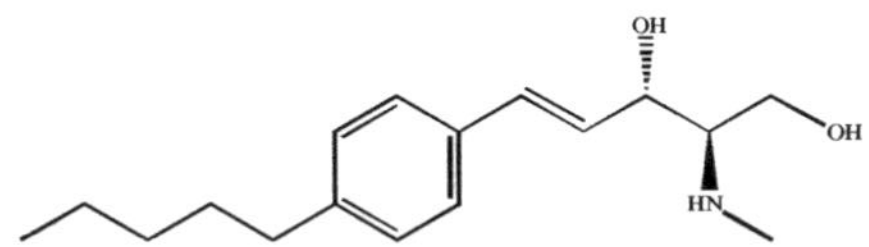

SKI-II	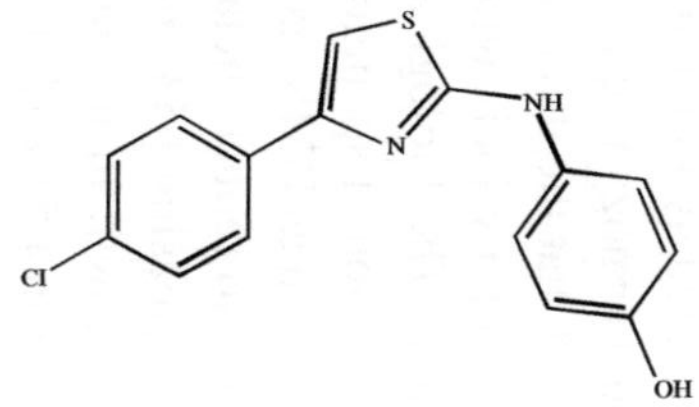
(s)-FTY720 vinylphosphonate	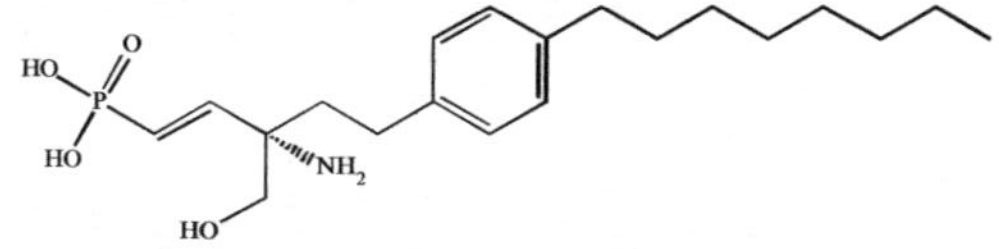
ABC294640	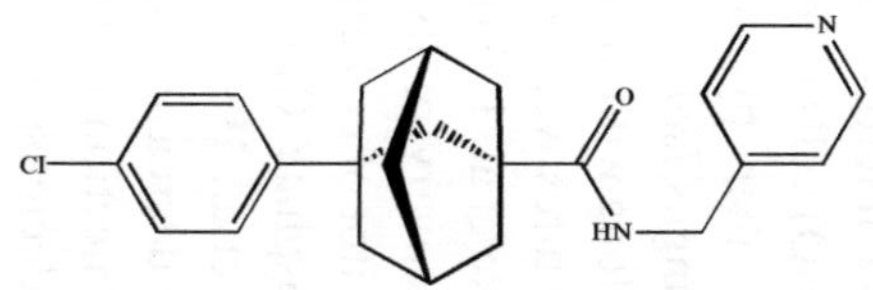

3.1 Serine-Palmitoyl Transferase

One of the first SPT inhibitors was beta-chloroalanine which was shown to inhibit SPT in vitro using rat liver microsomes as a source of SPT and in vivo, using Chinese hamster ovary (CHO) cells (Medlock and Merrill 1988). Nonetheless, this compound also inhibits other pyridoxal phosphate-dependent enzymes and transaminases. From the fungus *Isaria sinclairii*, (ISP-1) was isolated as an immunosuppressant compound along with the antibiotics myriocin and thermozymocidin, but once the chemical structures were resolved, it was appreciated that the three of them were identical (Miyake et al. 1995). The structural similarity of myriocin with sphingosine led to the discovery of its inhibitory effect on SPT. The biological effects of myriocin as inhibitor of cell growth were reverted by addition of sphingosine and sphingosine 1-phosphate. Other sphingosine analogues, such as DMS, had no effect on SPT (Miyake et al. 1995). Myriocin has been shown to reduce melanoma cell proliferation by decreasing sphingolipid levels and increasing p53 and p21 expression. In vivo, injection of myriocin reduced tumor growth in murine melanoma with a similar decrease in sphingolipids and increase in p53 and p21 showed in cells (Lee et al. 2012). L-cycloserine, another SPT inhibitor, has been shown to block taxol-induced ceramide, reducing apoptosis in breast cancer MCF-7 and MDA-MB-231 cell lines (Charles et al. 2001).

Recently, geranyl linalool, phytol, and farnesol have been also described as novel SPT inhibitors that reduce fumonisin B1-induced sphinganine accumulation and thus inhibit the first step of sphingolipid de novo synthesis (Shin et al. 2012).

3.2 Ceramide Synthases

Fumonisins were the first specific inhibitors described for the sphingolipid pathway. They are natural mycotoxins, with 15 different fumonisins, whereas FB1 is the most toxic. Fumonisins inhibit N-acylation of sphingoid bases by CerS, blocking the de novo synthesis of ceramide, and its recycling from sphingosine, causing an accumulation of free dihydrosphingosine and sphingosine resulting in a reduction of the total amount of complex sphingolipids and a total disruption of sphingolipid metabolism and eventually cellular metabolism failure.

3.3 Dihydroceramide Desaturase

The first DES inhibitor came from rational design. Based on the inhibitory mechanism of cyclopropene fatty acids on fatty acyl desaturase, the cyclopropene analogue of ceramide, *N*-[(1*R*, 2*S*)-2-hydroxy-1-hydroxymethyl-2-(2-tridecyl-1 cyclopropenyl)ethyl] octanamide, or briefly, GT11, was developed and found to inhibit DES (Triola et al. 2001). GT11 functioned as a competitive inhibitor with a Ki = 6 μM with *N*-octanoyl-sphinganine (Triola et al. 2003). In vivo studies showed GT11 to inhibit DES in different cultured cell lines. However, at higher concentrations (>5 μM), GT11 caused the accumulation of dihydrosphingosine 1-phosphate and sphingosine 1-phosphate in primary cerebellar neurons due to inhibition of sphingosine 1-phosphate lyase (Triola et al. 2004). From GT11 structure, another ceramide analogue, 5-thiadihydro- ceramide (XM462), containing a sulfur atom instead of the cyclopropene ring of GT11, also inhibited DES but with less potency than GT11, but was a more stable compound (Munoz-Olaya et al. 2008). From the work on GT11 and XM462, a new library of compounds was designed showing in vitro and in vivo inhibition of DES (Camacho et al. 2012).

Interestingly, although not designed and initially not used to inhibit DES, the drugs celecoxib and methylcelecoxib, which are COX-2 inhibitors, were shown to cause the accumulation of dihydroceramide species and dihydrosphingosine by inhibiting DES in HCT-116 cells, having no effect on reducing ceramide species (Schiffmann et al. 2009).

The most explored DES inhibitor is the synthetic retinoid *N*-(4-hydroxyphenyl) retinamide (fenretinide or 4-HPR) which has been shown to inhibit cell growth and induce cell death in several cancer cell lines. Moreover, 4-HPR has been approved for phase I and II in clinical trials. For example, 4-HPR is currently in clinical trials in pediatric patients with recurrent neuroblastoma (Villablanca et al. 2011), in premenopausal breast cancer risk (Macis et al. 2012), and in recurrent prostate cancer (Moore et al. 2010). 4-HPR was shown to cause accumulation of dihydroceramide and to act as an inhibitor of DES in cells (Wang et al. 2008). Mechanistically, the direct inhibition of DES by 4-HPR was demonstrated in in vitro studies using microsomes as a source of DES (Rahmaniyan et al. 2011). The cytotoxicity of 4-HPR has been related to accumulation of reactive oxygen species (ROS) and to inhibition of DES, although a recent study suggests that 4-HPR induction of cell death may occur independent of dihydroceramide and ROS accumulation (Apraiz et al. 2012).

3.4 Sphingomyelinases

In a screening amongst 10,000 microbial extract to find an nSMase2-specific inhibitor, a mycelial extract of *Trichopeziza mollissima* showed a micromolar-range inhibition for nSMase2 activity from rat liver microsomes. The active compound was scyphostatin, a water-insoluble and unstable compound when dried that can be stored at 20 °C in methanol (Nara et al. 1999a, b). The characterization of this inhibitor showed a reversible inhibitor, where the Km and Vmax were modified, and its specificity for neutral versus acid SMase was shown to be around 50-fold higher for nSMase, with IC_{50} for aSMase around IC_{50} = 49.3 μM, versus 1 μM for nSMase (Nara et al. 1999a, b). Of note, nSMase3 was inhibited by scyphostatin (Krut et al. 2006). Interesting synthetic analogues of scyphostatin have been developed looking for more potent nSMase inhibitors. Amongst them, kotylostatin, modified in its acyl chain, resulted in it being an irreversible inhibitor (Wascholowski et al. 2006). Not many biological studies have been carried out with scyphostatin or their analogues. Moreover, another natural and structural related inhibitor, manumycin A, is a reversible nSMase inhibitor with an affinity to nSMase comparable to the natural substrate, sphingomyelin. Some synthetic manumycin A analogues were shown to be irreversible inhibitors. Manumycin A has been shown to have antitumor activity. However, manumycin A has been shown to have Ras farnesyltransferase- and interleukin-1-converting enzyme inhibitory activities (Arenz et al. 2001). Manumycin A is currently used in cancer research, but its activity is attributed to Ras farnesyltransferase inhibition.

Probably, one of the most used nSMase2 inhibitors in research is GW4869. It is a potent, cell-permeable, noncompetitive, and specific nSMase inhibitor with an in vitro IC_{50} of 1 μM for nSMase, and it does not inhibit aSMase up to 150 μM. The inhibitor was shown to work in vivo as well by blocking the hydrolysis of sphingomyelin by nSMase induced by TNF (Luberto et al. 2002). GW4869 has been shown to reduce cellular ceramide levels and increase hyaluronic acid secretion (Qin et al. 2012), reducing secretion of miRNA (Kosaka et al. 2010) and counteracting the retinoic acid-induced growth inhibition (Somenzi et al. 2007).

Screening guanidinium derivatives, undecylidene-aminoguanidine (C11AG) was shown to inhibit nSMase and block HSV-1 replication by 50 % using 2.5 μM. It also blocked LPS-stimulated sphingomyelin hydrolysis (Amtmann et al. 2003) and enhanced cell death in Jurkat T-cell lymphoma cells.

There is not much literature on SMase3 in cancer progression, although it has been shown to be deregulated in several primary tumors (Corcoran et al. 2008), and it could play important roles in cellular homeostasis.

Similarly for nSMase, there are no selective inhibitors for this enzyme. However, aSMase was found to participate in evodiamine, a cytotoxic alkaloid, and induction of apoptosis in human gastric cancer SGC-7901 cells (Huang et al. 2011a, b), and it mediates apoptosis by ceramide production and radiosensibilization by the combination of the synergistic compounds sorafenib and vorinostat in cancer treatment (Park et al. 2010). The difluoromethylene sphingomyelin analogue SMA-7 is an SMase inhibitor that reduces levels of aSMase to basal in LPS-induced colon cancer cells, blocking the release of pro-inflammatory cytokines (Sakata et al. 2007). However, SMA-7 also inhibits nSMAse with a similar $IC_{50} = 3.3$ μM (Yokomatsu et al. 2001). The tricyclic group of compounds, such as desipramine, have been shown to indirectly inhibit aSMase by inducing its degradation.

3.5 *Ceramidases*

Ceramidases, and more exactly aCDase, has been a common target to drug development since they have a dual role in cancer progression. The first role is decreasing ceramide levels. Ceramide has been shown to drive cellular apoptosis, often in response of cytokines, or chemotherapeutic drugs. Thus activation of CDase results in decrease of ceramide levels, and failure of apoptosis. The second role is to participate in generation of sphingosine 1-phoshate, a pro-proliferative agonist. In that way, CDase activity hydrolyzes ceramide to form sphingosine, which in the presence of sphingosine kinase activity is converted to sphingosine 1-phosphate. This dual role makes CDase a strong player in cancer progression, and an attractive target for anticancer drug development. This opposite effect of ceramide and sphingosine 1-phosphate is not just limited to cell proliferation; for example, in our group we have recently described that the pro-migratory family of proteins ezrin, radixin, and moesin, which are overexpressed in several cancer cells, are also oppositely regulated by ceramide and sphingosine 1-phosphate (Canals et al. 2010).

Lyosomal aCDase and aSMase are inhibited by some amphiphilic tricyclic agents such as desipramine, and by other amphipathic amines such as chlorpromazine and chloroquine. These agents downregulate aCDase and aSMase protein levels, although they do not affect the RNA message levels. These compounds are not specific for SMases or CDases, affecting some, but not all, lysosomal enzymes (Canals et al. 2011). Looking for more specific inhibitors, and based on the sphingoid-base structure, synthetic chemistry has brought some families of CDase inhibitors. The first sphingolipid analogue-based CDase inhibitor was *N*-oleoylethanolamine (NOE) (Sugita et al. 1975). NOE has been broadly used as an aCDase inhibitor, increasing cellular ceramide, and inducing apoptosis in several cell lines, such as the mouse L929 fibroblast (Strelow et al. 2000), human glioma U87-MG cells, primary placenta trophoblast (Payne et al. 1999), and bone marrow-derived dendritic cells. However, NOE was found to also inhibit ceramide glycosylation and later, its aCDase inhibitor potency was found to be weak in vitro and in vivo (Grijalvo et al. 2006). Currently, NOE is not used as an aCDase inhibitor but as an endocannabinoid-related molecule. However, NOE has served as a scaffold to design other aCDase inhibitors.

Another attempt for CDase inhibitor was a ceramide analogue, (1*S*, 2*R*)-D-*erythro*-2-(*N*-Myristoylamino)-1-phenyl-1-propanol or D-e-MAPP, which was shown to accumulate intracellular ceramide in human promyelocytic HL-60 leukemia cells causing cell cycle arrest. It was shown to inhibit in vitro neutral and alkaline CDase at micromolar concentrations, but it had no effect on aCDase (Bielawska et al. 1996).

In a study of D-e-MAPP analogues, (1*R*, 2*R*)-2-(*N*-tetradecanoylamino)-1-(4-nitrophenyl)-1,3-propanediol (or B13) was shown to be a potent in vitro aCDase inhibitor (Bielawska et al. 2008). B13 also accumulated cellular ceramide in vivo, although the mechanism in vivo has been argued, since the neutral nature of the compound makes it difficult to be accumulated in the lysosome. To solve this problem, B13 was modified to enter in the lysosome, resulting in series of lysomotrophic molecules such as LCL204 (also known as AD 2646). Thus, LCL204 inhibits aCDase in vitro as well as in vivo, increasing ceramide levels in a myriad of cultured cells, and counteracting the resistance to apoptosis caused by overexpression of aCDase found in some cancers. LCL04 was found to reduce aCDase protein levels, and destroy the lysosomes and thus affect other lysosomal proteins, including aSMase (Bai et al. 2009).

3.6 Glucosylceramide Synthase

The ceramide analogue, D-*threo*-1-phenyl-2-decanoylamino-3-morpholino-1-propanol or D-*threo*-PDMP or just D-PDMP, is a competitive inhibitor of GSC. Interestingly, the L-PDMP isoform was found not to have an inhibitory effect on cells (Inokuchi et al. 1990). D-PDMP treatment of human leukemia HL-60 cells resulted in reduction of basal levels of glucosylceramide, lactosylceramide, and GM3. The reduction of glucosylceramide species and derivatives has been also observed in a myriad of cancer cells and it has been suggested as a chemotherapeutic agent, alone or to sensitize cells to other chemotherapeutic drugs. For example, Lewis lung carcinoma cells lost its lung-colonizing capacity after micromolar PDMP treatment. D-PDMP has been used in combination with imatinib to kill K562 leukemic cells (Baran et al. 2011), in combination with nanoliposomal C6-ceramide (Jiang et al. 2011), and with GNF-2- Bcr-Abl inhibitor to induce apoptosis in human chronic myeloid leukemia cell line (Huang et al. 2011a, b). Moreover, D-PDMP was seen to inhibit galactosylceramide synthase as well. The L-isomer of PDMP was not only found to have no inhibitory effect on GSC but also had the opposite effect of increasing the levels of glucosyl- and lactosylceramides and their respective metabolites (Chatterjee et al. 1996). However, L-PDMP was found to inhibit glycosylceramide glycanases, which are abnormally expressed in some tumor cancers such as colon cancer, neuroblastoma, and some breast cancer cell lines (Basu et al. 1999).

3.7 Sphingosine Kinases

There are two enzymes with sphingosine kinase activity to generate sphingosine 1-phosphate, SK1 (with three N-terminal variants Tonelli et al. 2010) and SK2. The participation of SK1 in cancer progression has been reported in many studies, and SK1 has been repeatedly considered as an oncogene. For example, overexpression of SK1 in NIH3T3 fibroblast cell line resulted in increase in cell proliferation, loss of cell growth inhibition by cell contact, increase in colony formation, and increase in the number of established tumor in mice injected with SK1-transfected cells.

Microarray databases, like oncomine (http://www.oncomine.org/), show statistically significant up-regulation of SK1 in a large number of cancers.

The *N*-methyl natural derivative of sphingosine, *N,N*-dimethylsphingosine (DMS), was found to inhibit SK activity in human platelets (Yatomi et al. 1996) and it has been reported to block chemotaxis towards growth factors, as well as inducing apoptosis in U937 human leukemic monocytes. As with other sphingosine analogues, DMS also inhibits PKC (Khan et al. 1990). Moreover, the quaternary ammonium derivative of DMS, TMS, which is a more stronger PKC inhibitor, has also been found to inhibit in vitro cancer cell growth, and in vivo growth of human tumor cells in nude mice, in a similar way than DMS (Endo et al. 1991). The similar results between DMS and TMS make it difficult to distinguish between SK and PKC effects using these compounds.

The synthetic sphingosine analogue L-*threo*-dihydrosphingosine (safingol) was first described as an inhibitor of PKC, and as with other sphingosine isomers and analogues was also found to inhibit SK activity in partially purified rat brain SK and from human platelet SK (Buehrer and Bell 1992). Moreover, safingol has been related to ceramide generation and induction of apoptosis (Noda et al. 2009), and in a sphingolipid-independent way inducing autophagy in human HCT-116 colon carcinoma cell line (Coward et al. 2009). Other protein kinases have been reported to be directly or indirectly inhibited by safingol such as the PI3k/Akt/mTOR pathway, and the ERK MAPK signaling upon bradykinin and PDGF stimulation (Tolan et al. 1996). Safingol has been shown to inhibit cancer cell growth (Schwartz et al. 1993) and sensitization of cancer cells to chemotherapeutic drugs, and it has been successfully used in animal models and successfully passed a phase I clinical trial in combination with doxorubicin (Schwartz et al. 1997). Safingol alone and in combination with cisplatin is ready to start clinical phase II trials (Dickson et al. 2011). From the safingol structure, an analogue library was constructed finding L-*threo*-*N*-(4-heptylbenzoyl)dihydrosphingosine derivative having a dramatic apoptotic effect in lung cancer A549 cells (Villorbina et al. 2007).

It is important to note that some of the effects of sphingoid analogues that were initially attributed to PKC inhibition could also be due to SK inhibition, or vice versa, or combination of the two enzymes.

Recently, another competitive inhibitor, PF-543, has been described to inhibit SK1 with a Ki of 3.6 nM, reducing the level of sphingosine 1-phosphate, and increasing the level of sphingosine in head and neck 1,483 carcinoma cell line. Surprisingly PF-543 did not affect the growth rate and total ceramide levels, but it affected the rate of newly synthesized ceramide (Schnute et al. 2012).

Other SK1 inhibitors are FTY720 (fingolimod) and (S)-FTY720 vinylphosphonate which were shown to inhibit SK1 activity (with a sphingosine competition inhibition of $K_{ic} = 2$ μM and uncompetitive inhibition of $K_{uc} = 17$ μM, respectively). The mechanism involves SK1 proteosomal degradation in smooth muscle, breast cancer MCF-7, and LNCaP prostate cancer cell lines. Thus, whereas FTY720 phosphate results in cell proliferation and migration, the FTY720 parent compound induces apoptosis. Moreover, the two compounds also showed opposite effects on ERK1/2, Akt, FAK, and caspase-3 (Tonelli et al. 2010). Of note, (s)-FTY720 vinylphosphonate acts as an antagonist of S1P receptors, increasing the potential therapeutical effect of the drug to induce cancer cell death. Interestingly, (R)-FTY720 methyl ester has been shown to inhibit selectively SK2 (Ki ~16 μM with sphingosine), implicating SK2 in cytoskeleton rearrangements in breast cancer MCF-7 cells (Lim et al. 2011).

Inhibition of SK using SKI-II (2-(p-hydroxyanilino)-4-(p-chlorophenyl) thiazole) ($IC_{50} = 0.5$ μM, for human recombinant SK1) typically employed a concentration of 10 μM, and this resulted in decreased multidrug-resistant breast cancer proliferation and viability (Antoon et al. 2012), blocked the activation of ezrin protein by S1P required for migration in cervical cancer HeLa cells (Canals et al. 2010), and resulted in SK1 degradation in MDA-MB-453 breast cancer cells (Ohotski et al. 2012). Indeed, this inhibitor has been suggested of not inhibiting directly SK1, but enhancing its lysosomal degradation (Ren et al. 2010). Recently, it has been used in combination with bortezomib to induce caspase-dependent apoptosis in leukemic cells (Li et al. 2011).

Specific for SK1, (2R,3S,4E)-*N*-methyl-5-(4-pentylphenyl)-2-aminopent-4-ene-1,3-diol (BML-258 or SK1-I) is a water-soluble ATP-competitive inhibitor with no activity towards SK2 and PKC. This inhibitor showed antiproliferative effects on human leukemia U937 cells inducing caspase- and BCL-2-dependent apoptosis. SK1-I showed an anti-proliferative effect and inhibition of migration and invasion in glioblastoma U373 and LN229 cells towards serum, EGF, and lysophosphatidic acid. Moreover, SK1-I inhibited the growth of acute myeloid leukemia xenograft tumors in mice and reduced tumor burden, serum S1P levels, metastasis, hemangiogenesis, and lymphangiogenesis in mouse metastasic breast cancer model (Nagahashi et al. 2012; Paugh et al. 2008).

Selective inhibitors for SK2 have been identified, and these include ABC294640 [3-(4-chlorophenyl)-adamantane-1-carboxylic acid (pyridin-4-ylmethyl)amide], a competitive inhibitor with respective sphingosine, Ki ~10 μM. In cell culture, ABC294640 inhibited cell proliferation, cell migration in several cancer cell lines (French et al. 2010), induced autophagy and apoptosis in PC-3 prostate and breast cancer MDA-MB-231 cancer cells, and reduced tumor incidence in an azoxymethane/dextran sulfate sodium mouse model (Beljanski et al. 2011; Chumanevich et al. 2010).

It is not the goal of this chapter to describe small molecules that inhibit sphingosine 1-phosphate receptors. However, it is important to point out that another set of inhibitors are available for these receptors, and they are also important targets in the current drug discovery research (Huwiler and Pfeilschifter 2008).

3.8 Antibodies as Chemotherapeutics

This chapter would not be complete if we do not mention that antibodies raised against sphingosine 1-phosphate and antibodies which recognize aberrant glycolipids expressed in tumors have become strong activators of the complement system, or directly induce cell death (Durrant et al. 2012). Amongst all the glycolipids, glycosphingolipids are the most up-regulated in tumor cells, and thus, appear to be the most important targets for cancer drug development. In this context, sphingomab, a murine monoclonal antibody that binds sphingosine 1-phosphate (the human version is known as sonepcizumab), has shown reduction in cancer progression in a murine model and in human cancers (Milstien and Spiegel 2006; Visentin et al. 2006). Sphingomab has already completed phase I clinical trials for cancer treatment and is being considered for phase II studies (Sabbadini 2011).

3.9 Chemotherapeutics that Modify Sphingolipid Pathway

Although the majority of therapeutics used in cancer are not directly targeted to sphingolipid enzymes, their action mechanism might require the participation of sphingolipids. For example resveratrol, which is an apoptotic compound used to induce cancer cells to die, may involve ceramide generation, since when breast cancer MDA-MB-231 cell lines are pretreated with myriocin, the cells do not accumulate ceramide in response to resveratrol and cells are rescued from the induction of apoptosis. Moreover, resveratrol has been reported to inhibit sphingosine kinase 1 in breast cancer MCF-7 cells (Lim et al. 2012). Other chemotherapeutic apoptotic agents such as fenretinide (Maurer et al. 1999), doxorubicin (Saad et al. 2007), ara-C (Grazide et al. 2002), etoposide (Perry et al. 2000), and Δ9-tetrahydrocannabinol have been shown to induce cell death by accumulation of ceramide (Galve-Roperh et al. 2000).

4 Perspectives

The identification of a role of sphingolipids in tumor behavior, development, and prognosis suggests a point of regulation in cancer progression, and new targets to block cancer growth or metastasis. While a few sphingolipid analogues have been already incorporated in clinical trials, other sphingolipid-regulating enzymes have poor- or low-efficiency inhibitors; other inhibitors are specific and potent in vitro but poor when used in vivo, or there is still a lack in their metabolism by the cell, or in blood. However, this is an active and exponentially growing field in cancer research, and we expect that in the near future chemotherapies involving inhibitors of sphingolipid-metabolizing enzymes, and other sphingolipid-regulated proteins, will be part of clinical trials and cancer patient treatments.

Acknowledgements We gratefully acknowledge Dr. Mike Airola and María Hernández-Corbacho for the careful reading of this manuscript. This work was supported by NIH grants CA87584 and CA97132.

References

Abe A, Shayman JA, Radin NS (1996) A novel enzyme that catalyzes the esterification of N-acetylsphingosine. Metabolism of C2-ceramides. J Biol Chem 271:14383–14389

Amtmann E, Baader W, Zoller M (2003) Neutral sphingomyelinase inhibitor C11AG prevents lipopolysaccharide-induced macrophage activation. Drugs Exp Clin Res 29:5–13

Antoon JW, White MD, Burow ME, Beckman BS (2012) Dual inhibition of sphingosine kinase isoforms ablates TNF-induced drug resistance. Oncol Rep 27:1779–1786

Apraiz A, Idkowiak-Baldys J, Nieto-Rementeria N, Boyano MD, Hannun YA, Asumendi A (2012) Dihydroceramide accumulation and reactive oxygen species are distinct and nonessential events in 4-HPR-mediated leukemia cell death. Biochem Cell Biol 90:209–223

Arenz C, Thutewohl M, Block O, Waldmann H, Altenbach HJ, Giannis A (2001) Manumycin A and its analogues are irreversible inhibitors of neutral sphingomyelinase. Chembiochem 2: 141–143

Bai A, Szulc ZM, Bielawski J, Mayroo N, Liu X, Norris J, Hannun YA, Bielawska A (2009) Synthesis and bioevaluation of omega-N-amino analogs of B13. Bioorg Med Chem 17: 1840–1848

Baran Y, Bielawski J, Gunduz U, Ogretmen B (2011) Targeting glucosylceramide synthase sensitizes imatinib-resistant chronic myeloid leukemia cells via endogenous ceramide accumulation. J Cancer Res Clin Oncol 137:1535–1544

Basu M, Kelly P, O'Donnell P, Miguel M, Bradley M, Sonnino S, Banerjee S, Basu S (1999) Ceramide glycanase activities in human cancer cells. Biosci Rep 19:449–460

Beljanski V, Knaak C, Zhuang Y, Smith CD (2011) Combined anticancer effects of sphingosine kinase inhibitors and sorafenib. Invest New Drugs 29:1132–1142

Bielawska A, Bielawski J, Szulc ZM, Mayroo N, Liu X, Bai A, Elojeimy S, Rembiesa B, Pierce J, Norris JS, Hannun YA (2008) Novel analogs of D-e-MAPP and B13. Part 2: signature effects on bioactive sphingolipids. Bioorg Med Chem 16:1032–1045

Bielawska A, Greenberg MS, Perry D, Jayadev S, Shayman JA, McKay C, Hannun YA (1996) (1S,2R)-D-erythro-2-(N-myristoylamino)-1-phenyl-1-propanol as an inhibitor of ceramidase. J Biol Chem 271:12646–12654

Birbes H, Bawab SE, Obeid LM, Hannun YA (2002) Mitochondria and ceramide: intertwined roles in regulation of apoptosis. Adv Enzyme Regul 42:113–129

Bourquin F, Capitani G, Grutter MG (2011) PLP-dependent enzymes as entry and exit gates of sphingolipid metabolism. Protein Sci 20:1492–1508

Buehrer BM, Bell RM (1992) Inhibition of sphingosine kinase in vitro and in platelets. Implications for signal transduction pathways. J Biol Chem 267:3154–3159

Camacho L, Simbari F, Garrido M, Abad JL, Casas J, Delgado A, Fabrias G (2012) 3-Deoxy-3,4-dehydro analogs of XM462. Preparation and activity on sphingolipid metabolism and cell fate. Bioorg Med Chem 20:3173–3179

Canals D, Jenkins RW, Roddy P, Hernandez-Corbacho MJ, Obeid LM, Hannun YA (2010) Differential effects of ceramide and sphingosine 1-phosphate on ERM phosphorylation: probing sphingolipid signaling at the outer plasma membrane. J Biol Chem 285:32476–32485

Canals D, Perry DM, Jenkins RW, Hannun YA (2011) Drug targeting of sphingolipid metabolism: sphingomyelinases and ceramidases. Br J Pharmacol 163:694–712

Charles AG, Han TY, Liu YY, Hansen N, Giuliano AE, Cabot MC (2001) Taxol-induced ceramide generation and apoptosis in human breast cancer cells. Cancer Chemother Pharmacol 47: 444–450

Chatterjee S, Cleveland T, Shi WY, Inokuchi J, Radin NS (1996) Studies of the action of ceramide-like substances (D- and L-PDMP) on sphingolipid glycosyltransferases and purified lactosylceramide synthase. Glycoconj J 13:481–486

Chumanevich AA, Poudyal D, Cui X, Davis T, Wood PA, Smith CD, Hofseth LJ (2010) Suppression of colitis-driven colon cancer in mice by a novel small molecule inhibitor of sphingosine kinase. Carcinogenesis 31:1787–1793

Corcoran CA, He Q, Ponnusamy S, Ogretmen B, Huang Y, Sheikh MS (2008) Neutral sphingomyelinase-3 is a DNA damage and nongenotoxic stress-regulated gene that is deregulated in human malignancies. Mol Cancer Res 6:795–807

Coward J, Ambrosini G, Musi E, Truman JP, Haimovitz-Friedman A, Allegood JC, Wang E, Merrill AH Jr, Schwartz GK (2009) Safingol (L-threo-sphinganine) induces autophagy in solid tumor cells through inhibition of PKC and the PI3-kinase pathway. Autophagy 5:184–193

Dickson MA, Carvajal RD, Merrill AH Jr, Gonen M, Cane LM, Schwartz GK (2011) A phase I clinical trial of safingol in combination with cisplatin in advanced solid tumors. Clin Cancer Res 17:2484–2492

Durrant LG, Noble P, Spendlove I (2012) Immunology in the clinic review series; focus on cancer: glycolipids as targets for tumour immunotherapy. Clin Exp Immunol 167:206–215

El Bawab S, Birbes H, Roddy P, Szulc ZM, Bielawska A, Hannun YA (2001) Biochemical characterization of the reverse activity of rat brain ceramidase. A CoA-independent and fumonisin B1-insensitive ceramide synthase. J Biol Chem 276:16758–16766

Endo K, Igarashi Y, Nisar M, Zhou QH, Hakomori S (1991) Cell membrane signaling as target in cancer therapy: inhibitory effect of *N*, *N*-dimethyl and *N*, *N*, *N*-trimethyl sphingosine derivatives on in vitro and in vivo growth of human tumor cells in nude mice. Cancer Res 51:1613–1618

Fabrias G, Munoz-Olaya J, Cingolani F, Signorelli P, Casas J, Gagliostro V, Ghidoni R (2012) Dihydroceramide desaturase and dihydrosphingolipids: debutant players in the sphingolipid arena. Prog Lipid Res 51:82–94

Farfel-Becker T, Vitner EB, Futerman AH (2011) Animal models for Gaucher disease research. Dis Model Mech 4:746–752

Flowers M, Fabrias G, Delgado A, Casas J, Abad JL, Cabot MC (2011) C6-Ceramide and targeted inhibition of acid ceramidase induce synergistic decreases in breast cancer cell growth. Breast Cancer Res Treat 133:447–458

French KJ, Zhuang Y, Maines LW, Gao P, Wang W, Beljanski V, Upson JJ, Green CL, Keller SN, Smith CD (2010) Pharmacology and antitumor activity of ABC294640, a selective inhibitor of sphingosine kinase-2. J Pharmacol Exp Ther 333:129–139

Galve-Roperh I, Sanchez C, Cortes ML, Gomez del Pulgar T, Izquierdo M, Guzman M (2000) Anti-tumoral action of cannabinoids: involvement of sustained ceramide accumulation and extracellular signal-regulated kinase activation. Nat Med 6:313–319

Granado MH, Gangoiti P, Ouro A, Arana L, Gonzalez M, Trueba M, Gomez-Munoz A (2009) Ceramide 1-phosphate (C1P) promotes cell migration Involvement of a specific C1P receptor. Cell Signal 21:405–412

Grazide S, Maestre N, Veldman RJ, Bezombes C, Maddens S, Levade T, Laurent G, Jaffrezou JP (2002) Ara-C- and daunorubicin-induced recruitment of Lyn in sphingomyelinase-enriched membrane rafts. FASEB J 16:1685–1687

Grijalvo S, Bedia C, Triola G, Casas J, Llebaria A, Teixido J, Rabal O, Levade T, Delgado A, Fabrias G (2006) Design, synthesis and activity as acid ceramidase inhibitors of 2-oxooctanoyl and N-oleoylethanolamine analogues. Chem Phys Lipids 144:69–84

Hakomori SI (2008) Structure and function of glycosphingolipids and sphingolipids: recollections and future trends. Biochim Biophys Acta 1780:325–346

Hannun YA, Loomis CR, Merrill AH Jr, Bell RM (1986) Sphingosine inhibition of protein kinase C activity and of phorbol dibutyrate binding in vitro and in human platelets. J Biol Chem 261: 12604–12609

Hannun YA, Obeid LM (2008) Principles of bioactive lipid signalling: lessons from sphingolipids. Nat Rev Mol Cell Biol 9:139–150

Hannun YA, Obeid LM (2011) Many ceramides. J Biol Chem 286:27855–27862

Heffernan-Stroud LA, Helke KL, Jenkins RW, De Costa AM, Hannun YA, Obeid LM (2012) Defining a role for sphingosine kinase 1 in p53-dependent tumors. Oncogene 31:1166–1175

Huang H, Zhang Y, Liu X, Li Z, Xu W, He S, Huang Y, Zhang H (2011a) Acid sphingomyelinase contributes to evodiamine-induced apoptosis in human gastric cancer SGC-7901 cells. DNA Cell Biol 30:407–412

Huang WC, Tsai CC, Chen CL, Chen TY, Chen YP, Lin YS, Lu PJ, Lin CM, Wang SH, Tsao CW, Wang CY, Cheng YL, Hsieh CY, Tseng PC, Lin CF (2011b) Glucosylceramide synthase inhibitor PDMP sensitizes chronic myeloid leukemia T315I mutant to Bcr-Abl inhibitor and cooperatively induces glycogen synthase kinase-3-regulated apoptosis. FASEB J 25:3661–3673

Huwiler A, Pfeilschifter J (2008) New players on the center stage: sphingosine 1-phosphate and its receptors as drug targets. Biochem Pharmacol 75:1893–1900

Inokuchi J, Jimbo M, Momosaki K, Shimeno H, Nagamatsu A, Radin NS (1990) Inhibition of experimental metastasis of murine Lewis lung carcinoma by an inhibitor of glucosylceramide synthase and its possible mechanism of action. Cancer Res 50:6731–6737

Jenkins RW, Canals D, Hannun YA (2009) Roles and regulation of secretory and lysosomal acid sphingomyelinase. Cell Signal 21:836–846

Jiang Y, DiVittore NA, Kaiser JM, Shanmugavelandy SS, Fritz JL, Heakal Y, Tagaram HR, Cheng H, Cabot MC, Staveley-O'Carroll KF, Tran MA, Fox TE, Barth BM, Kester M (2011) Combinatorial therapies improve the therapeutic efficacy of nanoliposomal ceramide for pancreatic cancer. Cancer Biol Ther 12:574–585

Khan WA, Dobrowsky R, el Touny S, Hannun YA (1990) Protein kinase C and platelet inhibition by D-erythro-sphingosine: comparison with *N*, *N*-dimethylsphingosine and commercial preparation. Biochem Biophys Res Commun 172:683–691

Kosaka N, Iguchi H, Yoshioka Y, Takeshita F, Matsuki Y, Ochiya T (2010) Secretory mechanisms and intercellular transfer of microRNAs in living cells. J Biol Chem 285:17442–17452

Krut O, Wiegmann K, Kashkar H, Yazdanpanah B, Kronke M (2006) Novel tumor necrosis factor-responsive mammalian neutral sphingomyelinase-3 is a C-tail-anchored protein. J Biol Chem 281:13784–13793

Ledesma MD, Prinetti A, Sonnino S, Schuchman EH (2011) Brain pathology in Niemann Pick disease type A: insights from the acid sphingomyelinase knockout mice. J Neurochem 116: 779–788

Lee YS, Choi KM, Lee S, Sin DM, Lim Y, Lee YM, Hong JT, Yun YP, Yoo HS (2012) Myriocin, a serine palmitoyltransferase inhibitor, suppresses tumor growth in a murine melanoma model by inhibiting de novo sphingolipid synthesis. Cancer Biol Ther 13:92–100

Li QF, Yan J, Zhang K, Yang YF, Xiao FJ, Wu CT, Wang H, Wang LS (2011) Bortezomib and sphingosine kinase inhibitor interact synergistically to induces apoptosis in BCR/ABl + cells sensitive and resistant to STI571 through down-regulation Mcl-1. Biochem Biophys Res Commun 405:31–36

Lim KG, Gray AI, Pyne S, Pyne NJ (2012) Resveratrol dimers are novel sphingosine kinase 1 inhibitors and affect sphingosine kinase 1 expression and cancer cell growth and survival. Br J Pharmacol 166:1605–1616

Lim KG, Sun C, Bittman R, Pyne NJ, Pyne S (2011) (R)-FTY720 methyl ether is a specific sphingosine kinase 2 inhibitor: effect on sphingosine kinase 2 expression in HEK 293 cells and actin rearrangement and survival of MCF-7 breast cancer cells. Cell Signal 23:1590–1595

Luberto C, Hassler DF, Signorelli P, Okamoto Y, Sawai H, Boros E, Hazen-Martin DJ, Obeid LM, Hannun YA, Smith GK (2002) Inhibition of tumor necrosis factor-induced cell death in MCF7 by a novel inhibitor of neutral sphingomyelinase. J Biol Chem 277:41128–41139

Maceyka M, Harikumar KB, Milstien S, Spiegel S (2012) Sphingosine-1-phosphate signaling and its role in disease. Trends Cell Biol 22:50–60

Macis D, Gandini S, Guerrieri-Gonzaga A, Johansson H, Magni P, Ruscica M, Lazzeroni M, Serrano D, Cazzaniga M, Mora S, Feroce I, Pizzamiglio M, Sandri MT, Gulisano M, Bonanni B, Decensi A (2012) Prognostic effect of circulating adiponectin in a randomized 2 × 2 trial of low-dose tamoxifen and fenretinide in premenopausal women at risk for breast cancer. J Clin Oncol 30:151–157

Mao C, Xu R, Bielawska A, Obeid LM (2000) Cloning of an alkaline ceramidase from Saccharomyces cerevisiae. An enzyme with reverse (CoA-independent) ceramide synthase activity. J Biol Chem 275:6876–6884

Mao C, Xu R, Szulc ZM, Bielawska A, Galadari SH, Obeid LM (2001) Cloning and characterization of a novel human alkaline ceramidase. A mammalian enzyme that hydrolyzes phytoceramide. J Biol Chem 276:26577–26588

Marchesini N, Luberto C, Hannun YA (2003) Biochemical properties of mammalian neutral sphingomyelinase 2 and its role in sphingolipid metabolism. J Biol Chem 278:13775–13783

Maurer BJ, Metelitsa LS, Seeger RC, Cabot MC, Reynolds CP (1999) Increase of ceramide and induction of mixed apoptosis/necrosis by N-(4-hydroxyphenyl)- retinamide in neuroblastoma cell lines. J Natl Cancer Inst 91:1138–1146

Medlock KA, Merrill AH Jr (1988) Inhibition of serine palmitoyltransferase in vitro and long-chain base biosynthesis in intact Chinese hamster ovary cells by beta-chloroalanine. Biochemistry 27:7079–7084

Milstien S, Spiegel S (2006) Targeting sphingosine-1-phosphate: a novel avenue for cancer therapeutics. Cancer Cell 9:148–150

Miyake Y, Kozutsumi Y, Nakamura S, Fujita T, Kawasaki T (1995) Serine palmitoyltransferase is the primary target of a sphingosine-like immunosuppressant, ISP-1/myriocin. Biochem Biophys Res Commun 211:396–403

Moore MM, Stockler M, Lim R, Mok TS, Millward M, Boyer MJ (2010) A phase II study of fenretinide in patients with hormone refractory prostate cancer: a trial of the Cancer Therapeutics Research Group. Cancer Chemother Pharmacol 66:845–850

Munoz-Olaya JM, Matabosch X, Bedia C, Egido-Gabas M, Casas J, Llebaria A, Delgado A, Fabrias G (2008) Synthesis and biological activity of a novel inhibitor of dihydroceramide desaturase. Chem Med Chem 3:946–953

Nagahashi M, Ramachandran S, Kim EY, Allegood JC, Rashid OM, Yamada A, Zhao R, Milstien S, Zhou H, Spiegel S, Takabe K (2012) Sphingosine-1-phosphate produced by sphingosine kinase 1 promotes breast cancer progression by stimulating angiogenesis and lymphangiogenesis. Cancer Res 72:726–735

Nara F, Tanaka M, Hosoya T, Suzuki-Konagai K, Ogita T (1999a) Scyphostatin, a neutral sphingomyelinase inhibitor from a discomycete, Trichopeziza mollissima: taxonomy of the producing organism, fermentation, isolation, and physico-chemical properties. J Antibiot (Tokyo) 52:525–530

Nara F, Tanaka M, Masuda-Inoue S, Yamasato Y, Doi-Yoshioka H, Suzuki-Konagai K, Kumakura S, Ogita T (1999b) Biological activities of scyphostatin, a neutral sphingomyelinase inhibitor from a discomycete, Trichopeziza mollissima. J Antibiot (Tokyo) 52:531–535

Noda T, Iwai S, Hamada M, Fujita Y, Yura Y (2009) Induction of apoptosis of detached oral squamous cell carcinoma cells by safingol. Possible role of Bim, focal adhesion kinase and endonuclease G. Apoptosis 14:287–297

Novgorodov SA, Wu BX, Gudz TI, Bielawski J, Ovchinnikova TV, Hannun YA, Obeid LM (2011) Novel pathway of ceramide production in mitochondria: thioesterase and neutral ceramidase produce ceramide from sphingosine and acyl-CoA. J Biol Chem 286:25352–25362

Ohotski J, Long JS, Orange C, Elsberger B, Mallon E, Doughty J, Pyne S, Pyne NJ, Edwards J (2012) Expression of sphingosine 1-phosphate receptor 4 and sphingosine kinase 1 is associated with outcome in oestrogen receptor-negative breast cancer. Br J Cancer 106: 1453–1459

Okazaki T, Bielawska A, Domae N, Bell RM, Hannun YA (1994) Characteristics and partial purification of a novel cytosolic, magnesium-independent, neutral sphingomyelinase activated in the early signal transduction of 1 alpha, 25-dihydroxyvitamin D3-induced HL-60 cell differentiation. J Biol Chem 269:4070–4077

Olsen I, Jantzen E (2001) Sphingolipids in Bacteria and Fungi. Anaerobe 7:103–112

Park MA, Mitchell C, Zhang G, Yacoub A, Allegood J, Haussinger D, Reinehr R, Larner A, Spiegel S, Fisher PB, Voelkel-Johnson C, Ogretmen B, Grant S, Dent P (2010) Vorinostat and sorafenib increase CD95 activation in gastrointestinal tumor cells through a Ca(2+)-de novo ceramide-PP2A-reactive oxygen species-dependent signaling pathway. Cancer Res 70: 6313–6324

Paugh BS, Paugh SW, Bryan L, Kapitonov D, Wilczynska KM, Gopalan SM, Rokita H, Milstien S, Spiegel S, Kordula T (2008) EGF regulates plasminogen activator inhibitor-1 (PAI-1) by a pathway involving c-Src, PKCdelta, and sphingosine kinase 1 in glioblastoma cells. FASEB J 22:455–465

Payne SG, Brindley DN, Guilbert LJ (1999) Epidermal growth factor inhibits ceramide-induced apoptosis and lowers ceramide levels in primary placental trophoblasts. J Cell Physiol 180:263–270

Perry DK, Carton J, Shah AK, Meredith F, Uhlinger DJ, Hannun YA (2000) Serine palmitoyl-transferase regulates de novo ceramide generation during etoposide-induced apoptosis. J Biol Chem 275:9078–9084

Pyne NJ, Pyne S (2011) Selectivity and specificity of sphingosine 1-phosphate receptor ligands: "off-targets" or complex pharmacology? Front Pharmacol 2:26

Qin J, Berdyshev E, Poirer C, Schwartz NB, Dawson G (2012) Neutral sphingomyelinase 2 deficiency increases hyaluronan synthesis by up-regulation of Hyaluronan synthase 2 through decreased ceramide production and activation of akt. J Biol Chem 287:13620–13632

Rahmaniyan M, Curley RW Jr, Obeid LM, Hannun YA, Kraveka JM (2011) Identification of dihydroceramide desaturase as a direct in vitro target for fenretinide. J Biol Chem 286: 24754–24764

Ren S, Xin C, Pfeilschifter J, Huwiler A (2010) A novel mode of action of the putative sphingosine kinase inhibitor 2-(p-hydroxyanilino)-4-(p-chlorophenyl) thiazole (SKI II): induction of lyso-somal sphingosine kinase 1 degradation. Cell Physiol Biochem 26:97–104

Saad AF, Meacham WD, Bai A, Anelli V, Elojeimy S, Mahdy AE, Turner LS, Cheng J, Bielawska A, Bielawski J, Keane TE, Obeid LM, Hannun YA, Norris JS, Liu X (2007) The functional effects of acid ceramidase overexpression in prostate cancer progression and resistance to chemotherapy. Cancer Biol Ther 6:1455–1460

Sabbadini RA (2011) Sphingosine-1-phosphate antibodies as potential agents in the treatment of cancer and age-related macular degeneration. Br J Pharmacol 162:1225–1238

Sakata A, Ochiai T, Shimeno H, Hikishima S, Yokomatsu T, Shibuya S, Toda A, Eyanagi R, Soeda S (2007) Acid sphingomyelinase inhibition suppresses lipopolysaccharide-mediated release of inflammatory cytokines from macrophages and protects against disease pathology in dextran sulphate sodium-induced colitis in mice. Immunology 122:54–64

Sanchez T, Hla T (2004) Structural and functional characteristics of S1P receptors. J Cell Biochem 92:913–922

Schaefer RM, Tylki-Szymanska A, Hilz MJ (2009) Enzyme replacement therapy for Fabry disease: a systematic review of available evidence. Drugs 69:2179–2205

Schiffmann S, Sandner J, Schmidt R, Birod K, Wobst I, Schmidt H, Angioni C, Geisslinger G, Grosch S (2009) The selective COX-2 inhibitor celecoxib modulates sphingolipid synthesis. J Lipid Res 50:32–40

Schnute ME, McReynolds MD, Kasten T, Yates M, Jerome G, Rains JW, Hall T, Chrencik J, Kraus M, Cronin CN, Saabye M, Highkin MK, Broadus R, Ogawa S, Cukyne K, Zawadzke LE, Peterkin V, Iyanar K, Scholten JA, Wendling J, Fujiwara H, Nemirovskiy O, Wittwer AJ, Nagiec MM (2012) Modulation of cellular S1P levels with a novel, potent and specific inhibitor of sphingosine kinase-1. Biochem J 444:79–88

Schwartz GK, Jiang J, Kelsen D, Albino AP (1993) Protein kinase C: a novel target for inhibiting gastric cancer cell invasion. J Natl Cancer Inst 85:402–407

Schwartz GK, Ward D, Saltz L, Casper ES, Spiess T, Mullen E, Woodworth J, Venuti R, Zervos P, Storniolo AM, Kelsen DP (1997) A pilot clinical/pharmacological study of the protein kinase C-specific inhibitor safingol alone and in combination with doxorubicin. Clin Cancer Res 3: 537–543

Shapiro D, Flowers HM (1962) Studies on sphingolipids VII. Synthesis and configuration of natural sphingomyelins. J Am Chem Soc 84:1047–1050

Shin KO, Park MY, Seo CH, Lee YI, Kim HS, Yoo HS, Hong JT, Jung JK, Lee YM (2012) Terpene alcohols inhibit de novo sphingolipid biosynthesis. Planta Med 78:434–439

Somenzi G, Sala G, Rossetti S, Ren M, Ghidoni R, Sacchi N (2007) Disruption of retinoic acid receptor alpha reveals the growth promoter face of retinoic acid. PLoS One 2:e836

Spiegel S, Milstien S (2011) The outs and the ins of sphingosine-1-phosphate in immunity. Nat Rev Immunol 11:403–415

Stoffel W (1970) Studies on the biosynthesis and degradation of sphingosine bases. Chem Phys Lipids 5:139–158

Strelow A, Bernardo K, Adam-Klages S, Linke T, Sandhoff K, Kronke M, Adam D (2000) Overexpression of acid ceramidase protects from tumor necrosis factor-induced cell death. J Exp Med 192:601–612

Sugita M, Willians M, Dulaney JT, Moser HW (1975) Ceramidase and ceramide synthesis in human kidney and cerebellum. Description of a new alkaline ceramidase. Biochim Biophys Acta 398:125–131

Tarabuso AL (2011) Fabry disease. Skinmed 9:173–177

Tolan D, Conway AM, Steele L, Pyne S, Pyne NJ (1996) The identification of DL-threo dihydrosphingosine and sphingosine as novel inhibitors of extracellular signal-regulated kinase signalling in airway smooth muscle. Br J Pharmacol 119:185–186

Tonelli F, Lim KG, Loveridge C, Long J, Pitson SM, Tigyi G, Bittman R, Pyne S, Pyne NJ (2010) FTY720 and (S)-FTY720 vinylphosphonate inhibit sphingosine kinase 1 and promote its proteasomal degradation in human pulmonary artery smooth muscle, breast cancer and androgen-independent prostate cancer cells. Cell Signal 22:1536–1542

Triola G, Fabrias G, Casas J, Llebaria A (2003) Synthesis of cyclopropene analogues of ceramide and their effect on dihydroceramide desaturase. J Org Chem 68:9924–9932

Triola G, Fabrias G, Dragusin M, Niederhausen L, Broere R, Llebaria A, van Echten-Deckert G (2004) Specificity of the dihydroceramide desaturase inhibitor N-[(1R,2S)-2-hydroxy-1-hydroxymethyl-2-(2-tridecyl-1-cyclopropenyl)ethyl]octanami de (GT11) in primary cultured cerebellar neurons. Mol Pharmacol 66:1671–1678

Triola G, Fabrias G, Llebaria A (2001) Synthesis of a cyclopropene analogue of ceramide, a potent inhibitor of dihydroceramide desaturase This work was supported by the Direccion General de Ensenanza Superior e Investigacion Cientifica (grant PB97-1171) and the Departament d'Universitats, Recerca i Societat de la Informacio, Generalitat de Catalunya (grant 1999-SGR 00187 and a Predoctoral fellowship to G.T.). We thank Dr. J. Casas, Dr. A. Delgado, and Dr. J. Joglar for their help in different aspects of this work. Angew Chem Int Ed Engl 40: 1960–1962

Villablanca JG, London WB, Naranjo A, McGrady P, Ames MM, Reid JM, McGovern RM, Buhrow SA, Jackson H, Stranzinger E, Kitchen BJ, Sondel PM, Parisi MT, Shulkin B, Yanik GA, Cohn SL, Reynolds CP (2011) Phase II study of oral capsular 4-hydroxyphenylretinamide (4-HPR/fenretinide) in pediatric patients with refractory or recurrent neuroblastoma: a report from the Children's Oncology Group. Clin Cancer Res 17:6858–6866

Villorbina G, Canals D, Carde L, Grijalvo S, Pascual R, Rabal O, Teixido J, Fabrias G, Llebaria A, Casas J, Delgado A (2007) Solid-phase synthesis of a combinatorial library of dihydroceramide analogues and its activity in human alveolar epithelial cells. Bioorg Med Chem 15:50–62

Visentin B, Vekich JA, Sibbald BJ, Cavalli AL, Moreno KM, Matteo RG, Garland WA, Lu Y, Yu S, Hall HS, Kundra V, Mills GB, Sabbadini RA (2006) Validation of an anti-sphingosine-1-phosphate antibody as a potential therapeutic in reducing growth, invasion, and angiogenesis in multiple tumor lineages. Cancer Cell 9:225–238

Wang H, Maurer BJ, Liu YY, Wang E, Allegood JC, Kelly S, Symolon H, Liu Y, Merrill AH Jr, Gouaze-Andersson V, Yu JY, Giuliano AE, Cabot MC (2008) N-(4-Hydroxyphenyl) retinamide increases dihydroceramide and synergizes with dimethylsphingosine to enhance cancer cell killing. Mol Cancer Ther 7:2967–2976

Wascholowski V, Giannis A, Pitsinos EN (2006) Influence of the scyphostatin side chain on the mode of inhibition of neutral sphingomyelinase. ChemMedChem 1:718–721
Wu BX, Clarke CJ, Hannun YA (2010) Mammalian neutral sphingomyelinases: regulation and roles in cell signaling responses. Neuromolecular Med 12:320–330
Yatomi Y, Ruan F, Megidish T, Toyokuni T, Hakomori S, Igarashi Y (1996) *N*, *N*-dimethylsphingosine inhibition of sphingosine kinase and sphingosine 1-phosphate activity in human platelets. Biochemistry 35:626–633
Yatomi Y, Yamamura S, Ruan F, Kume S, Ozaki Y, Igarashi Y (1997) *N*, *N*-dimethylsphingosine 1-phosphate activates human platelets. FEBS Lett 417:341–344
Yokomatsu T, Takechi H, Akiyama T, Shibuya S, Kominato T, Soeda S, Shimeno H (2001) Synthesis and evaluation of a difluoromethylene analogue of sphingomyelin as an inhibitor of sphingomyelinase. Bioorg Med Chem Lett 11:1277–1280

Pharmacology of the Sphingosine-1-Phosphate Signalling System

Dagmar Meyer zu Heringdorf, Katja Ihlefeld, and Josef Pfeilschifter

Contents

Abstract The recent success of FTY720 (Fingolimod, Gilenya®), which has been approved for the treatment of relapsing–remitting multiple sclerosis and is the first-in-class sphingosine-1-phosphate (S1P) receptor modulating drug, has boosted the interest in further drug development in this area. Several selective $S1P_1$ receptor-modulating drugs are being investigated in clinical trials for the treatment of diverse autoimmune disorders. Sphingosine kinase inhibitors are under development for the treatment of cancer, aberrant angiogenesis and inflammatory diseases; an inhibitor of SK2 with relatively low affinity is being analysed in patients with advanced solid tumours. While an indirect S1P lyase inhibitor has just failed the proof of concept in patients with rheumatoid arthritis, S1P lyase is still a promising target for the treatment of inflammatory and autoimmune diseases. Another approach is the development of S1P-scavenging or -clearing agents, including a monoclonal S1P antibody that has successfully passed phase I clinical trials and will be further developed for age-related macular degeneration.

D. Meyer zu Heringdorf (✉) • K. Ihlefeld • J. Pfeilschifter
Institut für Allgemeine Pharmakologie und Toxikologie, Klinikum der Johann Wolfgang Goethe-Universität, Theodor-Stern-Kai 7, 60590 Frankfurt am Main, Germany
e-mail: heringdorf@med.uni-frankfurt.de

E. Gulbins and I. Petrache (eds.), *Sphingolipids: Basic Science and Drug Development*, Handbook of Experimental Pharmacology 215, DOI 10.1007/978-3-7091-1368-4_13,

Keywords Sphingosine-1-phosphate • Sphingosine kinase • S1P lyase • Fingolimod • Sonepcizumab

1 Introduction

The bioactive lipid, sphingosine-1-phosphate (S1P), is a ubiquitous signalling molecule that regulates cell proliferation, migration and cell–cell contacts and thereby controls essential organ functions. S1P for example regulates immune cell trafficking, maintenance of the vascular barrier, angiogenesis, regulation of heart rate and many other functions (Chi 2011; Maceyka et al. 2012; also see other chapters of this book). Drugs that target the S1P signalling system are under development for the treatment of autoimmune diseases, inflammatory diseases, aberrant angiogenesis as well as cancer (Huwiler and Pfeilschifter 2008; Pyne and Pyne 2010; Edmonds et al. 2011). Established molecular targets within the S1P signalling system include the G-protein-coupled S1P receptors (S1P-GPCR) and the enzymes that catalyse the formation and degradation of S1P, the sphingosine kinases (SK1, SK2) and S1P lyase. The therapeutic potential of these targets and the presently available drugs will be discussed in this chapter, with a focus on compounds that are in clinical development or have been tested in animal models with a clear concept for further clinical development.

Notably, the S1P-GPCR are important regulators of cell migration during development and in immune and inflammatory responses (Chi 2011). These migratory activities are controlled by concentration gradients of S1P, for example by the blood-tissue gradient which directs lymphocyte migration from lymphatic tissues with low S1P concentrations to lymph and blood with high S1P concentrations (Hla et al. 2008). In fact, blocking the sensitivity of lymphocytes for this gradient or disturbing the gradient by different means are presently the most common mechanisms of drugs targeting the S1P signalling system.

The S1P signalling system also comprises the transporters that allow inside-out signalling by S1P and autocrine/paracrine activation of S1P-GPCR. Of these enzymes, the spinster-2 protein recently gained interest because it was shown that it contributes to the blood-tissue gradient of S1P (Fukuhara et al. 2012). Thus, the deletion of spinster-2 in mice led to a decrease in S1P plasma concentrations and defects in lymphocyte egress (Fukuhara et al. 2012). These effects were traced back to endothelial spinster-2, confirming the previously suggested important role of the vascular endothelium for maintenance of the S1P gradient (Fukuhara et al. 2012; Hla et al. 2008). Therefore, spinster-2 might be a target for inflammatory and autoimmune diseases.

2 S1P Receptor-Modulating Drugs

There are five high-affinity S1P-GPCR which are widely expressed and regulate a multitude of functions in a partly overlapping and partly antagonistic manner (Chun et al. 2010). The interest in the development of S1P receptor-modulating drugs has

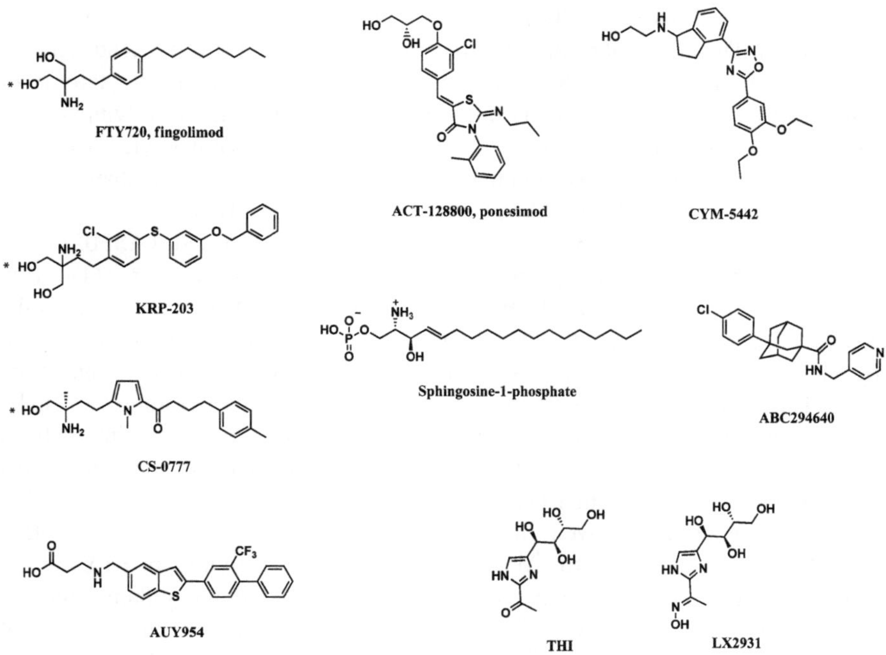

Fig. 1 Structures of selected modulators of the S1P signalling system. The *asterisks* mark compounds that need to be phosphorylated at the indicated sites to be able to interact with S1P receptors. For further information, see text

been initiated by the discovery that FTY720, an immunosuppressive drug with a unique influence on lymphocyte trafficking and homing, interacted with S1P-GPCR. The whole area is still strongly influenced by FTY720, and therefore, further drug development primarily focuses on selective functional antagonists at the $S1P_1$ receptor as immunosuppressive drugs. Several compounds have been synthesized that act as agonists or antagonists at the $S1P_{2/3/4/5}$ receptors, and they are used experimentally for studying the functional roles of these receptors. These rather experimental compounds will not be reviewed here.

FTY720 (fingolimod; Fig. 1) has recently been approved by FDA and EMEA for the treatment of relapsing–remitting multiple sclerosis (see Web sites http://www.accessdata.fda.gov/scripts/cder/drugsatfda/index.cfm?fuseaction=Search.DrugDetails and http://www.ema.europa.eu/docs/en_GB/document_library/EPAR_-_Product_Information/human/002202/WC500104528.pdf). Pharmacodynamics and pharmacokinetics of FTY720 as well as the pivotal clinical trials have been extensively reviewed (see, e.g., Brinkmann et al. 2010; Cohen and Chun 2011; David et al. 2012; Pelletier and Hafler 2012). FTY720 is the prodrug of FTY720-phosphate, which is produced by sphingosine kinases, preferentially by SK2, and acts as agonist with nanomolar affinity at the S1P-GPCR, $S1P_{1/3/4/5}$. Diverse studies have confirmed that FTY720's immunosuppressive action is based on activation of the $S1P_1$ receptor, which is subsequently internalized and may be degraded (Oo et al. 2007) or induce prolonged signalling (Mullershausen et al. 2009). This removal of the $S1P_1$ receptor from the cell surface, rendering the cells insensitive to external S1P, can be interpreted as superagonistic activity or functional antagonism. As a consequence, certain populations of T lymphocytes (mainly naïve and central memory T cells) as well as B lymphocytes lose their ability to sense the blood-tissue gradient of S1P and are trapped in lymphatic tissues (reviewed by Brinkmann et al. 2010; Cohen and Chun 2011). Lymphopenia can also be observed with $S1P_1$ receptor antagonists such as W146 and TASP0277308, provided the pharmacokinetic properties of the compounds are considered (Tarrasón et al. 2011; Fujii et al. 2012). Interestingly, other effects of FTY720, such as the $S1P_1$-mediated improvement of the endothelial barrier function, are not mimicked by $S1P_1$ receptor antagonists, in contrast, they induce a vascular leak (Sanna et al. 2006; Tarrasón et al. 2011; Fujii et al. 2012), suggesting that barrier improvement by FTY720 as agonistic effect may be due to prolonged signalling of the internalized ligand–receptor complex. Repeated administration of FTY720 and other $S1P_1$ receptor agonists, on the other hand, may exacerbate vascular leak, as observed in the mouse model of bleomycin-induced lung injury (Shea et al. 2010), indicating loss of $S1P_1$ on the long term. While the main in vivo activities of FTY720 are probably in large part due to impaired invasion of lymphocytes into inflamed tissues, the drug may also have direct effects on cells of the central nervous system that contribute to its clinical effectiveness in multiple sclerosis (for discussion, see Brinkmann et al. 2010; Cohen and Chun 2011). Interestingly, a specific knockout of $S1P_1$ in astrocytes alleviated the course of experimental autoimmune encephalitis (EAE) in mice and also abrogated the activity of FTY720, suggesting that the $S1P_1$ receptor plays an important role in astrocyte pathophysiology during EAE as well as for the activity of FTY720 (Choi et al. 2011).

Presently there are no clinical guidelines for the use of FTY720 in multiple sclerosis (Killestein et al. 2011; Pelletier and Hafler 2012). The most common serious adverse effects of FTY720 are bradycardia and atrioventricular block, which typically occur during initiation of the therapy and resolve within 24 h despite continued FTY720 treatment (Kappos et al. 2010; Cohen et al. 2010). The underlying mechanism is the activation of G-protein-gated inwardly rectifying potassium (GIRK) channels in the cardiac conduction system by S1P-GPCR. Notably, there is a report of a severe bradycardia and asystole even 21 h after the first dose of FTY720 (Espinosa and Berger 2011). Furthermore, a case of sudden death after the first dose of FTY720 has been reported in December 2011, and this case is being further investigated by FDA and EMEA (see Web sites http://www.fda.gov/Drugs/DrugSafety/ucm284240.htm and http://www.ema.europa.eu/docs/en_GB/document_library/Press_release/2012/01/WC500120703.pdf). Other side effects of FTY720 include mild infections as a consequence of immunosuppression, mild increases in systolic and diastolic blood pressure as well as a decrease in lung function probably due to S1P-GPCR modulation, asymptomatic increases in liver enzymes and macular oedema (Kappos et al. 2010; Cohen et al. 2010; Pelletier and Hafler 2012). FTY720 is being further developed for the treatment of primary progressive multiple sclerosis (Table 1).

Table 1 S1P modulating drugs in clinical development

Drug	Target(s)	Indication(s) tested in clinical trials	Developing company/sponsor
FTY720, fingolimod (Gilenya®)	$S1P_{1/3/4/5}$	• Relapsing–remitting multiple sclerosis (approved by FDA and EMEA) • Primary progressive multiple sclerosis (phase III)	Novartis
KRP-203	$S1P_{1>4}$	• Ulcerative Colitis (phase II) • Subacute cutaneous lupus erythematodes (phase II)	Novartis
BAF312	$S1P_{1/5}$	• Relapsing–remitting multiple sclerosis (phase II) • Polymyositis, dermatomyositis (phase II)	Novartis
ACT-128800, ponesimod	$S1P_1$	• Relapsing–remitting multiple sclerosis (phase II) • Severe chronic plaque psoriasis (phase II)	Actelion
ONO-4641	$S1P_1$	• Multiple sclerosis (phase II)	Ono Pharma
GSK2018682	$S1P_1$	• Relapsing–remitting multiple sclerosis (phase I)	GlaxoSmithKline
CS-0777	$S1P_1$	• Multiple sclerosis (phase I)	Daiichi Sankyo
ABC294640	SK2	• Advanced solid tumours (phase I)	#
LX3305 (=LX2931)	S1P lyase	• Rheumatoid arthritis (phase II)	Lexicon
Sonepcizumab	S1P	• Age-related macular degeneration (phase II)	Lpath, Pfizer

See http://www.clinicaltrials.gov. #, Medical University of South Carolina, Apogee Biotechnology Corporation, National Cancer Institute USA

Several S1P receptor-modulating drugs with selectivity for the $S1P_1$ receptor have been developed and some of them are being tested in clinical trials for the treatment of multiple sclerosis and other autoimmune diseases (Fig. 1 and Table 1). KRP-203 is the prodrug of KRP-203-phosphate, which induces $[Ca^{2+}]_i$ increases via the $S1P_1$ and $S1P_4$ receptors with EC_{50}s of ~1 and ~10 nM, respectively, and lacks activity at the $S1P_3$ receptor (Song et al. 2008). KRP-203 induced lymphopenia in mice and was active in rat organ transplantation, rat autoimmune myocarditis, chronic colitis in interleukin-10 deficient mice (reviewed by Huwiler and Pfeilschifter 2008) and autoimmune kidney disease in the MRL/lpr mouse model (Wenderfer et al. 2008). KRP-203 is being tested in phase II clinical trials for ulcerative colitis and subacute lupus erythematosus. AUY954 can activate the $S1P_1$ receptor without being phosphorylated. In a GTPγS binding assay, AUY954 activated the $S1P_1$ receptor with an EC_{50} of ~1 nM, while it activated $S1P_5$ and $S1P_3$ with EC_{50}s of 340 and 1,200 nM, respectively, and lacked activity at $S1P_2$ and $S1P_4$ (Pan et al. 2006). AUY954 induced lymphopenia and prolonged the survival of cardiac allografts in rats (Pan et al. 2006), reduced allergen-induced plasma leakage in mice (Blé et al. 2009) and attenuated paraparesis in a T cell-mediated autoimmune inflammatory demyelinating disease of the peripheral nervous system (Zhang et al. 2009). ACT-128800 (ponesimod) stimulated GTPγS binding with EC_{50}s of ~6 nM at $S1P_1$, 60 nM at $S1P_5$, 100 nM at $S1P_3$, and ~1,000 nM at $S1P_4$, while no activity at $S1P_2$ was detectable (Piali et al. 2011; Bolli et al. 2010). ACT-128800 attenuated the inflammatory responses in the skin of mice with delayed-type hypersensitivity and in rats with adjuvant-induced arthritis (Piali et al. 2011). The drug had a relatively short plasma half-life of ~3 h after a single oral dose in rats, which corresponded to a relatively rapid reversibility of the lymphopenia compared to FTY720, which has an elimination half-life of 6–9 days and requires 1–2 months for full recovery of the immunological effects (Piali et al. 2011). ACT-128800 is being tested in phase II clinical trials for relapsing–remitting multiple sclerosis and severe chronic plaque psoriasis. In agreement with the concept that the $S1P_5$ receptor contributes to the activity of FTY720 in multiple sclerosis, BAF312, an $S1P_{1/5}$ agonist (Gergely et al. 2009), is being tested in phase II clinical trials of relapsing–remitting multiple sclerosis and polymyositis/dermatomyositis. Indeed, in an organ model of de- and remyelination, in which the drugs' influence on lymphocyte trafficking is irrelevant, FTY720 and BAF312, but not the $S1P_1$-selective AUY954, increased markers of remyelination (Jackson et al. 2011). CS-0777 is another $S1P_1$ receptor modulator that is being tested in clinical trials. It again is a prodrug that needs to be phosphorylated, which is interestingly catalysed by fructosamine-3-kinase and fructosamine-3-kinase-related protein in erythrocytes (Yonesu et al. 2011). CS-0777 has a similar long half-life and slow reversibility of action as FTY720 (Moberly et al. 2012). CYM-5442 is a highly selective agonist at the $S1P_1$ receptor and inhibited forskolin-stimulated cAMP-responsive element transcription with an EC_{50} of ~1 nM (Gonzalez-Cabrera et al. 2008). This compound was able to activate the $S1P_1$ receptor without binding to the amino acid residues, R^{120} and E^{121} (Gonzalez-Cabrera et al. 2008), which form ionic interactions with the phosphate head group of S1P (Parrill et al. 2000; Hanson et al. 2012). CYM-5442 has a relatively short half-life in plasma (~3 h after oral application in rats; Gonzalez-Cabrera et al. 2008), but it strongly accumulates within the CNS (Gonzalez-Cabrera et al. 2012). In agreement, during once daily application of CYM-5442 in mice, lymphocyte counts recovered within each cycle, while the $S1P_1$ receptor was downregulated on astrocytes and neurons and EAE was attenuated (Gonzalez-Cabrera et al. 2012). These observations may support the hypothesis that the $S1P_1$ receptor on cells within the CNS is the major target in multiple sclerosis, but it may also be the case that a transient lymphopenia is sufficient for reducing the symptoms of EAE.

Taken together, the $S1P_1$ receptor modulators differ with regard to S1P receptor subtype selectivity and pharmacokinetics. Both the receptor profile and a potential organ accumulation will have an impact on the drugs' clinical activities. The hope that $S1P_1$ selective compounds will cause less cardiac arrhythmias, because GIRK activation is mediated by $S1P_3$ in mice (Sanna et al. 2004), has been challenged by reports showing that highly selective $S1P_1$ agonists can cause bradycardia in humans (Gergely et al. 2009). Compounds with shorter elimination half-life and quicker recovery from immunosuppression than FTY720 will have the advantage of rapid washout in case of side effects. A thorough comparison of the $S1P_1$ receptor modulators, however, requires more comprehensive data than publicly available. Therefore, access to original preclinical and phase I data is urgently needed.

3 Sphingosine Kinase Inhibitors

The two isoforms of sphingosine kinase are differentially expressed and regulated (Pitson 2011), and initiate diverse cellular responses depending on their subcellular localization (Maceyka et al. 2012). There is substantial evidence that sphingosine kinase inhibitors might be useful for the treatment of cancer, aberrant angiogenesis and inflammatory diseases. SK1, in particular, is regarded as a player in tumour cell growth and survival. In several human cancers, SK1 expression is higher than in adjacent normal tissue, and high SK1 expression correlates with malignancy and/or reduced patient survival (reviewed by Pyne and Pyne 2010). SK2, in contrast, has been associated with cell cycle arrest and apoptosis (Igarashi et al. 2003; Liu et al. 2003). However, observations from knockout mice, in which deletion of a single SK isoform was compensated for by the other isoform, and studies with SK2 inhibitors (see below) have challenged the view that SK1 and SK2 play antagonistic roles. Sphingosine kinases, with most studies focused on SK1, are furthermore involved in the regulation of inflammatory processes such as arthritis, colitis, asthma, and anaphylaxis, they contribute to the regulation of smooth muscle (vascular and bronchial) contraction, they stimulate angiogenesis and play a role in ischemia–reperfusion injury (for review, see Snider et al. 2010; Maceyka et al. 2012). Importantly, signalling by sphingosine kinases can either be due to secreted/exported S1P which acts on S1P-GPCR, a process that is called inside-out signalling, or to interaction of S1P with intracellular targets. Intracellular SK1/S1P signalling for example mediated NF-κB activation downstream of the TNF-α receptor, and nuclear SK2/S1P signalling inhibited histone deacetylase activity (reviewed by Maceyka et al. 2012). Thus, the functional roles of sphingosine kinases are even more complex than those of the S1P-GPCR.

The first nonlipid SK inhibitors were identified by screening of ~16,000 compounds for inhibition of purified recombinant human SK1 (French et al. 2003). Of these compounds, SKI-II (4-[[4-(4-Chlorophenyl)-2-thiazolyl]amino] phenol) was orally active and inhibited the growth of a mammary adenocarcinoma in a syngeneic mouse model (French et al. 2006). The mode of inhibition was non-competitive for ATP (French et al. 2003). Later, it was found that in intact cells, SKI-II caused an effective degradation of SK1 (Ren et al. 2010; Loveridge et al. 2010). There is some disagreement whether this occurs through lysosomal (Ren et al. 2010) or proteasomal (Loveridge et al. 2010) degradation.

While SKI-II also inhibits the SK2 isoform (Lim et al. 2011), the water-soluble sphingosine analog, SK1-I ((2R,3S,4E)-*N*-methyl-5-(4′-pentylphenyl)-2-aminopent-4-ene-1,3-diol), inhibited specifically SK1 with a K_i of ~10 μM, while it had no influence on SK2, ceramide kinase, and several protein kinases in a protein kinase activity screen (Paugh et al. 2008). It induced apoptosis in leukaemia cell lines and in primary blasts from patients with acute myelogenous leukaemia in vitro (IC_{50} ~7.5 μM), and inhibited the growth of U937 (human leukemic monocyte lymphoma) cell xenografts in vivo (Paugh et al. 2008). SK1-I furthermore induced cell death in glioblastoma cell lines and attenuated migration and invasion of these cells (Kapitonov et al. 2009). It also was active against glioblastoma xenografts and prolonged the survival of nude mice with intracranial glioblastoma cell xenografts (Kapitonov et al. 2009).

It is an irony in the history of SK inhibitor development that primarily SK1 has been regarded as the target for treatment of cancer, but an SK2 inhibitor became the first to be tested in clinical trials. ABC294640 (Fig. 1) inhibited recombinant SK2 with a K_i of ~10 μM but had no effect on SK1 up to at least 100 μM (French et al. 2010). Its mode of action was competitive to sphingosine. ABC294640 was inactive against 20 different protein kinases at a concentration of 50 μM. In cell lines, ABC294640 inhibited S1P formation with an IC_{50} of 26 μM and decreased cellular S1P levels. The compound inhibited the growth of various tumour cell lines in vitro, reduced migration and disrupted actin microfilament structure, suggesting that it might be able to reduce tumour metastasis. ABC294640 had a good oral bioavailability and tolerability in mice. Toxicity was associated with the haematopoietic system, with slight decreases in red blood cell number and haematocrit (French et al. 2010). At 35 and 100 mg/kg, ABC294640 reduced the growth of a mouse JC mammary adenocarcinoma cell line growing subcutaneously in immunocompetent Balb/c mice (French et al. 2010). The antitumor activity of ABC294640 has furthermore been shown in diverse xenograft models, including kidney, pancreatic and hepatocellular carcinoma (Beljanski et al. 2011a; Beljanski et al. 2011b). Interestingly, ABC294640 binds in the antagonist ligand-binding domain of the oestrogen receptor and has antiestrogenic effects, and therefore it has therapeutic potential for treatment of oestrogen receptor-positive breast cancer by inhibiting both SK and oestrogen signalling (Antoon et al. 2010). The low affinity to SK2 raises the suspicion that ABC294640 also has yet (an)other target(s). The drug is presently being tested in a phase I clinical trial involving patients with advanced solid tumours (Table 1). ABC294640 was furthermore active in diverse models of inflammatory diseases. The compound suppressed dextran sulphate sodium-induced and trinitrobenzene sulphonic acid-induced colitis in mice (models for ulcerative colitis and Crohn's disease, respectively), and reduced colitis-driven colon cancer development in mice (Maines et al. 2008, 2010; Chumanevich et al. 2010). ABC294640 also attenuated collagen- and adjuvant-induced arthritis in rats (Fitzpatrick et al. 2011a), although SK1 siRNA reduced and SK2 siRNA aggravated collagen-induced arthritis in mice (Lai et al. 2009), and even was beneficial in experimental osteoarthritis in rats (Fitzpatrick et al. 2011b).

A major problem of the above-described SK inhibitors is their low potency. Recently, the up to now most potent SK inhibitors were identified by in silico screening with a homology model of SK1 (Kennedy et al. 2011). Several amidine-based compounds were identified which inhibited SK1 and/or SK2 in the submicromolar concentration range (Kennedy et al. 2011). Compound "1a" inhibited SK1 with a K_i of 0.1 μM and SK2 with a K_i of 1.5 μM and effectively reduced S1P levels in cultured cells (Kharel et al. 2011). "1a" potently inhibited epidermal growth factor-stimulated S1P synthesis, ERK and Akt phosphorylation, while ~tenfold higher concentrations of "1a" were required to induce cytotoxicity, caspase-3 and PARP cleavage than for the reduction of S1P levels. Interestingly, intravenous application of "1a" to mice within 20 min reduced blood S1P levels in wild type, but not in $SK1^{-/-}$ mice, confirming that the drug acts as SK1 inhibitor in vivo and revealing that there is a rapid turnover of S1P in blood (Kharel et al. 2011). These data challenge the hypothesis that SK1 is a target for treatment of cancer, however, the novel compounds will help to redefine the roles of the two sphingosine kinases in health and disease.

4 S1P Lyase Inhibitors

S1P lyase catalyses the irreversible cleavage of the sphingosine backbone, therefore it is required for the ultimate degradation of sphingolipids (Kumar and Saba 2009). It was first reported in 2005 that the blood-tissue-gradient of S1P is dependent on the enzymatic activity of S1P lyase. This was discovered because the component of caramel colour III, 2-acetyl-4-tetrahydroxybutylimidazole (THI; Fig. 1), which was known to induce lymphopenia, elevated S1P concentrations in lymphoid tissues by inhibiting S1P lyase activity (Schwab et al. 2005). The high concentrations of S1P-induced internalization of the $S1P_1$ receptor in lymphocytes within lymphatic tissues and therefore had similar consequences as treatment with $S1P_1$ (functional) antagonists: decreased egress of lymphocytes from lymphatic tissues, lymphopenia and immunosuppression. The effects of THI were mimicked by downregulation of S1P lyase in hematopoietic cells and by 4-deoxypyridoxine (DOP), which inhibits pyridoxal phosphate-dependent enzymes including S1P lyase (Schwab et al. 2005). Thus, inhibition of S1P lyase emerged as a novel immunosuppressive principle. S1P lyase-deficient mice indeed have lymphopenia, but they suffer from many other pathologies in lung, heart, urinary tract and bone, have vascular defects and a strongly reduced life span (Vogel et al. 2009; Schmahl et al. 2007). While a full knockout of S1P lyase has features of a lipid storage disease, a partial restoration of S1P lyase activity was sufficient to protect from organ damages while lymphocyte trafficking remained impaired (Vogel et al. 2009). As a consequence, there might be a therapeutic window in which partial inhibition of S1P lyase could lead to clinical immunosuppression without causing significant organ damage. On this background, LX2931 (=LX3305; Fig. 1), which differs minimally from THI, was developed by Lexicon Pharmaceuticals (Princeton, NJ, USA) for immunosuppression. LX2931/3305 elevated S1P concentrations in lymphoid tissues, caused lymphopenia in multiple species and prevented the development of arthritis in the mouse collagen-induced arthritis model (Bagdanoff et al. 2010). In a phase I clinical trial in healthy subjects, the compound was well tolerated up to 180 mg daily and reduced lymphocyte counts (used as a surrogate marker for S1P lyase inhibition) by ~50 % at that dose (Bagdanoff et al. 2010). Recently, a phase II trial in patients with rheumatoid arthritis on stable methotrexate therapy has been completed and showed a favourable safety profile of LX2931/3305, but the compound failed the proof-of-concept (see Web sites http://clinicaltrials.gov/ct2/show/study/NCT00903383?term=lexicon&rank=3&sect=X36015 and http://www.lexpharma.com/images/

pdfs/EULAR%202011%205-20-11%20finalv3.pdf). The company presently opts for a scale-up in the hope to identify a therapeutically active dose http://www.lexpharma.com/images/pdfs/EULAR%202011%205-20-11%20finalv3.pdf.

An interesting fact about all S1P lyase inhibitors is that neither THI nor Lexicon's compounds directly inhibit the enzyme in vitro, and this cannot be explained by the occurrence of active metabolites (Bagdanoff et al. 2009, 2010). DOP inhibits all pyridoxal phosphate-dependent enzymes and therefore is highly unspecific. Thus, the run is open to find the first-in-class direct S1P lyase inhibitor.

5 S1P Scavenging or Clearing Agents

The effectiveness of S1P-scavenging agents to attenuate the pro-angiogenic and cancer promoting activities of S1P has first been shown with a monoclonal antibody to S1P. The anti-S1P antibody reduced tumour progression and tumour angiogenesis in murine xenograft and allograft models, and reduced the vessel formation in VEGF- and bFGF-containing Matrigel plugs in vivo (Visentin et al. 2006). A humanized version of this antibody, sonepcizumab, developed by Lpath Therapeutics (San Diego, CA, USA), reduced blood lymphocyte counts, indicating that it was able to affect the blood-tissue-gradient of S1P (O'Brien et al. 2009). Sonepcizumab inhibited choroidal and retinal neovascularization and sub-retinal fibrosis in ocular angiogenesis models when injected into the vitreous body (Xie et al. 2009; Caballero et al. 2009). From a screen of ~60 different lipids, the antibody also recognized dihydro-S1P and sphingosylphosphorylcholine and had a low affinity to D-*erythro*-sphingosine and sphingomyelin (O'c et al. 2009). Sphingomyelin is a highly abundant lipid (>1,000-fold excess over S1P; Hannun and Obeid 2008) that might be able to bind sonepcizumab in relevant amounts. Nevertheless, sonepcizumab passed several preclinical toxicology tests, and its safety and tolerability has been analysed in phase I clinical trials in patients with refractory advanced solid tumours (oncology formulation ASONEP®) and in patients with exudative age-related macular degeneration (ocular formulation iSONEP®) (reviewed by Sabbadini 2011). The favourable outcome led to its further development for treatment of age-related macular degeneration (Table 1).

Noxxon Pharma (Berlin, Germany) has chosen a different approach to S1P scavenging by developing an S1P-binding spiegelmer, NOX-S93. Up to now there are no published data regarding this spiegelmer. On its Web site http://www.noxxon.com/index.php?option=com_content&view=article&id=62&Itemid=101, the company states that NOX-S93 inhibited signalling of S1P through $S1P_1$ and $S1P_3$ with low nanomolar affinity and induced transient lymphopenia in mice. Target indications of NOX-S93 would be cancer angiogenesis, autoimmune disorders and ocular diseases.

Another approach to remove extracellular S1P in conditions such as cancer, inflammation, fibrosis and aberrant angiogenesis is the application of recombinant soluble S1P lyase. S1P lyase from the prokaryote *Symbiobacterium thermophilum* disrupted S1P signalling in cell culture models, inhibited tumour cell-induced angiogenesis in the chicken chorioallantoic membrane model, and effectively reduced S1P plasma levels when injected intravenously in mice (Huwiler et al. 2011). The advantage of the prokaryotic S1P lyase is its lack of an N-terminal transmembrane domain that makes it a soluble enzyme, while eukaryotic S1P lyases are integral membrane proteins.

6 Concluding Remarks

In conclusion, the $S1P_1$ receptor has been established as a target for immunosuppression, while drugs that modulate the other S1P receptors or inhibit SKs and S1P lyase still await the proof of their clinical usefulness. A possible advantage of the S1P cleaving or scavenging agents is their high specificity and the fact that their activity is restricted to extracellular pools of S1P, while small molecule inhibitors of SKs or S1P lyase would also modulate intracellular levels of S1P, and moreover, may have less specificity.

References

Antoon JW, White MD, Meacham WD, Slaughter EM, Muir SE, Elliott S, Rhodes LV, Ashe HB, Wiese TE, Smith CD, Burow ME, Beckman BS (2010) Antiestrogenic effects of the novel sphingosine kinase-2 inhibitor ABC294640. Endocrinology 151:5124–5135

Bagdanoff JT et al (2009) Inhibition of sphingosine-1-phosphate lyase for the treatment of autoimmune disorders. J Med Chem 52:3941–3953

Bagdanoff JT et al (2010) Inhibition of sphingosine 1-phosphate lyase for the treatment of rheumatoid arthritis: Discovery of (E)-1-(4((1R,2S,3R)-1,2,3,4-tetrahydroxybutyl)-1H-imidazol-2-yl)ethanone oxime (LX2931) and (1R,2S,3R)-1-(2-(Isoxazol-3-yl)-1H-imidazol-4-yl)butane-1,2,3,4-tetraol (LX2932). J Med Chem 53:8650–8662

Beljanski V, Knaak C, Zhuang Y, Smith CD (2011a) Combined anticancer effects of sphingosine kinase inhibitors and sorafenib. Invest New Drugs 29:1132–1142

Beljanski V, Lewis CS, Smith CD (2011b) Antitumor activity of sphingosine kinase-2 inhibitor ABC294640 and sorafenib in hepatocellular carcinoma xenografts. Cancer Biol Ther 11:524–534

Blé F, Cannet C, Zurbruegg S, Gérard C, Frossard N, Beckmann N, Trifilieff A (2009) Activation of the lung $S1P_1$ receptor reduces allergen-induced plasma leakage in mice. Br J Pharmacol 158:1295–1301

Bolli MH et al (2010) 2-Imino-thiazolidin-4-one derivatives as potent, orally active $S1P_1$ receptor agonists. J Med Chem 53:4198–4211

Brinkmann V, Billich A, Baumruker T, Heining P, Schmouder R, Francis G, Aradhye S, Burtin P (2010) Fingolimod (FTY720): discovery and development of an oral drug to treat multiple sclerosis. Nat Rev Drug Discov 9:883–897

Caballero S, Swaney J, Moreno K, Afzal A, Kielczewski J, Stoller G, Cavalli A, Garland W, Hansen G, Sabbadini R, Grant MB (2009) Anti-sphingosine-1-phosphate monoclonal antibodies inhibit angiogenesis and sub-retinal fibrosis in a murine model of laser-induced choroidal neovascularization. Exp Eye Res 88:367–377

Chi H (2011) Sphingosine-1-phosphate and immune regulation: trafficking and beyond. Trends Pharmacol Sci 32:16–24

Choi JW, Gardell SE, Herr DR, Rivera R, Lee C, Noguchi K, Teo ST, Yung YC, Lu M, Kennedy G, Chun J (2011) FTY720 (fingolimod) efficacy in an animal model of multiple sclerosis requires astrocyte sphingosine 1-phosphate receptor-1 ($S1P_1$) modulation. Proc Natl Acad Sci U S A 108:751–756

Chumanevich AA, Poudyal D, Cui X, Davis T, Wood PA, Smith CD, Hofseth LJ (2010) Suppression of colitis-driven colon cancer in mice by a novel small molecule inhibitor of sphingosine kinase. Carcinogenesis 31:1787–1793

Chun J, Hla T, Lynch KR, Spiegel S, Moolenaar WH (2010) International Union of Basic and Clinical Pharmacology LXXVIII. Lysophospholipid receptor nomenclature. Pharmacol Rev 62:579–587

Cohen JA, Barkhof F, Comi G, Hartung H, Khatri BO, Montalban X, Pelletier J, Capra R, Gallo P, Izquierdo G, Tiel-Wilck K, de Vera A, Jin J, Stites T, Wu S, Aradhye S, Kappos L (2010) Oral fingolimod or intramuscular interferon for relapsing multiple sclerosis. N Engl J Med 362:402–415

Cohen JA, Chun J (2011) Mechanisms of fingolimod's efficacy and adverse effects in multiple sclerosis. Ann Neurol 69:759–777

David OJ, Kovarik JM, Schmouder RL (2012) Clinical pharmacokinetics of fingolimod. Clin Pharmacokinet 51:15–28

Edmonds Y, Milstien S, Spiegel S (2011) Development of small-molecule inhibitors of sphingosine-1-phosphate signaling. Pharmacol Ther 132:352–360

Espinosa PS, Berger JR (2011) Delayed fingolimod-associated asystole. Mult Scler 17:1387–1389

Fitzpatrick LR, Green C, Frauenhoffer EE, French KJ, Zhuang Y, Maines LW, Upson JJ, Paul E, Donahue H, Mosher TJ, Smith CD (2011a) Attenuation of arthritis in rodents by a novel orally-available inhibitor of sphingosine kinase. Inflammopharmacology 19:75–87

Fitzpatrick LR, Green C, Maines LW, Smith CD (2011b) Experimental osteoarthritis in rats is attenuated by ABC294640, a selective inhibitor of sphingosine kinase-2. Pharmacology 87:135–143

French KJ, Schrecengost RS, Lee BD, Zhuang Y, Smith SN, Eberly JL, Yun JK, Smith CD (2003) Discovery and evaluation of inhibitors of human sphingosine kinase. Cancer Res 63:5962–5969

French KJ, Upson JJ, Keller SN, Zhuang Y, Yun JK, Smith CD (2006) Antitumor activity of sphingosine kinase inhibitors. J Pharmacol Exp Ther 318:596–603

French KJ, Zhuang Y, Maines LW, Gao P, Wang W, Beljanski V, Upson JJ, Green CL, Keller SN, Smith CD (2010) Pharmacology and antitumor activity of ABC294640, a selective inhibitor of sphingosine kinase-2. J Pharmacol Exp Ther 333:129–139

Fujii Y, Hirayama T, Ohtake H, Ono N, Inoue T, Sakurai T, Takayama T, Matsumoto K, Tsukahara N, Hidano S, Harima N, Nakazawa K, Igarashi Y, Goitsuka R (2012) Amelioration of collagen-induced arthritis by a novel $S1P_1$ antagonist with immunomodulatory activities. J Immunol 188:206–215

Fukuhara S, Simmons S, Kawamura S, Inoue A, Orba Y, Tokudome T, Sunden Y, Arai Y, Moriwaki K, Ishida J, Uemura A, Kiyonari H, Abe T, Fukamizu A, Hirashima M, Sawa H, Aoki J, Ishii M, Mochizuki N (2012) The sphingosine-1-phosphate transporter Spns2 expressed on endothelial cells regulates lymphocyte trafficking in mice. J Clin Invest 122:416–1426

Gergely P, Wallstroem E, Nuesslein-Hildesheim B, Bruns C, Zecri F, Cooke N, Traebert M, Tuntland T, Rosenberg M, Saltzman M (2009) Phase I study with the selective $S1P_1/S1P_5$ receptor modulator BAF312 indicates that $S1P_1$ rather than $S1P_3$ mediates transient heart rate reduction in humans. Mult Scler 15:S125–S126, Abstract

Gonzalez-Cabrera PJ, Cahalan SM, Nguyen N, Sarkisyan G, Leaf NB, Cameron MD, Kago T, Rosen H (2012) $S1P_1$ receptor modulation with cyclical recovery from lymphopenia ameliorates mouse model of multiple sclerosis. Mol Pharmacol 81:166–174

Gonzalez-Cabrera PJ, Jo E, Sanna MG, Brown S, Leaf N, Marsolais D, Schaeffer M, Chapman J, Cameron M, Guerrero M, Roberts E, Rosen H (2008) Full pharmacological efficacy of a novel $S1P_1$ agonist that does not require S1P-like headgroup interactions. Mol Pharmacol 74:1308–1318

Hannun YA, Obeid LM (2008) Principles of bioactive lipid signalling: lessons from sphingolipids. Nat Rev Mol Cell Biol 9:139–150

Hanson MA, Roth CB, Jo E, Griffith MT, Scott FL, Reinhart G, Desale H, Clemons B, Cahalan SM, Schuerer SC, Sanna MG, Han GW, Kuhn P, Rosen H, Stevens RC (2012) Crystal structure of a lipid G protein-coupled receptor. Science 335:851–855

Hla T, Venkataraman K, Michaud J (2008) The vascular S1P gradient-cellular sources and biological significance. Biochim Biophys Acta 1781:477–482

Huwiler A, Bourquin F, Kotelevets N, Pastukhov O, Capitani G, Grütter MG, Zangemeister-Wittke U (2011) A prokaryotic S1P lyase degrades extracellular S1P in vitro and in vivo: implication for treating hyperproliferative disorders. PLoS One 6:e22436

Huwiler A, Pfeilschifter J (2008) New players on the center stage: sphingosine 1-phosphate and its receptors as drug targets. Biochem Pharmacol 75:1893–1900

Igarashi N, Okada T, Hayashi S, Fujita T, Jahangeer S, Nakamura S (2003) Sphingosine kinase-2 is a nuclear protein and inhibits DNA synthesis. J Biol Chem 278:46832–46839

Jackson SJ, Giovannoni G, Baker D (2011) Fingolimod modulates microglial activation to augment markers of remyelination. J Neuroinflammation 8:76

Kapitonov D, Allegood JC, Mitchell C, Hait NC, Almenara JA, Adams JK, Zipkin RE, Dent P, Kordula T, Milstien S, Spiegel S (2009) Targeting sphingosine kinase-1 inhibits Akt signaling, induces apoptosis, and suppresses growth of human glioblastoma cells and xenografts. Cancer Res 69:6915–6923

Kappos L, Radue E, O'Connor P, Polman C, Hohlfeld R, Calabresi P, Selmaj K, Agoropoulou C, Leyk M, Zhang-Auberson L, Burtin P (2010) A placebo-controlled trial of oral fingolimod in relapsing multiple sclerosis. N Engl J Med 362:387–401

Kennedy AJ, Mathews TP, Kharel Y, Field SD, Moyer ML, East JE, Houck JD, Lynch KR, Macdonald TL (2011) Development of amidine-based sphingosine kinase-1 nanomolar inhibitors and reduction of sphingosine-1-phosphate in human leukemia cells. J Med Chem 54:3524–3548

Kharel Y, Mathews TP, Gellett AM, Tomsig JL, Kennedy PC, Moyer ML, Macdonald TL, Lynch KR (2011) Sphingosine kinase type-1 inhibition reveals rapid turnover of circulating sphingosine-1-phosphate. Biochem J 440:345–353

Killestein J, Rudick RA, Polman CH (2011) Oral treatment for multiple sclerosis. Lancet Neurol 10:1026–1034

Kumar A, Saba JD (2009) Lyase to live by: sphingosine phosphate lyase as a therapeutic target. Expert Opin Ther Targets 13:1013–1025

Lai W, Irwan AW, Goh HH, Melendez AJ, McInnes IB, Leung BP (2009) Distinct roles of sphingosine kinase-1 and -2 in murine collagen-induced arthritis. J Immunol 183:2097–2103

Lim KG, Sun C, Bittman R, Pyne NJ, Pyne S (2011) (R)-FTY720 methyl ether is a specific sphingosine kinase-2 inhibitor: effect on sphingosine kinase-2 expression in HEK-293 cells and actin rearrangement and survival of MCF-7 breast cancer cells. Cell Signal 23:1590–1595

Liu H, Toman RE, Goparaju SK, Maceyka M, Nava VE, Sankala H, Payne SG, Bektas M, Ishii I, Chun J, Milstien S, Spiegel S (2003) Sphingosine kinase type-2 is a putative BH3-only protein that induces apoptosis. J Biol Chem 278:40330–40336

Loveridge C, Tonelli F, Leclercq T, Lim KG, Long JS, Berdyshev E, Tate RJ, Natarajan V, Pitson SM, Pyne NJ, Pyne S (2010) The sphingosine kinase-1 inhibitor 2-(p-hydroxyanilino)-4-(p-chlorophenyl)thiazole induces proteasomal degradation of sphingosine kinase-1 in mammalian cells. J Biol Chem 285:38841–38852

Maceyka M, Harikumar KB, Milstien S, Spiegel S (2012) Sphingosine-1-phosphate signaling and its role in disease. Trends Cell Biol 22:50–60

Maines LW, Fitzpatrick LR, French KJ, Zhuang Y, Xia Z, Keller SN, Upson JJ, Smith CD (2008) Suppression of ulcerative colitis in mice by orally available inhibitors of sphingosine kinase. Dig Dis Sci 53:997–1012

Maines LW, Fitzpatrick LR, Green CL, Zhuang Y, Smith CD (2010) Efficacy of a novel sphingosine kinase inhibitor in experimental Crohn's disease. Inflammopharmacology 18:73–85

Moberly JB, Rohatagi S, Zahir H, Hsu C, Noveck RJ, Truitt KE (2012) Pharmacological modulation of peripheral T and B lymphocytes by a selective sphingosine-1-phosphate receptor-1 modulator. J Clin Pharmacol 52:996–1006

Mullershausen F, Zecri F, Cetin C, Billich A, Guerini D, Seuwen K (2009) Persistent signaling induced by FTY720-phosphate is mediated by internalized $S1P_1$ receptors. Nat Chem Biol 5:428–434

O'Brien N, Jones ST, Williams DG, Cunningham HB, Moreno K, Visentin B, Gentile A, Vekich J, Shestowsky W, Hiraiwa M, Matteo R, Cavalli A, Grotjahn D, Grant M, Hansen G, Campbell M, Sabbadini R (2009) Production and characterization of monoclonal anti-sphingosine-1-phosphate antibodies. J Lipid Res 50:2245–2257

Oo ML, Thangada S, Wu M, Liu CH, Macdonald TL, Lynch KR, Lin C, Hla T (2007) Immunosuppressive and anti-angiogenic sphingosine 1-phosphate receptor-1 agonists induce ubiquitinylation and proteasomal degradation of the receptor. J Biol Chem 282:9082–9089

Pan S, Mi Y, Pally C, Beerli C, Chen A, Guerini D, Hinterding K, Nuesslein-Hildesheim B, Tuntland T, Lefebvre S, Liu Y, Gao W, Chu A, Brinkmann V, Bruns C, Streiff M, Cannet C, Cooke N, Gray N (2006) A monoselective sphingosine-1-phosphate receptor-1 agonist prevents allograft rejection in a stringent rat heart transplantation model. Chem Biol 13:1227–1234

Parrill AL, Wang D, Bautista DL, van Brocklyn JR, Lorincz Z, Fischer DJ, Baker DL, Liliom K, Spiegel S, Tigyi G (2000) Identification of Edg1 receptor residues that recognize sphingosine-1-phosphate. J Biol Chem 275:39379–39384

Paugh SW, Paugh BS, Rahmani M, Kapitonov D, Almenara JA, Kordula T, Milstien S, Adams JK, Zipkin RE, Grant S, Spiegel S (2008) A selective sphingosine kinase-1 inhibitor integrates multiple molecular therapeutic targets in human leukemia. Blood 112:1382–1391

Pelletier D, Hafler DA (2012) Fingolimod for multiple sclerosis. N Engl J Med 366:339–347

Piali L, Froidevaux S, Hess P, Nayler O, Bolli MH, Schlosser E, Kohl C, Steiner B, Clozel M (2011) The selective sphingosine 1-phosphate receptor-1 agonist ponesimod protects against lymphocyte-mediated tissue inflammation. J Pharmacol Exp Ther 337:547–556

Pitson SM (2011) Regulation of sphingosine kinase and sphingolipid signaling. Trends Biochem Sci 36:97–107

Pyne NJ, Pyne S (2010) Sphingosine 1-phosphate and cancer. Nat Rev Cancer 10:489–503

Ren S, Xin C, Pfeilschifter J, Huwiler A (2010) A novel mode of action of the putative sphingosine kinase inhibitor 2-(p-hydroxyanilino)-4-(p-chlorophenyl) thiazole (SKI II): induction of lysosomal sphingosine kinase-1 degradation. Cell Physiol Biochem 26:97–104

Sabbadini RA (2011) Sphingosine-1-phosphate antibodies as potential agents in the treatment of cancer and age-related macular degeneration. Br J Pharmacol 162:1225–1238

Sanna MG, Liao J, Jo E, Alfonso C, Ahn M, Peterson MS, Webb B, Lefebvre S, Chun J, Gray N, Rosen H (2004) Sphingosine 1-phosphate (S1P) receptor subtypes $S1P_1$ and $S1P_3$, respectively, regulate lymphocyte recirculation and heart rate. J Biol Chem 279:13839–13848

Sanna MG, Wang S, Gonzalez-Cabrera PJ, Don A, Marsolais D, Matheu MP, Wei SH, Parker I, Jo E, Cheng W, Cahalan MD, Wong C, Rosen H (2006) Enhancement of capillary leakage and restoration of lymphocyte egress by a chiral $S1P_1$ antagonist in vivo. Nat Chem Biol 2:434–441

Schmahl J, Raymond CS, Soriano P (2007) PDGF signaling specificity is mediated through multiple immediate early genes. Nat Genet 39:52–60

Schwab SR, Pereira JP, Matloubian M, Xu Y, Huang Y, Cyster JG (2005) Lymphocyte sequestration through S1P lyase inhibition and disruption of S1P gradients. Science 309:1735–1739

Shea BS, Brooks SF, Fontaine BA, Chun J, Luster AD, Tager AM (2010) Prolonged exposure to sphingosine 1-phosphate receptor-1 agonists exacerbates vascular leak, fibrosis, and mortality after lung injury. Am J Respir Cell Mol Biol 43:662–673

Snider AJ, Orr Gandy KA, Obeid LM (2010) Sphingosine kinase: role in regulation of bioactive sphingolipid mediators in inflammation. Biochimie 92:707–715

Song J, Matsuda C, Kai Y, Nishida T, Nakajima K, Mizushima T, Kinoshita M, Yasue T, Sawa Y, Ito T (2008) A novel sphingosine 1-phosphate receptor agonist, 2-amino-2-propanediol hydrochloride (KRP-203), regulates chronic colitis in interleukin-10 gene-deficient mice. J Pharmacol Exp Ther 324:276–283

Tarrasón G, Aulí M, Mustafa S, Dolgachev V, Domènech MT, Prats N, Domínguez M, López R, Aguilar N, Calbet M, Pont M, Milligan G, Kunkel SL, Godessart N (2011) The sphingosine-1-phosphate receptor-1 antagonist, W146, causes early and short-lasting peripheral blood lymphopenia in mice. Int Immunopharmacol 11:1773–1779

Visentin B, Vekich JA, Sibbald BJ, Cavalli AL, Moreno KM, Matteo RG, Garland WA, Lu Y, Yu S, Hall HS, Kundra V, Mills GB, Sabbadini RA (2006) Validation of an anti-sphingosine-1-phosphate antibody as a potential therapeutic in reducing growth, invasion, and angiogenesis in multiple tumor lineages. Cancer Cell 9:225–238

Vogel P, Donoviel MS, Read R, Hansen GM, Hazlewood J, Anderson SJ, Sun W, Swaffield J, Oravecz T (2009) Incomplete inhibition of sphingosine 1-phosphate lyase modulates immune system function yet prevents early lethality and non-lymphoid lesions. PLoS One 4:e4112

Wenderfer SE, Stepkowski SM, Braun MC (2008) Increased survival and reduced renal injury in MRL/lpr mice treated with a novel sphingosine-1-phosphate receptor agonist. Kidney Int 74:1319–1326

Xie B, Shen J, Dong A, Rashid A, Stoller G, Campochiaro PA (2009) Blockade of sphingosine-1-phosphate reduces macrophage influx and retinal and choroidal neovascularization. J Cell Physiol 218:192–198

Yonesu K, Kubota K, Tamura M, Inaba S, Honda T, Yahara C, Watanabe N, Matsuoka T, Nara F (2011) Purification and identification of activating enzymes of CS-0777, a selective sphingosine 1-phosphate receptor-1 modulator, in erythrocytes. J Biol Chem 286:24765–24775

Zhang Z, Zhang Z, Zug C, Nuesslein-Hildesheim B, Leppert D, Schluesener HJ (2009) AUY954, a selective $S1P_1$ modulator, prevents experimental autoimmune neuritis. J Neuroimmunol 216:59–65

Index

E. Gulbins and I. Petrache (eds.), *Sphingolipids: Basic Science and Drug Development*,
Handbook of Experimental Pharmacology 215, DOI 10.1007/978-3-7091-1368-4,

MIX
Papier aus verantwortungsvollen Quellen
Paper from responsible sources
FSC® C105338

If you have any concerns about our products,
you can contact us on
ProductSafety@springernature.com

In case Publisher is established outside the EU,
the EU authorized representative is:
Springer Nature Customer Service Center GmbH
Europaplatz 3, 69115 Heidelberg, Germany

Printed by Libri Plureos GmbH
in Hamburg, Germany